Olivier Dangles

Climate Change on Mountains

Reviving Humboldt's Approach to Science

 Springer

Olivier Dangles
French National Research Institute for Sustainable Development
Montpellier, France

ISBN 978-3-031-39530-7   ISBN 978-3-031-39528-4   (eBook)
https://doi.org/10.1007/978-3-031-39528-4

Cover illustration: **The true mountain that inspired Humboldt**. Artistic adaptation of Humboldt's Tableau Physique, a groundbreaking document of the altitudinal ranges of tropical mountain vegetation. The contour of Mt. Antisana, Ecuador, is a reproduction of a sketch made by Humboldt in his travel diaries (Ette & Maier 2018, p. 330). Note the signs of glacier retreat and the upslope shift of plants as a response to climate change. The Andean fox is included as a symbol of the transdisciplinarity and 'nomadic thinking' that characterizes Humboldt's work (Ette, 2014). Concept: Olivier Dangles; illustration: Paula Terán Ospina.

This Springer imprint is published by the registered company Springer Nature Switzerland AG
The registered company address is: Gewerbestrasse 11, 6330 Cham, Switzerland

Paper in this product is recyclable.

For the true naturalist, the one who has experienced a scientific awakening, there are, up there [on mountain peaks], important questions to be addressed to nature and wonderful answers to be gathered.

Daniel Dollfus-Ausset[1]

Natural history is nothing less than a way to live in the world poetically.

Patrick Drevet[2]

Today's science continues to find its legitimacy upon its recourse to the data, which are repeatedly checked and rechecked in a never-ending search for truth through the elimination of error. [...] With our hopes and dreams suffused in the ether of illusion, life itself appears diminished. Shorn of its creative impulse, it no longer gives cause for wonder or astonishment.

Tim Ingold[3]

---

[1] Dollfus-Ausset (1863–1870), t.4, vol. 7 (Ascencions), p. 393. In 1843, Daniel Dollfus-Ausset (1797–1870) joined Louis Agassiz's expeditions and became fascinated by the study of glaciers, which he continued to visit and investigate every summer until his death. Dollfus-Ausset drew on these works to publish "Materials for the study of glaciers" in 15 volumes and numerous daguerre-otypes. See Frochot (2020).

[2] Le Corps Du Monde by Patrick Drevet, Editions Seuil, Copyright © 1997, p. 318.

[3] Ingold (2013), pp. 734–735. Used with permission of John Wiley & Sons - Books, from Dreaming of dragons: on the imagination of real life, The Journal of the Royal Anthropological Institute, 19(4); permission conveyed through Copyright Clearance Center, Inc.

*En profonde gratitude pour le temps passé
dans les montagnes et pour tous les
non-humains avec lesquels j'ai fait
connaissance.*[4]

*A Paola,
Llenó mi vida de amor y cariño sin límites.*[5]

---

[4]In deep gratitude for the time spent in the mountains and for all the non-humans I got to know.
[5]To Paola, she filled my life with love and affection without limits.

Preface

If the 20th century killed Humboldt, the 21st century must revive him.

Laura Dassow Walls[6]

The idea for this book started forming in my mind in 2015, halfway through co-writing the textbook *Ecology of High-altitude Waters* with my friend Dean Jacobsen. While I was well versed in the dry style of scientific papers, the textbook allowed a creative freedom that stimulated my desire to write about scientific practices, discoveries, and beyond. A crossover book with a wider audience in mind seemed an appealing next step. And so I started working on a book about the ecological effects of climate change in the tropical Andes—my scientific focus over the last decade—that would blend compelling scientific findings with personal memoir. My objective was to share my first-hand experience on the accelerating impacts of climate change in tropical mountains, effects caused by rising temperatures, melting glaciers, and changing precipitation patterns. In the field, I have observed the wide-ranging impacts these changes are having on ecosystems—the shifts in plant and animal populations, the increased risk of natural disasters. Working with South American researchers and students in the natural outdoor laboratory of the Andes, we investigated how insects, plants, and vertebrates are coping with altered temperatures and water availability. To better understand how nature here will face the challenges created by rapidly warming conditions, we measured the physical environment using the latest technologies, collected and surveyed biological communities in remote places, conducted lab and field experiments, worked with local communities, and developed tools and models to analyze the data. In its early stage, this book was distantly inspired by the geographer and polymath Alexander von Humboldt, who spent 5 years exploring the forests and

---

[6]The Search for Humboldt, Laura Dassow Walls, (2006), Geographical Review, p. 477. Copyright © 2006 American Geographical Society of New York, https://americangeo.org, reprinted by permission of Taylor & Francis Ltd., http://www.tandfonline.com on behalf of 2006 American Geographical Society of New York, https://americangeo.org.

mountains of tropical South America where he observed the interconnectedness of mountain ecosystems and their vulnerability to environmental changes.

Then came 2019, the 250th anniversary of Humboldt's birth (he was born in 1769), and an explosion of international publications and events paying tribute to him. While venerated in his time, Humboldt's holistic approach, combining science, the humanities, and the arts—the rational and the sensory—had been lost over the twentieth century. The issues of the twenty-first century brought it back into fashion, and for this 250th anniversary hundreds of symposiums, special issues, books, articles, new editions, and translations of his works appeared, many of which argued that Humboldt's worldview would help achieve better science and a more sustainable future. Laura Dassow Wall, who wrote *The Passage to Cosmos: Alexander von Humboldt and the Shaping of America* in 2009, arguing for Humboldt's modern importance, seemed to see her wish fulfilled.

But there was something that bothered me in all this buzz around Humboldt: these contributions preached for Humboldt's model, but none really explained how to put it into practice. Does Humboldt have a place in science today? If so, how? How can we learn from his writings and philosophy to do better science, to get a broader understanding of the world, to protect the Earth? I could not find these answers in what I read or heard about Humboldt. Yet, all my years working on the impacts of climate change in the tropical Andes had given me a sort of "universal" perspective not unlike his that I felt would be worth sharing. So rather than a distant mentor, Humboldt became a major protagonist in this book. While the focus is on the effects of climate change on mountains, the story weaves in anecdotes that illuminate Humboldt's approach to science, which I believe is crucial to tackle the challenges we face today.

As a practicing scientist, for me "reviving Humboldt" is not about celebrating his memory as a historical figure, but about a careful reading of his texts, his extensive footnotes, his complex drawings and figures, his endless data tables; it is about drawing links from his expeditions and records to contemporary studies; it is about embracing his way of thinking by not being restricted to an academic silo but integrating different disciplines in science, the arts, and humanities; it is about merging the rational and the sensory, logic and the imagination, the textual and the visual. Above all, it is a way of perceiving the world in which everything is connected: people, disciplines, places, historical eras. I hope that this book contributes to "reviving Humboldt" in this sense.

Montpellier, France   Olivier Dangles

Acknowledgments

Elise Bradbury was my faithful editorial companion over the years of working on this project. Elise offered her editing skills and suggested stylistic improvements to help a non-native English writer find the right words. She also pulled me out of the pit of discouragement in the rollercoaster process of finding a publisher. Thank you, Elise, I couldn't have done it without you.

Paula Terán Ospina's illustrations immeasurably add to the text. She is the embodiment of the book's merging of science and art, using her incredible skill at drawing to convey the finest details, and was always open to suggestions.

The invisible archivists who scanned and put online the hundreds of documents I read for this book are not possible to name but their work is invaluable. Among them, the staff of Gallica, the digital library of the Bibliothèque Nationale de France; Calames, the online catalog of higher education archives; the Internet Archive; the Biodiversity Heritage Library; the Archives de Paris; CNRS Images; Bibliothèque du Muséum National d'Histoire Naturelle, Rare and Manuscript Collections of Cornell University; and the Centre de Recherche du Château de Versailles. Without access to these materials, it would not have been possible to find the links between the many historical and contemporary figures that appear in the book.

Several colleagues deserve a special mention. Pierre Moret introduced me to the history of science and the art of navigating archives. He also provided many insightful comments on the manuscript. The reviewer and historian of science Michael Reidy saw much more in this book than I had seen myself; his invaluable advice led me to key references and improved the structure of the book. Antoine Rabatel, Jean-Mathieu Nocquet, Dean Jacobsen, and Denis Torres also offered valuable feedback and ideas for revision for several chapters of the book. Lastly, Jérôme Casas has been an intellectual mentor and friend for over two decades, and his breadth of knowledge has been a never-ending source of inspiration for me.

Several friends and colleagues allowed me to use their photos: Roger Calvez, Sophie Cauvy-Fraunié, Dean Jacobsen, Ricardo Jaramillo, Vincent Jomeli, Juan Diego Pérez Arias, Matthieu Perrault, Pierre Moret, and Juan Sebastián Rodríguez. Many shared documents and opinions that shaped the topics in this book: Matthias

Abram, Christiana Borchart de Moreno, Segundo E. Moreno Yánez, Eleanor Harvey, Ottmar Ette, Aaron Sachs (Alexander Humboldt); Nelson Hairston and Alexander Flecker (James Needham); Jean Vacher and Salomé Ketabi (Alcide d'Orbigny); Edward C. Dickinson (Darwin's ñandou); Clara Poirier (Tim Ingold); Juan Diego Pérez Arias (Arturo Eichler), Claire Nicklin, Matias Cortese, Jonhatan Miller, Karen Merikangas Darling (writing); Paula Terán Ospina, Belén Mena, François Nowicki (art); Thomas Condom, Bernard Francou, David Whiteman, Luis Daniel Llambi, Antoine Rabatel, Jean-Emmanuel Sicard (glaciers and rivers); Anne-Gaël Bilhaut, Alden Yépez, Elizabeth Allison (anthropology/culture); Hervé Jourdan (New Caledonia), and Armando Castellanos (bears).

As I've worked in four countries in the last 15 years—France, the United States, Ecuador and Bolivia—any attempt to list everyone who has helped me can only fail by serious omission. Many students, postdoctoral fellows, and colleagues collaborated on the studies described in this book: Patricio Andino, Fabien Anthelme, Laura Arcos, Álvaro Barragán, Santiago Burneo, Roger Calvez, Ivan Cangas, Rafael Cárdenas, Tatiana Cárdenas, Sophie Cauvy-Fraunié, Thomas Condom, Bruno Condori, Verónica Crespo-Pérez, Daniela Cueva, Antonio Daza, Bert De Bièvre, the Delgado family, Andrea Encalada, Agathe Frochot, Santiago García Lloré, Jere Gilles, Martin Kraemer, Rubén Lamas, Pablo Lloret, Rosa Isela Meneses, Paula Iturralde-Pólit, Jacques Gardon, Karina Gonzalez, Dunia González-Zeas, Harry W. Greene, Ladislvav Hamerlik, Jesper Kuhn, Xavier Lazzaro, Marie-Pierre Ledru, Philip Madsen, Luis Maisincho, Florencio Maza, Diego Mina, Rommel Montúfar, Jorge Molina-Rodriguez, Andrés Morabowen, Priscilla Muriel, Kazuya Naoki, Estefania Quenta-Herrera, Frédérique Rolandone, Daniela Rosero-López, the community of Sajama, Álvaro Soruco, Quentin Struelens, Esteban Suárez, Renato Valencia, Todd Walter, David Winkler, Gabriel Zeballos, and Kabil Zerouali.

Maria L. Eisner, Jay Hart, and Carol Pearson Ralph took the time to answer my questions via email, and Rebecca Finnel dared go into her local post office during the pandemic to send me two original drawings by Francois Vuilleumier: an invaluable gift.

My research institute, the French Research Institute for Sustainable Development (IRD), has provided unconditional support for my work in South America since 2006. Special thanks to Bernard Dreyfus, Thomas Changeux, Marie-Noëlle Favier Thomas Mourier, Marie-Lise Sabrié, Jean-François Silvain, and Jean Vacher. My warmest thanks goes to Valérie Verdier for making possible my stay at Cornell University. I am also grateful to the Pontificia Universidad Católica del Ecuador and its staff for embracing me with open arms and supporting my projects.

I am grateful to those at Springer who have shepherded this book throughout its production: Ramon Khanna, Antje Endemann, Neelofar Yasmeen, and Barbara Amorese. I consider it fate that Ramon Khanna, the executive editor of the astronomy and cosmology list, born in Berlin and frequenter of the Humboldthain Park as a child, supported the publication of this book that pays tribute to Humboldt.

Finally, because everything begins with childhood, I am grateful beyond words to my beloved parents, Solange and Roger, my grandparents, Louise and Paul, and my sister Christine, who instilled the passion for traveling in me. My *querida suegrita* María Eugenia Bucheli and late *querido suegro* Alvaro Ponce, lover of the páramos, accepted me as one of the family. And my bottomless sources of inspiration, my wife Paola and our two sons Nicolas and Matias, thank you for your boundless love.

Contents

Chapter 1
Introduction: The Subterranean Networks of Humboldt

The kind of attention to life [that can be traced] in Alexander Humboldt's work [...] is not a 'school of thought' but a subterranean network, clandestine links that run through and connect different places, people, mediums, works, practices, fields, historical monuments.

Estelle Zhong Mengual (2021, p. 240)

The main protagonists of this chapter are shown in Fig. 1.1.

1.1 Ecuadorian Andes

It's a cold February predawn in the high **sierra** of Ecuador. I stick my head through the tent flap and in the absence of light can barely discern the surrounding landscape, which is lightly sprinkled with snow. A piercing wind chills my face. I didn't get much sleep and have a lancing pain in my head along with a general lack of energy. Nothing out of the ordinary after a first night at an altitude of 4500 m. I'm tempted to stay in the warmth of my sleeping bag, but I know I have to get out of my tent and go to the stream to collect samples. Since our arrival at the study site yesterday afternoon, our small team of ecologists have taken turns every three hours to monitor the insect fauna of a glacier-fed stream. The study is part of a long-term international research project supported by Denmark, France and Ecuador with the objective of documenting the effect of glacier retreat on freshwater life. Several previous studies have addressed this issue in temperate regions, but it is a world first in the tropics. There is no question of renouncing my shift; I must collect the 5:00 a.m. sample.

© The Author(s), under exclusive license to Springer Nature Switzerland AG 2023
O. Dangles, *Climate Change on Mountains*,
https://doi.org/10.1007/978-3-031-39528-4_1

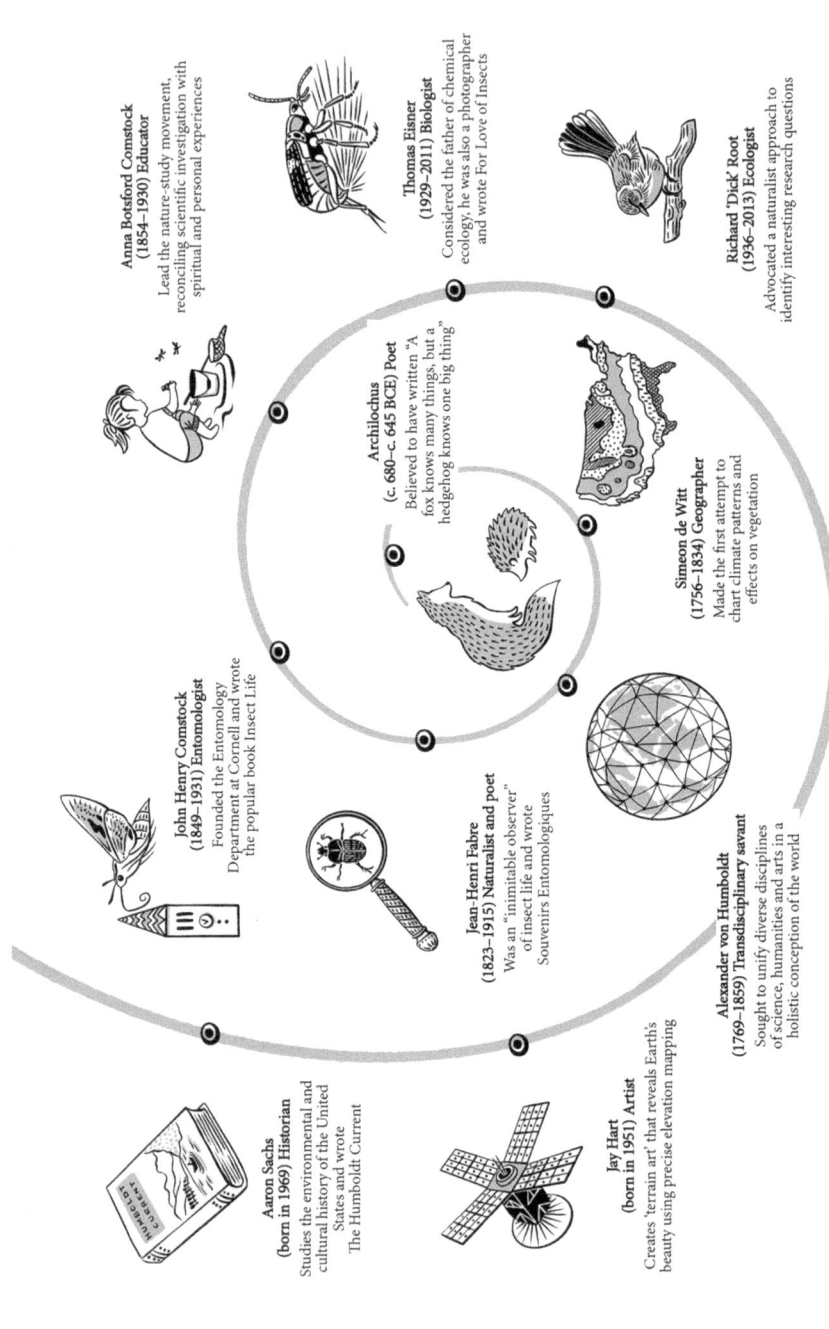

Fig. 1.1   Central figures in this chapter. This spiral timeline shows ten historical or contemporary figures who contributed important ideas mentioned in this section. Drawings by Paula Terán Ospina

Fig. 1.2 The páramo of Mt Antisana with Mt Cotopaxi in the background. Above 3500 m in the Ecuadorian Andes, the green carpeted hills of the páramo punctuated by icy summits are a naturalist's paradise, both rich in endemic species and one of the most scenic places on Earth. Photo by Olivier Dangles

I grab my woolly hat, gloves, headlamp and rubber boots and step out into the bitter darkness. Guided by the sound of the babbling stream nearby, I waddle like a penguin to avoid falling over on the sloping, slippery terrain. After a short walk, I reach the site and go into the usual routine. First, I take a long-exposure photo with a camera in a fixed position on the bank, then I measure a few water-quality parameters, and finally I retrieve the drift nets. Only 300 m downstream from the glacier mouth, the water is just above freezing point, poor in oxygen and loaded with glacial silt; yet, life abounds in the stream. Within a couple of hours, our nets have captured hundreds of tiny insect larvae, most unknown to science. As glaciers continue to vanish, most of these may disappear in a few decades, before entomologists have the time to describe them.

Collecting the samples takes me about an hour, and by the time I get back to the camp, dawn breaks, revealing the landscape. Just in front of me, as if I could touch it, stands the imposing Cotopaxi volcano with its nearly perfect conical ice-capped peak—a breathtaking view (Fig. 1.2). It's time to put science aside and engage the other side of my brain: the immediate, intuitive, emotional right side, which contrasts to the slower, more deliberative and logical left side. As a rule, the right brain is the "forgotten half of scientific thinking."[1] Yet for me, these modes of thinking are

---

[1] Scheffer (2014).

complementary. They combine aesthetic appreciation of nature and its scientific understanding; childlike curiosity for the living world and meticulous measurements; time for deep receptivity and observation and reflecting on acquired scientific ideas; physical experiences and mental concepts. Wildlife photography is my passport to my right brain, a sphere of focused attentiveness, imagination, intuition and being attuned to the beauty of nature's composition.[2] I am convinced that my pleasure in photographing nature strengthens my ability to observe the living world and, ultimately, leads to a broader understanding of how nature works. It feeds back into my left brain.

Facing Cotopaxi, I unfold my tripod, mount the camera to take a couple of pictures, and sit on the edge of a rock. Behind the mountain, a plume of smoke wafts from the summit of Tungurahua, another active volcano some 80 km southeast. To the west, the tip of the colossal Mount Chimborazo emerges from the clouds. Like islands in the sky, each of these magnificent volcanoes has its own unique appearance: shape, color, ice cover, wildlife and 'eruptive personality,' which make them so special to photograph. And each has its own set of characteristics: altitude, climate, size, isolation and biogeographical history, which make them so fascinating to study. On Mount Antisana, the 'island' where I embarked with my team yesterday, I vividly feel this merging of the aesthetic and the scientific. Looking down from my observation point, I take in the diversity of forms and colors of the grassland landscape that surrounds me: the pinkish-grey meandering waters of glacial streams, the shamrock green of spongy cushion plants, the yellowish tufts of tussock grasses, the pale blue bumpy ice of the glacier, and the dark green of montane forest patches. All these habitats have a graphic beauty to capture. And, as I will discover over the next ten years living in the tropical Andes, all have important scientific secrets to reveal.

Through my binoculars I scan the mosaic of habitats, trying to grasp their interrelationship and the challenges associated with their study. Over the last decade, I have relished viewing the natural world through the lens of different disciplines, from physics to ecology to photography. To borrow a quote from the Ancient Greek poet **Archilochus**, I am a fox rather than a hedgehog. Foxes are generalists, wandering from one object of study to another and drawing on a wide variety of experiences and disciplines in their work. Hedgehogs are more focused on a narrow field of research, delighting in the details of specific issues to solve problems one at a time. [3] The two strategies are complementary and, while I'm more inclined to be a fox, I have many hedgehog colleagues. The risk is that as scholars continue to race down separate pathways, converting them into pure hedgehogs, knowledge becomes increasingly fragmented, and communication across disciplines more the exception than the rule. These boundaries obstruct our ability to address global challenges such

---

[2] For a thoughtful discussion on the visual character of ecological knowledge and the role played by photography to engage naturalists (around the turn of the nineteenth to twentieth centuries) in both scientific and subjective explorations of nature, see Hughes (2022), pp. 271–420.

[3] Gould (2011). For a parallel analogy with birds and frogs, see also Dyson (2009).

as preserving nature and adapting to climate change. The mosaic of habitats in the Ecuadorian highlands seems to offer up the opportunity to venture into the *terra nova* of interdisciplinary research that bridges different scientific communities. It may be a coincidence, but hedgehogs are absent from the high Andes, and foxes are rather common, wandering between glaciers, cliffs and wetlands. And sure enough, it is the silhouette of an Andean fox that I spot while scanning the landscape. I grab my camera and tripod, change my wide-angle for a telephoto lens and head down in its direction. A fox, chasing another fox.

I enjoy wandering in nature. It helps me develop a feeling for the intricacy of organisms in their environment, to gather impressions and intuitions beyond the reach of automatic data logging. For me, moving through the landscape is a way of understanding as well as discovering scientific questions.[4] After a while walking in the tussock grass, I've lost track of the fox and realize that I am close to the lowest study site along our glacier-fed stream. We call it the 'Humboldt site,' the name our hydrologist colleagues gave to the gauging station installed here. They chose this name as the site is located near a well-known hut where the Prussian physical geographer, naturalist and explorer **Alexander von Humboldt** and his company had supposedly camped when they visited Mount Antisana (Fig. 1.3).[5] At the time, I sort of know Humboldt, or, I should say, I know certain things that have been named after him: the Humboldt Current, squid and penguin. But his achievements and importance in the history of science are hazy in my mind. As I approach the entrance of the hut, I see a plaque attached to the white adobe wall: "*200 years ago, on a day like today, Alexander von Humboldt, celebrated scientist and the 'true discoverer of America,' visited and climbed Antisana to unravel its secrets.*" As I will discover, Humboldt is a legend in South America, where he is still the most influential European figure of the nineteenth century. So why do I have only a vague idea who this man was? To find the answer, I need to go back to my country of birth, France. That is where this story starts.

1.2 France

I got hooked on nature as a kid. While virtually all kids enjoy contact with the natural world, most lose their interest in nature as they grow up. I did not. For some reason, the time spent in nature that I experienced as a boy made its way to the deepest part

---

[4] Social anthropologist Tim Ingold has repeatedly argued that the movement of walking itself entails making knowledge, even when it appears undirected. See Ingold (2000), p. 230, Ingold and Vergunst (2008), pp. 1–19.

[5] In fact, it seems more likely that Humboldt and his company spent the night not in the hut, but in the main hacienda of Antisana, a larger building across the stream. This hacienda was also visited by the volcanologist Alphons Stübel and the alpinist Edward Whymper and was painted by Rafael Troya around 1872. The hut with the plaque can be considered more as the symbolic start from which Humboldt began his exploration at the Antisana. See Salzer and Nöbauer (2021).

Fig. 1.3 **Plaque commemorating Humboldt on the slopes of Mt Antisana, Ecuador.** In March 1802, Humboldt and his company spent 4 days on the slopes of Antisana. This sojourn is commemorated by a plaque (**b**) on the wall of the so-called 'Humboldt's hut' (**a**) (shown here with a glacial stream in the foreground and Mt Antisana in the background). The plaque was put on the front wall of the hut in 2002 by the Antisana Foundation, the Delgado family (former owners of the area), the mountain climbing club of the Pontifical Catholic University of Ecuador (PUCE), the French Research Institute for Sustainable Development (IRD), and the National Institute of Meteorology and Hydrology (INHAMI). Photos by Olivier Dangles

of me, so that my curiosity and passion to understand nature have never faded. French writer Patrick Drevet depicts the importance of childhood experiences to scientific naturalists very well:

> Like words that resonate far beyond the object they designate, or that enrobe it with imagination, the term natural history evokes less a discipline than memories. Of all the sciences, it is certainly the only with a sentimental connotation. Its domain brushes against wonder, sometimes merging with it. It goes back to a child's first emotions as its senses develop and then its first adventures reveal the world. It is linked to these initial adventures, to the child's first frights and greatest amazements. There is no doubt that the botanist, the entomologist, the zoologist, the geologist, the astronomer continue to be driven by this child's impulse—the original thirst to discover. The naturalist remains as struck as he was by the first objects that widened his eyes and his senses, however far he may have come behind the sophisticated instruments that he wields.[6]

Indeed, for me there is no doubt that there is an invisible connection linking my boyhood nature spots and the field study sites where I work today.

* * *

My introduction to nature was in the Aveyron, a sparsely populated region in southwestern France that is squeezed between the highlands of the ancient volcanic Plateau of Aubrac and the gorges of the River Tarn. Like all regions with high environmental heterogeneity, in particular along altitudinal and climatic gradients, Aveyron hosts a diverse array of habitats, plants and animals. The many hedgerows still tracing lines across the countryside provide both a refuge for wildlife and a pattern and texture that enhance the beauty of the landscape (Fig. 1.4a). It is a paradise for naturalists. But for a child, Aveyron was simply the place where my grandparents lived, and where I spent all my holidays. On their farm they had ducks, rabbits, hens, rock piles, a well with a green chain pump feeding a water tank, a vegetable garden, and a big oak tree. When I was feeling adventurous, I 'traveled' to immerse myself in the forest down the hill where there was a brook full of crayfish. My grandparents' farm and the countryside around it offered a nearly infinite variety of things to interest me. This may explain why I became a fox. As the American psychologist Susan Engel explains in her book *The Hungry Mind*: "*Some [kids] want to know more about everything they encounter, while for others the urge to find out is focused on a few topics about which they have unwavering and infinite interest.*"[7] What differentiates a fox from a hedgehog may originate from how we acquire, as kids, particular kinds of curiosity.

But mere contact with nature is not enough to see, feel and learn about natural things. For this, one needs to interact with a place's non-human inhabitants, to 'experience' nature. Experience affects how you feel in ways that can ultimately be integrated into your individual identity. My grandparents' house was a formative

---

[6]Le Corps Du Monde by Patrick Drevet, Editions Seuil, Copyright © 1997, p. 317

[7]The Hungry Mind: The Origins of Curiosity in Childhood by Susan L. Engel, Cambridge, Mass.: Harvard University Press, Copyright © 2015 by the President and Fellows of Harvard College. Used by permission. All rights reserved, p. 19.

Fig. 1.4 Childhood connections with nature. Like many naturalists, my interest in nature and entomology began with experiences during childhood: (**a**) the Aveyron countryside near Saint-Léons, the village where entomologist Jean-Henri Fabre was born and near where my grandparents lived; (**b**) a Laguiole knife with its signature fly on the handle. All Laguiole knives feature either a bee or a fly, but these come in all shapes and patterns. Photos (**a**) Olivier Dangles, (**b**) https://commons.wikimedia.org/wiki/File:Laguiole_(Messer)_jm120846.jpg © Jörgens.mi

place to develop my own experience of nature. I would try to catch one of the many lizards sunbathing on the old mossy stone wall behind the well. I collected snails, sorted them by their shapes and colors, and kept them in a bucket filled with lettuce and twigs. I walked through the woods with my grandfather, Paul, to collect chestnuts, mushrooms, *respounchous* and earthworms (for fishing, not eating!). On our walks, my grandfather would use the pocketknife from which he was inseparable, for trimming plants, digging into the soil, carving a stick from a chestnut tree branch or cutting off a hunk of Roquefort cheese for our snack. I was totally fascinated by this knife. It was a Laguiole, famous for their beautifully handcrafted handles—my grandfather's was made of marbled cow horn—with a forged fly on the spring where the knife folds (Fig. 1.4b). This very attractive fly (resembling the horseflies that commonly congregate on the backs and flanks of Aveyron cattle) on the 'magic' knife of my beloved grandfather perhaps lies at the origin of my particular curiosity for insects and other bugs. Even today, seeing the fly on a Laguiole knife has the same effect on me as madeleines did for Proust, leading me to relive my childhood memories of the time I spent in nature. Insects were beautiful and were everywhere around me, so I started spending hours observing them, training my eyes to really look. There was magic in every observation. This was the time when I also became interested in books, which led me to discover that Aveyron had been home to one of the greatest field observers of insects who ever lived.

* * *

The French naturalist and entomologist **Jean-Henri Fabre** was born a couple of days before Christmas in the village of Saint-Léons, which lies only 50 km east of my grandparents' house. Throughout his childhood he was fascinated by natural history, collecting stones and insects in the Aveyron countryside. Later, Fabre would devote his life to exploring the secrets of the insect world, using the countryside around his home as his laboratory. His special talent was making precise, detailed observations in the field, to such an extent that Charles Darwin described him as an *inimitable observer.*[8] But Fabre was also a skilled experimentalist. In one of his most famous experiments, he forced a group of pine processionary caterpillars to follow a continuous head-to-tail circle around the rim of a large plant pot. As each caterpillar instinctively followed the silken track of the one ahead of it, the formation went on marching in an endless circle for 7 days. Fabre calculated that the caterpillars did the loop 335 times, covering a distance of more than 450 m! Fabre compiled his discoveries in his classic ten-volume masterpiece *Souvenirs Entomologiques*, which at the age of 12 would cement my passion for natural history, entomology and art. Several singularities of Fabre's work strongly influenced me as a budding entomologist. First, he wrote beautifully in lively prose, addressing the reader as if speaking to an old friend. In Fabre's poetic description, the processionary caterpillar tale becomes:

> Very much more sumptuous than ours, their system of road-making consists of upholstering with silk instead of macadam. We sprinkle our roads with broken stones and level them with the pressure of a heavy steamroller; they lay over their path a soft satin rail, a work of general interest to which each contributes his thread.[9]

Second, his *Souvenirs Entomologiques* was illustrated by dozens of pictures taken by his youngest son Paul (Fig. 1.5a). While I would not buy my first camera until my twenties, much younger I was moved by Paul Fabre's macrophotography, in

---

[8]Darwin (1866), p. 100. "... *the males of certain hymenopterous insects have been frequently seen by that inimitable observer M. Fabre, fighting for a particular female, who sits by an apparently unconcerned beholder of the struggle, and then retires with the conqueror.*" Fabre was a Christian and never accepted the theory of evolution. However, his work was respected by Darwin and the two exchanged a couple of letters.

[9]Fabre (1925) Tome 6, p. 358. Translation by A. Teixeira de Mattos and B. Miall in Fabre J.-H. (1918) *The Wonders of Instinct.* New York: The Century Co. p. 120.

Fig. 1.5 Early entomological photography. As an educator, entomologist Jean-Henri Fabre
placed much importance on ensuring his books reached a wide public through an engaging writing
style and visuals. (**a**) Jean-Henri Fabre (seated, age 89) and his son Paul in a photo session with live
insects in a terrarium, probably one of the first attempts at macrophotography (written on the back of
the photo: "Celebrations in honor of Fabre: The famous entomologist, having had to give up his
personal work because of his old age, naturally works with his son to capture scenes of insect life on
film."); (**b**) one of the stunning photographs in Fabre's book: a spider wasp (*Cryptocheilus
alternatus*) discovering the hole of a tarantula wolf spider (*Lycosa tarantula*). Photos (**a**) Albert
Harlingue (1912) and (**b**) Paul Fabre in *Souvenirs Entomologiques* (1925), tome 2, Fig. IX

particular the images catching nature in the act. The spider wasp attacking the
tarantula wolf spider was my favorite (Fig. 1.5b). Lastly, although *Souvenirs
Entomologiques* focused on insect instinct, Fabre attached particular importance to
the environments where insects live, describing their interactions with other plants
and animals, their lifecycles and food habits, their response to environmental
conditions, and even their role in nature (as when he recorded the speed with
which carrion beetles bury a dead mole). Through his work I realized that, beyond
natural history and entomology, I was fascinated by ecology.

I visited my grandparents' farm every holiday until the age of 14. Then a new
chapter of my naturalist life opened as I began to spend more time in Paris. My
favorite neighborhood was the fifth arrondissement, the Latin Quarter; it was full of
naturalist bookshops, libraries, and, most of all, home to the *Jardin des Plantes*. I
loved making my way through the palm trees, creepers, giant ferns and orchids of the
tropical greenhouse as well as visiting the snakes at the Menagerie vivarium. I spent
hours reading books in the library of the adjacent National Museum of Natural
History, visiting zoologists and botanists in their labs. Progressively, I started to
connect the books, disciplines and theories to the famous French naturalists behind
them, whose names were everywhere around me, on busts and statues and the street
names around the gardens: Buffon, Lamarck, Cuvier, Jussieu, Geoffroy Saint-
Hilaire. It was only some 20 years later, after my stay in South America, that I
realized that one name was missing among all the others. The name of a naturalist
who was once, after Napoleon Bonaparte, the most famous man in Paris. And as I
would discover, that name had been deliberately erased from French collective
memory.

* * *

Born in Berlin in September 1769, under a comet that passes Earth every 2090 years, Alexander von Humboldt descended on his mother's side from a French Huguenot (Protestant) family who had sought refuge in Prussia to escape religious persecution. In 1790, Humboldt traveled for the first time to Paris, which at that time was the intellectual capital of the world. He arrived when preparations were underway for the first anniversary of the storming of the Bastille. From that moment he fell in love with Paris, committed to the three fundamental concepts of the new French republic: liberty, equality and fraternity. Humboldt reported: *"The sight of the Parisians, their rallying together as a nation, of their still unfinished temple of liberty [. . .]: all that floats in my soul like a dream."*[10]

In 1798, Humboldt came to Paris again, this time to study at the *Jardin des Plantes* and the *Observatoire de Paris* in the company of the most famous botanists, chemists and mathematicians of his time. There he met Aimé Bonpland, who was studying medicine and botany at the National Museum of Natural History. A few months later, Humboldt and Bonpland left Paris for Marseille, and then traveled along the Mediterranean coast to Spain and then across the Atlantic to South America. From 1799 to 1804 they traversed about 10,000 km, journeying through the Spanish American colonies (modern-day Venezuela, Colombia, Ecuador, Peru, Mexico and Cuba) to observe nature. On his return to France, Humboldt lived continuously in Paris from 1804 to 1827 and wrote extensively, composing several volumes relating to his South American journey, as well as thousands of personal letters. Later, in the years between 1842 and 1847, he would frequently return to Paris. In total, Humboldt lived in Paris for more than 30 years—a third of his lifetime. His brother Wilhelm found regrettable that he "has ceased being German and, in almost every detail, has become Parisian."[11] He was such a Francophile that he thought in French[12] and wrote most of his books in French rather than German; some of his early works had to be translated into his native language.

Beyond his writings, his contributions significantly enriched the collections of the National Museum of Natural History in Paris. In 1815, during the Prussian invasion of Paris, Humboldt negotiated to prevent thousands of soldiers from invading the museum's grounds, and may have helped to avoid the destruction of the Iena Bridge that spans the river Seine. Humboldt was also a member of the French Academy of Sciences and a co-founder of the influential Society of Arcueil (1806–22).[13] What is indisputable is that he had a profound influence on nineteenth-century French scientists, writers, artists, educators, explorers and politicians, as well as the public at large. On 8 May 1859, 2 days after Humboldt's death, the front page of the French

---

[10] In Duviols and Minguet (1994), p. 15.

[11] Letter to Caroline von Humboldt, 24 Aug. 1813, In Meinhardt (2018), p. 233.

[12] In Nelken (1980), p. 32.

[13] When in 1810 Napoleon sought to have Humboldt (a supposed Prussian spy) expelled from Paris, the Society of Arcueil promptly defended its member, showing how France had become Humboldt's adopted country. Chemist Jean-Antoine Chaptal (1756–1832), in close contact with Napoleon, pointed out that without Humboldt, "science in Paris would be at a complete standstill;" "when he travels it is like the entire academy on tour." In Crosland (1967), p. 76.

newspaper *La Presse* celebrated his *"immortal memory and unforgettable example."* Yet in less than half a century, Humboldt's status in France would change from popular hero to unknown. As the French Humboldt specialist Charles Minguet wrote: *"Alexander de Humboldt did not have in our country the audience he deserves. If our compatriots know the name of Humboldt, it is mainly thanks to the philological and philosophical works of Wilhelm, his elder brother, whose glory long eclipsed Alexander's. [...] While the Germans have never ceased to pay him just tribute, the French have shown themselves to be very ungrateful for this genius who combined Germanic seriousness with Latin warmth."*[14]

How could someone so highly regarded by artists and intellectuals, who had become such a fixture in the popular imagination and an "idol" in French society,[15] simply vanish from view? I decided to investigate this mystery and started by looking for street names and statues paying tribute to Humboldt in Paris. I did find a Humboldt Street, but it is a short and grim lane, lost in the far northeastern part of the French capital. Built in the early 1980s, the uninteresting street was first named 'road BX/19' and was renamed Rue Humboldt in 1988, marked with the terse plaque "Alexandre de Humboldt (1769–1859) *Naturaliste et voyageur allemand* [German naturalist and traveler]." Not a description that would seem fitting for a national hero. I did discover in the Paris archives that there was once a pleasant Rue Humboldt, located a stone's throw from the prestigious *Observatoire de Paris* and parallel to the Boulevard François Arago, a mathematician who was a very close friend of Humboldt's for 40 years (Fig. 1.6). However, today this street is named after Jean Dolent, an obscure twentieth-century French writer and art critic. What happened? I continued to investigate and learned that the street name was likely changed in 1914, a period that saw the rise of strong anti-German sentiment in France, following the French defeat in the Franco–Prussian War of 1870. The desire for revenge against Germany, particularly to recover the 'lost provinces' of Alsace and Lorraine, and a concurrent push to 'de-Germanize' France probably explains Humboldt's disappearance from Paris, and from our collective memory. There is still a statue of Humboldt on the façade of the University Palace in Strasbourg, the main city in Alsace, the easternmost region of France, which belonged to Germany until the end of the First World War. Yet even this statue is forgotten now, disfigured by 130 years of erosion.[16]

---

[14] In Minguet (1969), p. 7.

[15] Humboldt was so called by George Ticknor (1822–66), an American academician specialized in languages and literature; see Walls (2009), p. 115.

[16] There is another statue of Humboldt near Paris, in Versailles. Shortly after receiving the news of Humboldt's death, French Minister of State Achille Fould (1800–67) proposed that Emperor Napoleon III "honors the memory of M. von Humboldt and resolve that his statue be placed in the Gallery of Versailles." On 9 May 1859, a 1.8-m marble statue was commissioned for 12,000 francs to be made by French sculptor Augustin-Alexandre Dumont (1801–84), who knew Humboldt well. At the request of the government, the statue remained in the artist's studio until 1884, when the executor of Dumont's will brought it to the Palace of Versailles where it can still be seen

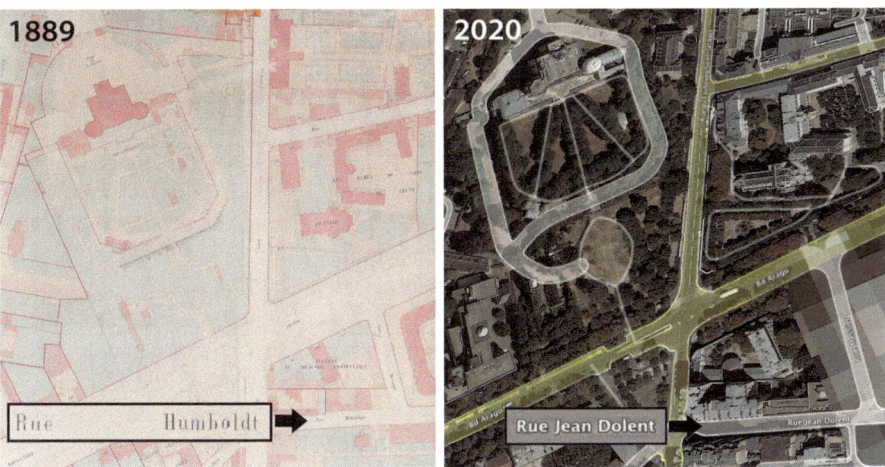

Fig. 1.6 Few traces of Humboldt left in Paris. The most famous scientist of his time, Humboldt lived for about 30 years in Paris, where a street was named after him near the Observatoire de Paris. Perpendicular to the Rue du Faubourg Saint Jacques, the street is still visible on a map of 1889, but today (2020) has become Rue Jean Dolent, a poorly known French writer and art critic who died in 1909. Illustrations: 1889, Archives de Paris, PP/11859/E and PP/11859/F, 2020, Map data: Google

In fact, French scholars never nurtured the kind of admiration for Humboldt that was to be found in the German, Hispanic and Anglo-American world. The same can be said of Fabre as compared to his popularity in Russia, China and, most of all, in Japan, where he is something of a cult figure. The reason for this may date back to the beginning of the seventeenth century and the profound influence that the great philosopher René Descartes had on French thought.[17] In the Cartesian tradition, emotions and perceptions of reality are thought to be the source of untruth and illusions. In contrast, Humboldt and Fabre were empirical thinkers who emphasized sensory experience as the source of knowledge about the natural world. Driven by a restless, never-ending curiosity for everything they came across, they were the epitome of naturalists rather than scientists. They combined right brain and left brain modes of thinking despite a surrounding scientific community that was increasingly cold and discipline-based and more inclined to rule on what was interesting to study or not. The more I learned about Humboldt and Fabre, the more their approach to scientific research fascinated me. Yet as their memory had vanished from most streets, universities and minds in France, I would need to find the spirit of my two muses roaming around somewhere else. And, by chance, I found the place.

---

today (although somewhat hidden in the lower gallery). See Nelken (1980), p. 40; Blankenstein (2014), p. 168.

[17] For an analysis of the special characteristics of French thought, see Hazareesingh (2015).

1.3   Ithaca

The Bombardier Dash 80 was buffeted by a terrible gale. The small turboprop plane convulsed and jerked to the side. It was noisy, bumpy, shaken by air pockets: everything a passenger doesn't want. I had faced unsettling turbulence before, flying over the Andes or approaching windy cities like Quito, Loja, Uyuni or La Paz. But the intensity of the storm combined with the darkness surrounding the matchbox-sized plane made the situation quite intimidating. My wife was holding back panic, although our two boys were peacefully sleeping, exhausted by over 10 h of travel since we left Quito. It was July 2017, and after eleven years doing research in the tropical Andes, my role in the field study had come to an end. Instead of flying straight back to France, I decided to stay on this side of the Atlantic for a sabbatical at Cornell University. I was not sure what I aimed to achieve during my stay there, yet, unconsciously, I was continuing to link my vocation of research to the pleasure of travel, of movement, of the unexpected. While tensely gripping my seat, I remembered that more than 200 years earlier, in May 1804, Humboldt's ship, the *Concepción*, braved a terrible storm in the strait between Cuba and the Bahamas. At the end of his stay in South America, before returning to Europe, Humboldt too made a stop in the United States. On the way, his ship was hit so violently by a storm that he worried for his life and that of his shipmates, as well as for his maps, manuscripts and collected specimens, representing, as he wrote, *"all the fruits of my labor."*[18] While I had traveled to South America without knowing Humboldt, this time I was following in his footsteps, and the storm seemed to confirm I was heading to the right destination. And so we arrived in Ithaca.

* * *

It may be no coincidence that **Aaron Sachs**, a history professor at Cornell and a specialist in American environmental and cultural history, chose the Café De Witt in downtown Ithaca for our breakfast meeting. **Simeon De Witt** was an American geographer who drew some of the first maps of the area. In 1792, he made perhaps the first attempt to chart climate patterns and their effects on agriculture and vegetation,[19] a subject that Humboldt would later conceptualize with his famous *Tableau Physique*. There is no evidence that Humboldt ever read De Witt's publication, nor that Aaron chose the café as a tribute to past geographers—as the Café De Witt is known for serving the best locally sourced, organic breakfasts in Ithaca, this was likely the reason for Aaron's choice. I had arranged to meet him, as a few years before I had voraciously read his book *The Humboldt Current,* an account of the lives and work of four nineteenth-century North American travelers who followed Humboldt's ideals. The book argues that Humboldt was a pervasive influence on American cultural perspectives on both nature and science, so over breakfast we discussed the difference between Humboldt's legacy in the United States and in

---

[18]Humboldt (2003), p. 397.

[19]See De Witt (1802).

France. While Humboldt visited the United States only once, for about a month, in 1869 a centennial commemoration was proclaimed across the entire North American continent to mark the 100th anniversary of his birth. In contrast, Humboldt lived in Paris for 30 years, but in the same year, only a few supporters from Paris's German community marked the occasion, which fell just 10 years after his death. At the time the French were busy celebrating the centennial of Napoleon's birth instead. Unlike France, no country has named more places in Humboldt's honor than the United States. However, even in the United States Humboldt disappeared from collective memory a few decades after his death;[20] it is thanks to the work of modern historians like Aaron that his importance is once again gaining recognition.

Aaron and I ranged over a variety of topics: the links between art and science, tips for writing books, spots to photograph owls, why storms are so frequent in the Ithaca region, and how delicious Café De Witt's omelets are. Just before parting, Aaron asked me why I had chosen Cornell for my sabbatical. I clumsily tried to form a sensible response with my main areas of interest, 'natural history,' 'ecology,' 'entomology,' 'sustainability science,' 'climate change,' 'tropical Andes,' 'Humboldt,' 'Fabre,' but it just showed how unfocused things were in my head. Aaron reminded me that Humboldt had much clearer ideas. He came to the United States to meet President Thomas Jefferson, who was also at that time President of the American Philosophical Society, an eminent organization of the finest minds in the arts and sciences in America. I answered that, though my reasons for coming to the United States may be unclear, it was certainly not to meet then president, Donald Trump. Aaron affirmed that neither would have Humboldt. While I was still unsure about the goal of my sabbatical, our discussion reassured me that Humboldt's spirit was at the root of American environmentalism, and perhaps more present in American academia than in France. Particularly at Cornell, a university renowned for its long legacy in natural history and ecology. With Humboldt onboard, would I also get the company of Jean-Henri Fabre's ghost for my sabbatical?

* * *

My quest for Fabre at Cornell started by encountering another Jean-Henri, **John Henry Comstock**. I discovered his portrait in the entrance of the Entomology building, Comstock Hall, as I was heading to my first Jugatae Entomology seminar.[21] Comstock began teaching entomology as a sophomore in 1871, a few years after Cornell was founded. At a time when entomology barely existed as a discipline in the United States, Comstock created the university's Entomology Department; his research, reviews, innovations in pest management, and outreach books would leave

---

[20] A thoughtful analysis of the reasons why Humboldt dropped out of public consciousness in the US is given by Nichols (2006).

[21] In the early 1900s, Comstock and his wife, Anna Botsford Comstock, launched the Cornell Jugatae Club, an entomological club that welcomed both faculty and students. John Henry Comstock divided Lepidoptera into two suborders, Jugatae (moths) and Frenatae (butterflies), each with different wing morphology. While Lepidoptera classification has undergone many changes since Comstock, Jugatae is still the name for Cornell's entomology seminars.

an indelible mark on American entomology. Comstock and Fabre were contemporaries and, despite contrasting professional environments, had much in common. They both started out as penniless self-made entomologists and were both gifted educators dedicated to raising public interest in entomology and the study of nature. Both also received letters from Darwin and made entomological discovery a family business. While Fabre enlisted his son Paul for photography, Comstock worked closely with his wife **Anna Botsford Comstock**, an outstanding scientific illustrator, educator and writer of popular science books. In 1897, the Comstocks produced a popular book entitled *Insect Life*, which, like Fabre's *Souvenirs Entomologiques* at the same period, paved the way to a new appreciation of insects and their kin. On two different continents, as if linked by an invisible spiritual thread, Comstock and Fabre independently engendered convergent ideas and approaches in entomology.

But Comstock is not the only link at Cornell that summons up Fabre's pioneering vision and methodological approach to entomology. Between 1957 and 2011, a *"modern Fabre,"* as his friend Edward O. Wilson, the Harvard biologist, called him,[22] enlightened the Cornell Entomology Department and later the Neurobiology and Behavior Department. This man was **Thomas Eisner**, a brilliant, visionary biologist whose scientific achievements made him an entomological legend. Fabre (born in rural France in 1823) and Eisner (born in Berlin in 1929) had very different lots in life. Fabre's parents were modest plowmen; he spent much of his life without a *franc*, in various rural locations in southern France. Eisner's father was a chemist, a colleague of the Nobel Prize-winning Fritz Haber at the Kaiser Wilhelm Institute of Electrochemistry, and his mother was an artist. At Cornell, one of the top universities for insect studies, Eisner was surrounded by many cooperative researchers and had access to substantial grant money. In his lifetime, Fabre published some 100 popular science books and school textbooks on entomology and other scientific fields. Eisner published more than 200 peer-reviewed papers: 44 in the Proceedings of the National Academy of Sciences, 2 in *Nature* and 43 in the journal *Science*—for which he holds the record number of covers![23] But beyond their distinct times and places, an Eisner–Fabre connection is clear. Like Fabre, Eisner was fascinated by the complexity and beauty of insects and spiders, dedicating his life to deciphering the mysteries of their behavior, in particular their defensive strategies against predators. Like Fabre, Eisner's discoveries were invariably triggered by sharp-eyed observations during outdoor walks. Like Fabre, his approach merged detailed descriptions of elusive behaviors, the experimental study of mechanisms, and detours into the fields of chemistry and physics. Like Fabre, Eisner was a wonderful storyteller, combining sparkling prose with captivating images. His book *For Love of Insects*, a first-person account of his groundbreaking discoveries, is a kind of modernist, rocket-science version of Fabre's *Souvenirs Entomologiques*; both share the same curiosity for nature and enthusiasm for discovery. While describing his personal odyssey to escape the Nazi regime (the first step of which was his family moving from Germany

---

[22] Eisner (2003), p. ix.

[23] For articles on Eisner's life and legacy, see Steele (1987), Baldwin (2004), Raguso (2011).

to Barcelona in 1933 when he was four[24]) Eisner said, *"The outbreak of the Spanish Civil War in 1936 forced us to flee to France, where we briefly established residence in Paris. I learned French, but hardly anything new about insects."*[25] It is regrettable that during his stay he did not come across a French edition of Fabre's *Souvenirs Entomologiques*, though he would discover it later.[26]

Aside from his stature as an entomologist, I was fascinated by Eisner the naturalist. He published several articles in which he described the importance of contemporary natural history to modern science and to nature conservation.[27] In Eisner's time, the vertiginous rise of molecular biology and theoretical ecology meant natural history was falling out of vogue among scientists. Yet with his riveting insect stories published in top journals, Eisner ensured natural history had a prominent place in the scientific arena. An avid photographer, toward the end of his life, diminished by Parkinson's disease, Eisner traded his camera for ... a color copier! Using this, he created artistic works of a new kind, capturing his original vision of nature.[28] Of these, Eisner's arrangement of shells that compose '*The naturalist*' is truly beautiful (Fig. 1.7). In it, a Lake Titicaca-frog-like creature is pursued by a naturalist wearing a Victorian top hat. I can't help but see it as Humboldt! I like the idea of linking Humboldt with Eisner. Both had observed a myriad of strange creatures in the field, returned home and wrote about science in ways that inspired future generations to take up the quest; an approach that would inspire my research philosophy.

The spirits of former field naturalists are very much alive at Cornell, embodied in the unifying philosophy of a new generation of entomologists, ecologists and evolutionary biologists. The Ecology and Evolutionary Biology faculty still offers field ecology courses with a similar structure to the 'Exploration, Discovery and Follow-up' courses initiated by Eisner, in which students make discoveries in the field and then evaluate their research potential by following up with field or laboratory experiments. During my meetings, seminars and hallway chats with bird, coral, fish, frog, reptile and insect specialists at Cornell, I could see how expertise in organismal biology was used to address fundamental questions in

---

[24]With unrest in Europe spreading, the Eisner family left France in 1937, emigrating first to Argentina and then to Uruguay, where they spent the next ten years. It is in this *"entomological paradise"* that Eisner nurtured his interest in natural history and insects. In 1947, the Eisners moved to the United States.

[25]For Love of Insects by Thomas Eisner, Cambridge, Mass.: The Belknap Press of Harvard University Press, Copyright © 2003 by the President and Fellows of Harvard College. Used by permission. All rights reserved.

[26]I wrote to Maria Eisner, Thomas Eisner's widow, to ask whether he had known of Fabre. She answered that Thomas Eisner was fluent in French, *"had a French edition of Fabre's* Souvenirs Entomologiques *(from 1924) and regarded Fabre highly."* p. 3.

[27]See, for example, Eisner (1982).

[28]Other natural compositions by Eisner using a color copier can be seen in Gorman, J. (2006). *Eye-Catching Images of Nature, Made with a Common Machine*, The New York Times, October 10, 2006. https://www.nytimes.com/2006/10/10/science/10eisn.html

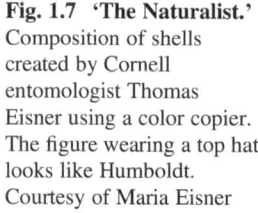

Fig. 1.7 'The Naturalist.' Composition of shells created by Cornell entomologist Thomas Eisner using a color copier. The figure wearing a top hat looks like Humboldt. Courtesy of Maria Eisner

ecology and evolution. Maybe one reason for this is that Cornell lies in the midst of forests, hills, lakes and rivers, making it possible to quickly switch, physically or mentally, from your computer to nature.[29] One day, I spotted a Cooper's hawk from my office window and after that kept my camera and tripod right next to my desk. During our weekly lab meetings in the Lamont Cole room, a framed image of a scholar gazed out at us. This was an announcement of a symposium in honor of **Richard "Dick" Root**, an influential Cornellian ecologist and entomologist and a super field naturalist.[30] His teaching model was to encourage students to first observe in the field and then to use those observations to initiate further inquiry. The title of the homage sounds like a credo: "*Let nature tell you what's interesting.*" Humboldt, Fabre, Comstock and Eisner would certainly have agreed.

1.4   Connections

As a naturalist, I think I am a reasonably good observer. Yet it was only two months after I started my sabbatical at Cornell that I really noticed the imposing, vibrantly colored map hanging in the atrium of my office building, Corson Hall. It took some effort to recognize the place on the map. A place where I had nonetheless spent a quarter of my life: the South American Andes. The confusing orientation was the reason for my obliviousness. The 6-m-long map was horizontal, with the east at the

---

[29] During a discussion with Cornell ornithologist David W. Winkler, he argued that another reason why natural history and ecology are still so strong at Cornell may be Cornell medical school's move to New York City in the 1950s. This helped give prominence to organismal biology and the science of plants and animals at the Ithaca campus.

[30] See the biography of Dick Root by Levin et al. (2013).

Fig. 1.8 'Andes.' A work by the artist Jay Hart in the atrium of Corson Hall at Cornell University (part of the 'Earth Pattern' exhibition). The painting combines scientific information (physical geography) and artistic composition (colors, orientation and graphics). Photo by Olivier Dangles

top (Fig. 1.8). Was this a reference to medieval times, when European cartographers oriented maps toward the east, where the sun rose and Paradise was thought to lie? Or was it because the Andean cordillera is so long—at 7000 km, the longest mountain range on Earth—that only a horizontal representation would allow its details to be fully appreciated? Or because **Jay Hart**, the local artist who created the map, wanted to challenge the observer? Equally confusing were the map's colors. The cordillera looked like a giant layer cake: grey-green for low-altitude Andean forests and grasslands, grey for rocks, white for salt pans, and purple icing the summits. Purple? At altitudes above 4800 m, purple represented the oxygen-deficient zone. In this map, art met data.

Schooled in image science, geomorphology, cartography and plant ecology, Jay Hart is a purveyor of 'terrain art,' through which he reveals the beauty of Earth's landforms using a combination of precise elevation mapping and coloring. Following the path forged by the pioneering American eco-artists Helen and Newton Harrison, he reworks maps by digitally altering, unlabelling, painting over and reorienting the original images so that familiar landmarks are upended. His belief is that "landscape is a major muse for humans' higher ambitions" and that creative cartography is a privileged form of communication to connect viewers' personal experiences with the world. Jay's art visualizes how mountain landscapes, from a distance, are a subject for both the scientist and the artist, giving complementary perspectives of the same reality. In me, Jay's artistic work triggered a psychological zoom effect, connecting my personal experiences to something much bigger. On the map, I could identify the position of Mt Antisana, where I had gathered scientific data and natural history observations that I could link, through other scientists, to different places and times: Humboldt and Fabre, Comstock and Eisner, the Andes, Aveyron, Paris and Cornell. It was there, facing this map, that I realized that many of my experiences were connected to the Andes. While I was physically in one place, my imagination could roam in many places and many epochs at once.

* * *

The tropical Andes lay at the confluence of the historical figures, disciplines and geographical areas that have shaped my hybrid approach to research, and brought into play both left-brain and right-brain modes of responding to the world. In this natural outdoor laboratory that ranges 25° in latitude, from the equator to the southern tip of Bolivia, I have tried over the last 15 years to uncover how nature in the tropical Andes is facing the challenge created by rapidly warming conditions. Tropical mountains, of which 90% lie in the Andes, are one of the most sensitive regions to climate change, which is having already observable consequences on temperature extremes, glaciers, the water cycle, species distribution, and the resilience of local livelihoods.

In this book I present recent scientific advances on the ecology of warming mountains and their wild inhabitants, both plants and animals, on land and in water. While the book focuses on the tropical Andes, it weaves in related information from other parts of the world: other mountain ranges, ocean islands and continents, where key discoveries—including those of past naturalists—have laid the foundations for climate change science. It incorporates approaches from other disciplines, from different branches of science, from history and the arts to offer a more comprehensive approach. Hard data builds our understanding of the world, but appealing to the emotions makes us want to understand it. In the spirit of Humboldt and other naturalists of his time and since, the book makes the argument for the importance of storytelling and bridging science with the arts and humanities to raise awareness and concern about climate change. A way of doing science that speaks to our humanity as well as increases our knowledge about nature.

This book touches on my own development as a scientist, interweaving this with historical accounts about naturalists and their findings that are relevant to my research approach and to the central theme of understanding mountains in a warming world. The narrative moves in space and time, from the present to the past, from continent to continent, from laboratory to field, from archives to mathematical models, from behind the camera to in indigenous communities. The intertwined histories of past and modern naturalists reveal that many recent breakthroughs in climate change ecology were preceded by the kind of deep observational studies that have fallen out of fashion in the past half-century. The development of natural history created a common language and methodology whereby scientists with different expertise could compare observations, formulate patterns and begin to make predictions. The broad, eclectic approach of nineteenth-century naturalists may appear old-fashioned and unadapted to the reality of modern scientific research.

However, in an era of big data, resulting in a plethora of predictions at global scales, the book makes the argument for different types of thinking. I argue that a naturalist approach, which sees the world from the subject's point of view, is crucial to a true understanding of the ecological consequences of global warming. Scientific specialization in the twentieth and twenty-first centuries has sidelined the quest for connection and integration common in Humboldt's time. And yet the spirits of past naturalists are found in many of the facts, techniques and perspectives that form the basis of state-of-the-art climate change ecology. Indeed, many of these early discoveries are even more relevant today than they were in their own time.

* * *

Each section in this book takes a different angle to understand the impact of climate change in the tropical Andes. Chapter 2, *Islands*, focuses on some of the early naturalists' most fascinating and strange subjects of study, creatures such as the dodo and giant tortoise, which were fertile ground for the most influential theories in ecology and evolution. Yet ecologically speaking, islands do not have to be surrounded by water, but are any habitats separated from other similar habitats. Seen from above, the hundreds of high-altitude peaks distributed along the spine of South America resemble islands of an archipelago. Today, these islands in the sky are embedded in a sea of crops, pastures, roads and villages, which are inevitably rising upwards, thus shrinking the available habitat for wildlife. The island concept does not only apply at the scale of isolated mountain tops. Ecosystem 'islands' appear repeatedly at different spatial levels in the Andean landscape, from oasis-like wetlands fed by glaciers to plants that act as islands on a micro-scale. I use this island approach to examine the vulnerability to climate change of three key life forms in the high Andes: ground beetles, birds and cushion plants.

Chapter 3, *Time capsules*, takes a deep dive into the central theme of this book to explain how scientists can predict the ecological effects of climate change. For this, we basically need a time machine that allows us to assess the ecological situation in these mountains hundreds of years from now. The best we can do to achieve this is to take a voyage back in time in one of two ways. The first option is to truly travel back in time by analysing historical data that may have hidden ecological relevance. Humboldt's *Tableau Physique* (1807) is one of the most tempting historical documents in our case, as it is by far the oldest existing dataset on altitudinal ranges of tropical mountain vegetation, representing a unique data source to assess vegetation shifts in response to climate change. Working with a historian colleague to uncover the secrets of Humboldt's diagram, we were able to avoid the traps previous ecologists had fallen into and quantify the altitudinal range shift of high-altitude Andean vegetation over the last two centuries. The second option is to substitute time with space. Near a glacier, knowledge collected by glaciologists allows estimating the distance from the terminus of a glacier as a proxy for site age, which ecologists can use to observe how plant communities were structured over time.

Chapter 4, *Underwater flies*, zooms in further on the book's central theme with a focus on my specific research for more than a decade: the ecology of high-altitude waters. It discusses the amazing adaptations of aquatic invertebrates, fish, frogs and other mountain organisms and describes observational and experimental studies that have tried to decipher the respective role of physical factors such as water temperature, oxygen and turbidity on their survival. I describe the high-tech and low-tech approaches and methodologies we adopted or invented to deal with this, from thermal cameras and the mathematical technique of the Fourier transform, to a rock-piling intervention. These studies allowed a new understanding of how aquatic systems respond to melting ice: glaciers not only provide the water that makes aquatic life possible on mountain peaks, they also give rise to a mosaic of unique environmental conditions, detectable kilometres downstream from the glacier.

Chapter 5, *Telling stories*, starts with the observation that while climate change affects everyone, it is often difficult for scientists to transmit their knowledge in ways that speak to the public. I argue that we have much to learn from the past naturalists who were excellent storytellers. Although this aspect lies outside of 'hard' science, natural science is also based on the senses, on perception, on emotions that have the potential to make a profound impact on science communication: few have read Audubon's writing on birds, but his images speak volumes. Based on two personal experiences—an expedition to the top of Mt Chimborazo and a photographic study of the Andean bear—I argue that reviving the approaches of past naturalists to communicate scientific knowledge can be effective in conveying the reality of climate change to a broad audience.

In this sense, the conclusion argues that adopting a more naturalist philosophy as was practiced in the eighteenth century might help academia to better address the entrenched obstacles to achieving sustainable development goals. A more truly transdisciplinary and multicultural approach that cross-pollinates science, the humanities and the arts may be our best hope for tackling the urgent challenges we face.

* * *

The journey of writing this book allowed me to discover Humboldt and his way of connecting living things, places and people. The more I got to know Humboldt, the more I felt he held the key to a more holistic understanding of the ecological effects of climate change. The book is not an attempt to follow in his footsteps in wonder at the places and landscapes where he traveled, but ultimately a call to embrace Humboldt's philosophy in scientific thinking and try to reveal the interconnectedness of our world. I discovered Humboldt's "*subterranean network, clandestine links*" running through the places that had been crucial to my own research: France, the Andes and the northeastern United States. Humboldt was an inspiration in making these connections and in joining the triangle that forms my personal cosmos. Let's start the journey.

Chapter 2
Islands

The snowy, rounded summit of the mountain stands out like an island in the middle of the plain.

Alexander von Humboldt (1854, p. 25)

The new main protagonists of this chapter are shown in Fig. 2.1.

2.1 Sea Islands

I had never looked closely at a map of New Caledonia until I prepared a trip there in late March 2016. I had been invited to sit on a PhD committee and give a series of seminars on my work in the tropical Andes—geographically distant, but in fact mountaintop and island ecologists have much to share. Annexed by France in 1853, New Caledonia hosts the second-largest reef system in the world after Australia's Great Barrier Reef, which lies some 1200 km to the west. The crystal-clear lagoons surrounding the main island, Grande Terre, are the outstanding feature in a satellite map of New Caledonia, yet my attention was focused on the mountain range running the 350-km length of the interior. Reading the names and altitudes of the mountains, I spotted an unexpected name in a French territory: Mt Humboldt, 1618 m. While Humboldt had been erased from the collective memory of my country, I was delighted that there was still one major 'French' place commemorating him. The island was being explored when Humboldt died in 1859, and the mountain then considered its highest peak was named after him. A few years later, when things named after Humboldt were being renamed throughout mainland France, New Caledonia's remoteness probably preserved this tribute. A few kilometers north of Mt Humboldt, another mountain, Mt Arago, was named after the mathematician François Arago, one of the dearest friends of Humboldt. The two had also been commemorated in streets in close proximity in Paris until Humboldt's was renamed (see Fig. 1.6). It is consoling to know that the two scientists, who shared a common vision regarding the importance of knowledge for social progress, remain side by side in the mountains of New Caledonia.

O. Dangles, *Climate Change on Mountains*,
https://doi.org/10.1007/978-3-031-39528-4_2

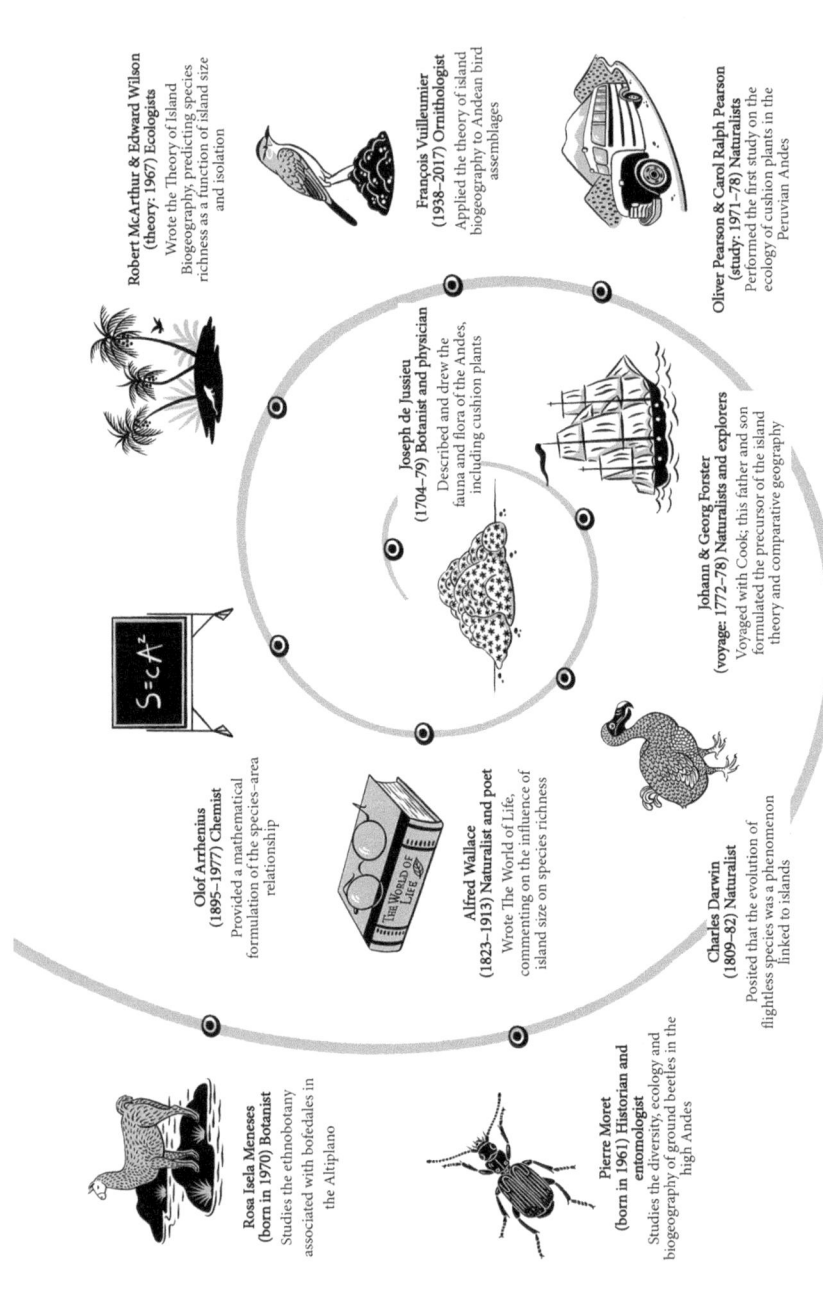

Fig. 2.1   Central figures in this chapter. This spiral timeline shows ten historical or contemporary figures who contributed important ideas mentioned in this section. Drawings by Paula Terán Ospina

Robert McArthur & Edward Wilson (theory: 1967) Ecologists
Wrote the Theory of Island Biogeography, predicting species richness as a function of island size and isolation

François Vuilleumier (1938–2017) Ornithologist
Applied the theory of island biogeography to Andean bird assemblages

Oliver Pearson & Carol Ralph Pearson (study: 1971–78) Naturalists
Performed the first study on the ecology of cushion plants in the Peruvian Andes

Joseph de Jussieu (1704–79) Botanist and physician
Described and drew the fauna and flora of the Andes, including cushion plants

Johann & Georg Forster (voyage: 1772–78) Naturalists and explorers
Voyaged with Cook; this father and son formulated the precursor of the island theory and comparative geography

Olof Arrhenius (1895–1977) Chemist
Provided a mathematical formulation of the species–area relationship

$S = cA^z$

Alfred Wallace (1823–1913) Naturalist and poet
Wrote The World of Life, commenting on the influence of island size on species richness

Charles Darwin (1809–82) Naturalist
Posited that the evolution of flightless species was a phenomenon linked to islands

Rosa Isela Meneses (born in 1970) Botanist
Studies the ethnobotany associated with bofedales in the Altiplano

Pierre Moret (born in 1961) Historian and entomologist
Studies the diversity, ecology and biogeography of ground beetles in the high Andes

Yet the connection between Humboldt and the Pacific Islands is not limited to a toponym in New Caledonia. Humboldt grew up during the time of the major overseas discovery expeditions of Cook, Commerson, Poivre, and Bernardin de Saint Pierre, and the literature that emerged around these adventures fascinated him from an early age. He was particularly enthralled by *A Voyage Round the World*, published in 1777 by the German traveler **Georg Forster** to whom Humboldt was "indebted for the lively interest which prompted [him] to undertake distant travels." [1] In 1772, at the age of 17, Forster joined his father, who had been appointed ship's scientist on board the HMS Resolution, to accompany James Cook on his second journey around the world in search of the great southern hemisphere *Terra Australis Incognita*, a hypothetical land appearing on maps in antiquity and based on the theory that a southern landmass must exist to 'balance out' known lands in the north. While this mythical continent eluded Cook,[2] his voyage through innumerable Pacific Islands (among them New Caledonia, which he was the first European to discover) yielded a wealth of new geographic, scientific, and ethnographic knowledge.

The young Forster compiled this vast knowledge in a highly readable book that combined scientific observations, paintings and travelogue. This bestseller would greatly influence Humboldt's passion for the visual arts and inspire his romantic view of tropical regions, but it also had an even more transformative impact. As Forster had visited so many different islands with distinct latitudes, sizes and degrees of isolation, *A Voyage Round the World* laid the foundation of comparative ethnology and regional geography. For Humboldt, this book would initiate *"a new era of scientific traveling, having for its object the comparative knowledge of nations and of nature in different parts of the Earth's surface."*[3] Humboldt and Forster became friends and traveled together across Western Europe in 1790. Humboldt's determination to travel grew—he hoped to do for continental interiors what Forster had done for islands.[4] To some extent, the research outlook Humboldt adopted during his travels to the tropical Andes originated in Foster's works in Oceania.

Today, we associate **Charles Darwin** with the Galápagos and **Alfred Wallace** with the Malay Archipelago. Humboldt's legacy is not linked to any specific islands,

---

[1] Humboldt (1846–1858), vol 1, p. 317.

[2] At the end of February 1773, Cook (1728–79) almost encountered mainland Antarctica but, facing the Antarctic winter, he was finally forced to sail north to Tahiti and resupply his ship. On 30 January 1774, the Resolution reached 71°S latitude. No European had come this close to the South Pole before. But an immense ice field kept him from continuing.

[3] In Humboldt (1846–1858), vol. 2, p. 70.

[4] Humboldt argued: "I have enjoyed one advantage which few scientific travelers have shared to an equal degree, in having seen not merely coasts, and districts little removed from the edge of the ocean, as in voyages of circumnavigation, but in having, moreover, traversed, both in the new and the old world, extensive continental districts presenting the most striking contrasts; [...] Such opportunities could not fail to encourage the tendencies of a mind predisposed to generalization, and were well fitted to animate me to the attempt of treating in a special work our present knowledge of the sidereal and terrestrial phenomena of the universe in their empirical connection." In Humboldt (1846–1858), vol. 1, pp. xviii–xix

although he visited a few, such as Cuba and the Canaries.[5] But what are islands exactly? Can any environments spaced far apart from each other be considered islands, whether separated by land or sea? Certainly. Cool alpine regions are, in fact, essentially high-altitude islands separated from the surrounding lowland regions by a physical barrier that prevents the dispersion of animal and plants between the two. Climbing many mountains during his travels, Humboldt, like ocean island naturalists, would also be confronted with questions related to landscape isolation, specificities of fauna and flora in specific places, and differences in plant communities between summits. Although his approach lacked an evolutionary perspective, Humboldt's Forster-inspired comparative geography was fundamental in creating the conditions for Darwin and Wallace to advance evolutionary science. To better understand the effects of warming on life in mountains, we need to look more closely at the island-like configuration of the Andes.

2.2   Sky Islands

Faults and fire. The Andes, the backbone of South America, is the youngest, longest, and most geologically active mountain range on Earth: forced up by the collision of tectonic plates, it started to rise only 10 million years ago[6]—a time when humans began to split off from great apes. The range extends from the warm white sand shores of the Caribbean to the icy storm-tossed waters of Cape Horn on the doorstep of Antarctica. It is shot through with a dense fractal network of active seismic faults and over 150 active volcanoes, part of the Pacific Ring of Fire, testaments that the chain is still in the full throes of formation. This reality became irrefutable as my son Matias was blowing out his seven birthday candles when the 7.8 Pedernales earthquake struck on 16 April 2016 at 7:00 pm. Although the epicenter was on the Pacific coast, over 150 km from our home in Quito, and occurred at a depth of 20 km, the 50 s of tremors seemed endless. At least 676 people were killed and 16,600 people injured. A few months earlier, on 14 August 2015, we had nervously observed the dense cloud of smoke and ash that suddenly erupted from Mt Cotopaxi, barely 70 km from Quito. No lava flowed from the peak and no landslides thundered down its flanks, still, the eruption made thousands of people flee the area around the volcano for safer ground. As Humboldt put it, daily life in Ecuador is intimately related to the

---

[5] Humboldt loved the Canary Islands and would happily have stayed there for months, but the captain argued that after four or five days they had to move on (Walls 2009). As pointed out by Williams (2015), naturalists' interests were frequently at odds with the commanders of discovery expeditions. On terrestrial islands, Humboldt would have had more freedom in making his own schedule.

[6] While the mountain-building process of the Andes started some 70 million years ago, it is thought that the cordillera as we see it today started rising in the second half of that period, in particular during the last 10 million years, though this age varies depending on what part of the cordillera is considered.

rumblings of the Earth's crust, "which alternately falls asleep and wakes up."[7] Local people are "accustomed to sleeping in peace on the brink of catastrophe"[8] while travelers become "almost as familiar with the somewhat abrupt movements of the ground as we are, in Europe, with the sound of thunder."[9]

At the heart of this highly active range, the tropical Andes stretch from 11° North to 23° South—the largest cool region in the Earth's tropics, covering more than 1.7 million km², an area about the size of Alaska (see Fig. 1.8). In elevation, its peaks range from 1000 m to over 6700 m (Mt Huascarán, Peru). From the equator looking northward to Colombia and Venezuela, the cordillera runs roughly parallel to the Pacific Ocean and then swings eastward, while branching into three distinct ranges separated by deep valleys. It ends with the colossal 5710-m Santa Marta in Colombia, which stands alone just 40 km from the Caribbean coast. Further south, in Peru and Bolivia, both the shape and the climate of the tropical Andes change dramatically. Here, a vast, dry, high-mountain plain, the **Altiplano**, is bordered on every side by mountains such as the Cordillera Real and volcanoes such as Mt Sajama, which stands more than 3400 m above sea level. The Altiplano once held a vast inland sea, which turned to lakes, which were then transformed by the gradually drying climate into a series of saltpans. From a bird's eye view, the area appears as a massive uninterrupted mountain range stretching on either side of the equatorial line. There are no islands in sight.

In order to see an island pattern emerge from the Andes, it is necessary to draw contour lines around increasingly higher altitudes. If you do this with the Andean cordillera in Ecuador (Fig. 2.2), at 3000 m the main cordillera appears as one unit, with the two parallel chains barely differentiated. At 4000 m two cordilleras appear, separated by a 80–160-km-wide gap, which is a fertile central valley lying at about 2800 m. At 4200 m an archipelago of sky islands, or as Humboldt more poetically wrote "islands spread in the atmospheric ocean,"[10] clearly appears: a range of high-elevation habitats, geographically isolated within different mountain ranges, with a variety of shapes and forms. At 4400 m, each of these habitats contracts to a specific volcano, whose island-like shape is probably best represented by Mt. Cotopaxi. Each volcano has its own geological history and climate, which in turn determines its own set of environmental characteristics—its unique fingerprint: area, topographic ruggedness and profile, temperature gradient, precipitation pattern, rock and soil type, erosion and runoff processes, volcanic activity, etc. Yet while every Andean sky

---

[7] Humboldt (1854), p. 47.

[8] Cited in Terra (1955), p. 125. The original text is from a letter in French that Humboldt wrote to his brother in Lima, on 25 November 1802 (Humboldt 1803, p. 326): "C'est ainsi que l'homme s'accoutume à s'endormir paisiblement sur le bord d'un précipice." Terra's translation of the words "on the edge of a precipice" as "on the brink of catastrophe" captures the diversity of dangers (volcanic and seismologic) faced by Quito inhabitants, which are described by Humboldt in his letter.

[9] Humboldt (1852–1853), vol 1, p. 349.

[10] Humboldt (1854), p. 119.

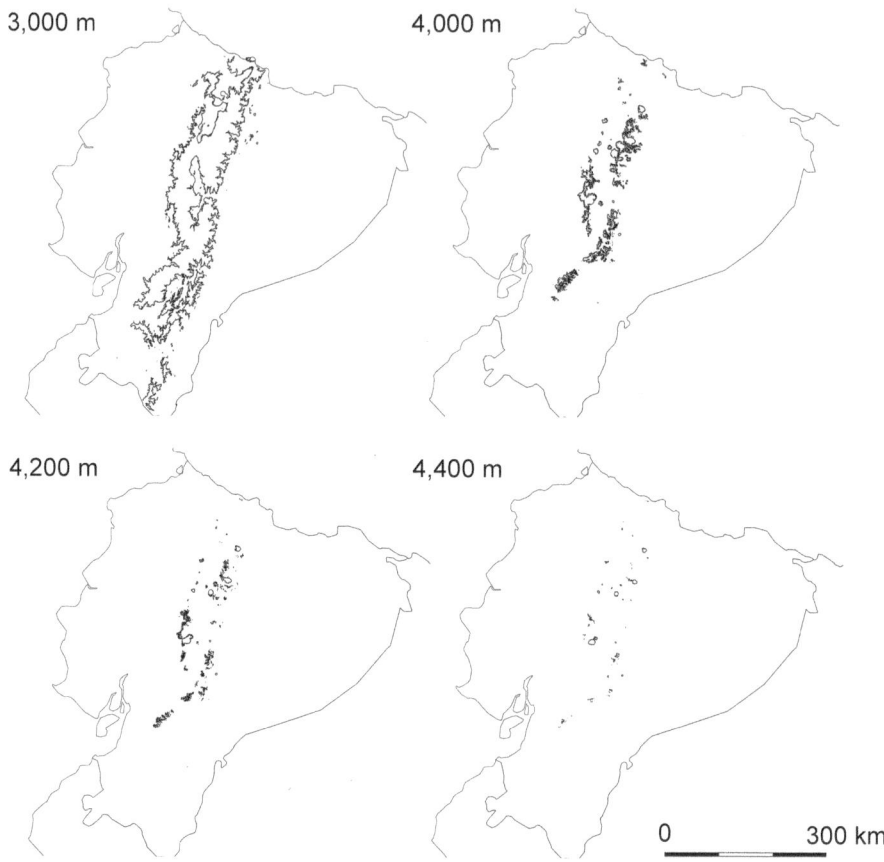

Fig. 2.2 Andean sky islands. Contour lines drawn around increasingly higher altitudes (from 3000 to 4400 m) in Ecuador. While at 3000 m the Andean cordillera appears as a continuous mountain range, increasingly isolated islands of páramos progressively appear at higher altitudes. The interconnectedness between the islands had been dynamic over the last millennia, following climate oscillations. Illustration by Olivier Dangles

island is unique, in Ecuador, Colombia and Venezuela, most host the same singular ecosystem, the páramo.

The Andean páramo is one of the most scenic landscapes on Earth, with green carpets of spongy cushion plants punctuated by colorful shrubs, run through by meandering streams and scattered with glaciers (Fig. 1.2). They are clearly demarcated from lower-elevation habitats of montane forest remnants, crops, pastures, roads and villages. It is the relative spatial arrangement of these páramo sky islands within an archipelago that makes them most interesting for ecologists and evolutionary biologists. Analogous to the water between oceanic archipelagos, low-elevation habitats act as a barrier to dispersal between sky islands, facilitating the **divergence** of isolated plant and animal populations. Humboldt himself noticed the uniqueness of these ecosystems from both a scientific and aesthetic point of view: *"No zone of alpine vegetation in the temperate or cold parts of the globe can be compared with that of the Páramos in the tropical Andes."*[11] *"Nowhere, perhaps, can be found collected together, in so small a space, such beautiful compositions, and so remarkable in regard to the geography of plants."*[12] His delight in the patterns of plant diversity did not distract him from musing about the processes behind this:

> Alpine plants offer a curious example of species similarity, despite the great distance that separates the mountains. I have observed elsewhere than at the Silla de Caracas the same Befaria [a plant in the heather family] whose purple flowers adorn mountainsides in the kingdom of New Grenada. I will not ask how the seed of this beautiful plant came to this lofty peak, the only one in the entire coastal range that, because of its elevation, enjoys a climate cold enough to sustain the Befaria; I will not ask, because in good philosophy, the first origin of things can be neither a problem of history nor an object of research for a naturalist.[13]

Humboldt may not have dared ask, but we will.[14]

The diversity of life in specific mountain regions varies markedly across the globe. In the Arctic and temperate regions, mountains usually have a low number of species, slightly exceeding that of neighboring lowlands, and often do not host their own endemic species, i.e., those restricted to a specific location. In contrast, tropical mountains are unusually diverse: some, such as the tropical Andes, are exceptionally so. Created relatively recently, around 3–5 million years ago when the Andes underwent a major uplifting event, the páramo offered new habitats for plants and animals to flourish. Within only a few million years—a mere flap of a hummingbird's wing on the scale of species evolution—the páramo has been the setting for a dramatic diversification of lifeforms and is one of the fastest evolving and most densely diverse ecosystems in the world. For example, with more than 3400 endemic species of plants over an area only the size of Maryland, the average net diversification rate of páramo plant **lineages** is faster than any of the documented recent **speciation** (the formation of new and distinct species) processes in any other ecosystem[15]—faster even than the textbook examples of fruit flies and honeycreepers in the Hawaiian archipelago. Remarkably, speciation in the páramo has occurred more or less over the same period of time in multiple unrelated lineages, in contrast to, for example, the single rapid diversification of cichlid fishes that

---

[11] Humboldt (1854), p. 392.

[12] Humboldt (1852–1853), vol. 1, p. 422.

[13] Humboldt (1811–1833), vol. 1, p. 38. The genus name *Befaria* has since been changed to *Bejaria*.

[14] Humboldt was reluctant to venture into the field of evolutionary processes that led to the current state of ecosystems. While his global vision made him masterful in comparing different places at a specific point of time (isotherms, the correspondence between landscapes under similar conditions in very distant regions, etc.), he was less bold in comparing different epochs. His ideas would soon be marginalized by Darwin's theory of evolution, even though Darwin owed much to Humboldt for the development of his theory. See Gould (2000), Lansley (2018).

[15] Madriñán et al. (2013).

occurred in a restricted area in East African lakes. No doubt the island-like structure of the high tropical Andes is the key to explaining these evolutionary processes. But to better understand the complex history of high Andean species, many questions need to be answered. Are páramo species completely restricted to their sky islands, or can they disperse? Are they confined because of physiological inability to survive in the lowlands, or due to competition from species better adapted to lowland conditions? Did they reach their islands by dispersing directly from summit to summit, by trickling through the lowlands, by migrating over vanished bridges between páramos, or by evolving from lowland ancestors? One biologist had a particular interest in solving these enigmas, and has been a long-lasting influence on subsequent generations of sky island ecologists, including myself.

* * *

The son of a sculptor and amateur naturalist, **François Vuilleumier** was naturally curious about wildlife. Frustrated by the nocturnal habits of most wild mammals in Switzerland, his native country, and inspired by the Swiss ornithologist Paul Géroudet, with whom he spent hours in the field, his interest would come to focus on birds.[16] In 1963 he went to the United States to do his graduate studies at the University of Illinois, and 1 year later he made his first field trip to the Andes, where he started his research on Andean birds. Vuilleumier observed that páramos had a distinct bird community, with each patch supporting species distinct from those of lower-elevation habitats, soon leading him to conceptualize the páramos as an archipelago of sky islands. In 1963, the biologists **Robert MacArthur** and **Edward Wilson** had published their seminal paper 'An equilibrium theory of insular zoogeography' that established the conceptual foundation of the study of island biotas. Over the next 4 years, MacArthur and Wilson refined their biological arguments and mathematical proofs to publish, in 1967, *The Theory of Island Biogeography*, one of the most influential books in the history of ecology. Their theory built on the basic principles of population ecology and genetics to explain how isolation and land area combine to regulate a balance between immigration and extinction in island populations. The model was simple and, for an empirical ornithologist like Vuilleumier, attractive, as it offered several robust predictions that could be tested with field observations. Vuilleumier collected data on 180 high Andean bird species and found the patterns of speciation and distribution the equilibrium theory predicted: bird faunas tend to show a positive correlation with the area of a sky island, and an inverse correlation with the distance from this source. In this way, the páramo islands seem to behave like a true oceanic archipelago.

Exploring this issue occupied Vuilleumier for the rest of his career. He would conduct fieldwork throughout the Andes attempting to decipher the complex history of its bird fauna, using specific bird groups such as sierra finches and ground tyrants as study models (Fig. 2.3). In 1987, Vuilleumier summarized much of the knowledge he and other scholars had gained in *High Altitude Tropical Biogeography*, a

---

[16]For details on F. Vuilleumier's life, see Brush (2017).

Fig. 2.3 Sketches by an ornithologist. Four Andean species of ground tyrant (*Muscisaxicola* spp.) drawn by ornithologist François Vuilleumier. An avid illustrator of birds, F. Vuilleumier was influenced by Swiss naturalist–artists such as Robert Hainard and Paul Geroudet. These drawings were gifted to me by Rebecca Finnell, F. Vuilleumier's wife

book he co-edited with Maximina Monasterio, the Venezuelan pioneer in tropical mountain ecology research. This book, describing the geographical distribution patterns for a variety of plants, birds and insects and exploring the historical processes that explain these patterns, was what led me to view the Andean páramos as sky islands.

While field data on species distribution provides a snapshot of present-day conditions, the diversity of life on sky islands bears the signature of deep-time evolutionary and ecological processes. Digging for fossils, taking mud cores or looking at evolutionary information stored in DNA are powerful tools to understand the history of species. But another approach is to look at the distribution of endemic species. The level of **endemism** in an area can provide an indication of the age of flora and fauna and the relative importance of immigration and local speciation. When endemics evolve in a place, isolation is a strong contributing factor as this reduces immigration and gene flow. Isolated groups of species evolve into a wide variety of types adapted to specialized modes of life, a type of local speciation called **radiation**. In Vuilleumier's footsteps, a group of colleagues and I decided to analyze endemism patterns in birds, plants and ground beetles in the sky islands of the northern Andes.[17]

Using data we collected in the field as well as from the literature, we found that, depending on the location, between one- and two-thirds of all plants, birds and ground beetles occurring in Andean sky islands were endemic to a specific place. In the case of ground beetles, the proportion of endemism could even reach 90%. We further observed that the smallest and most isolated sky islands had the highest proportion of endemism (Fig. 2.4). This made sense concerning the effect of isolation, but not of area. We would have expected larger islands to contain a higher proportion of endemics deriving from previously present lineages that had then undergone local speciation. More surprisingly, when we ran some statistics to better grasp the effect of isolation and altitude, we found that these only explained a mere third of the total variance in our data. The páramo of Santa Marta in Colombia, just 90 km north of the main range, had a much higher proportion of endemic plants and birds than more isolated páramos. We observed the same pattern with the soggy páramo of Llanganates, in Ecuador, a few kilometers north of the city of Riobamba, which contained an unexpectedly high proportion of endemic ground beetles considering its small area. In fact, the proportion of endemics in páramos of a similar size was highly variable. The equilibrium theory could not satisfactorily explain these patterns. Was there something wrong with the theory?

While MacArthur and Wilson had advanced scientific knowledge with the concept of the equilibrium of species diversity on islands, their theory had a rather static view of islands themselves. It has become increasingly evident that in fact they are dynamic: changing in size, configuration and isolation as the Earth's crust and the climate change over geological timescales. This is particularly true with Andean páramos. Fossil pollen records have revealed that over the past 2.5 million years there have been almost 100 alternating periods of warming and cooling! As a consequence, the altitude of páramo belts has changed markedly. Some 20,000 years ago, páramos descended hundreds of meters below the altitude where they lie today and were at least three times larger. As the climate fluctuated, páramo island contour lines moved up and down the slopes of the Andes, resulting in configurations that shifted between being fragmented and connected.[18] Imagine the contour lines in Fig. 2.2 that represent the páramos moving up and down from one panel to the next. Consequently, sky islands formed, separated and reconnected through repeated phases of expansion and contraction. And, as each volcano with its páramo ecosystem has a specific topography, the size and shape of these sky islands can change at different amplitudes, frequencies and moments through time in the same region.

This fascinating interplay between climate and geology can cause the isolation or fusion of plant and animal populations, facilitating the intermixing of once single-island endemics into archipelago endemics and increasing the overall species

---

[17] Anthelme et al. (2014b).

[18] Flantua et al. (2019).

Fig. 2.4 Effect of sky island size and isolation on Andean species endemism. These graphs show the relationship between the size and isolation of páramos and the percentage of endemic species of plants, birds and ground beetles in the tropical Andes. As a general pattern, the smallest and most isolated páramos contain the highest proportion of endemic species, but some places— like the páramo of Santa Marta in Colombia or Llanganates in Ecuador (*arrows*)—deviate from the predictions. These deviances are explained by changes in a páramo's size and shape due to past climate oscillations. From Anthelme et al. (2014b), Alpine, Arctic and Antarctic Research ©2014 Regents of the University of Colorado

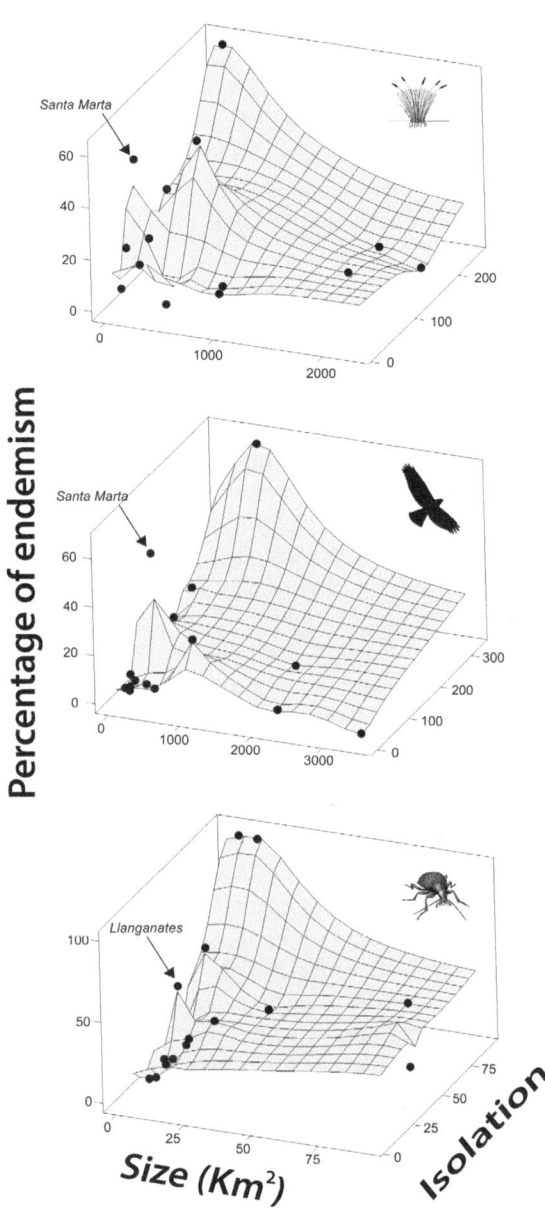

richness of connected islands. This effect of historical connectivity on species diversity explains our unexpected data. Small islands located between two larger islands, or that were once part of one continuously connected island, can retain the legacy of the historic connection, and thus depart from the expected species–area–isolation patterns. In general, previously connected islands have higher total species richness and lower single-island endemism than predicted by their present-day size and isolation. Effectively, every sky island is the remaining fragment of a larger lost island, which was once less isolated than today. These lost islands have shaped current species diversity and, in some cases, even strongly affected species appearance and lifestyle. The best place to look for these effects is in a group known for its rapid evolution and adaptation to new places: insects. We had seen that there were a huge number of endemic ground beetles in the páramos. So, from a ground beetle perspective, what's a sky island?

* * *

It may not seem obvious that the best person to answer this question would be an archaeologist, specializing in Iron Age architecture and town planning in Iberia and southern Gaul. However, beetles—with nearly half a million described species— have always been a favorite not only of prominent biologists, such as Carl Linnaeus, Wallace and Darwin, but many amateur collectors. **Pierre Moret** is one of them (Fig. 2.5a, b). Pierre is an internationally known archaeologist at Toulouse University, but insects are his first passion. From the age of nine, in the days when it was still common to give children an insect net, he chased after butterflies, but soon switched to beetles, fascinated by their amazingly diverse forms and lifestyles. Many years later, during his first trip to Ecuador in 1984, Pierre would sample ground beetles at different altitudes on the slope of Cotopaxi. Compared to the colorful species prized by most beetle lovers, these species were not spectacular at all (see Fig. 2.5c), but Pierre felt they might have the potential to answer interesting scientific questions. At the time, he did not even know the genus names of the Cotopaxi beetles he collected—a challenging taxonomic issue in the days before the internet! Over the three decades since, Pierre has collected over 20,000 ground beetles and named 240 new species out of the 264 that are now known in the high Andes of Ecuador. In his view, shifting between archaeology and entomology helps him to "*aerate his brain*" and "*avoid the routine that leads to intellectual sclerosis.*" Eclecticism has its advantages.

My first contact with Pierre was in 2009, when I was editing a special issue on entomology in Ecuador in the *Annales de la Société Entomologique de France*. Pierre contributed an elegant paper on the biogeography of páramo ground beetles.[19] One key finding of his studies that particularly intrigued me was that despite the high level of beetle diversification in the páramo, these species shared the same evolutionarily novel adaptation: flightlessness. All 264 species collected by Pierre in Ecuadorian páramos are unable to fly (with no or drastically reduced wings), except eight that live occasionally in the lower zone of the páramos (and are also found at lower altitudes). The loss of dispersion capacity is a classic **island syndrome**, as exemplified by many island bird species: think of New Zealand's kiwi and the extinct Mauritius dodo, among the most iconic examples. But such generalized flight loss in a group of island animals as was the case with these ground beetles

---

[19]Moret (2009).

Fig. 2.5 Beetle collecting in sky islands. (**a**, **b**) Pierre Moret collecting tiny ground beetles at 5000 m on the slope of Mt Antisana in Ecuador. Having turned over thousands of stones in dozens of páramos, Pierre seems to have a sixth sense where to find the beetle he is looking for. (**c**) *Dyscolus danglesi*, a flightless ground beetle described in 2020 by Pierre, wingless and endemic to a montane forest in a páramo near Loja, southern Ecuador (Like many entomologists, Pierre Moret is confronted with the challenge of finding names for the hundreds of new species he describes. Some ground beetle enthusiasts, such as the famous American entomologist Terry Erwin (1940–2020), chose humor: for example, naming new species in the genus Agra—his favorite—*Agra cadabra*, *Agra memnon*, or *Agra vation*. Pierre prefers to dedicate species to his colleagues. The illustration shows the ground beetle named after me: *D. danglesi*. Pierre's reason? "*I found that it had a funny or at least original head, different from the others!*", which I chose to take as a compliment.). This is the only known individual of this species, hence the missing leg, which was removed for DNA analysis. Photos: (**a**, **b**) Olivier Dangles, (**c**) Pierre Moret

was new to me. Researching this island syndrome, I discovered that the first theories had not been based on easily observed wingless birds, but on ... ground beetles! And this theory had been put forward by none other than one of the most famous beetle lovers: Charles Darwin.

A few years before publishing *The Origin of Species* (1859), Darwin heard about a surprising discovery: more than one-third of the 550 beetles native to the Madeira archipelago, a group of four tiny islands 160 km west of Morocco, were flightless, with atrophied wings.[20] He found it *"a very curious point in the astounding proportion of the Coleoptera that are apterous* [wingless]."[21] Darwin posited that the evolution of flightless species from ancestors capable of flight was a phenomenon linked to islands, suggesting that a winged beetle would stand less chance of surviving on an island where it would be subject to the action of winds carrying it to sea, than one which flew less or was incapable of flight. Later, the question of wing atrophy in insects living in isolated places intrigued a number of entomologists, who observed that wing reduction was common at both extremes of Earth's verticality, in sites such as caves and mountaintops.[22]

Might winglessness on oceanic islands and sky islands be due to the same reasons? While high-altitude ridges and mountaintops are windy, it is commonly accepted that this syndrome in these habitats is more related to cold than to wind, with all due respect to Darwin. Cold temperatures make the energy costs of flying high, so mountain insects that do not have to rely on flight in their daily activity may be selected for flightlessness. As their name suggests, ground beetles live almost exclusively low to the ground, often along stream banks, lakes or bogs or in the moss of trees. If you lift a stone in the páramo, you might see ground beetles living in dense local concentrations, which eases the problem of finding mates. Since their food (mostly springtails) is also very localized, they have no reason to wander widely. Moreover, tropical mountains have a range of **microclimates**, so ground beetles do not need to move great distances in search of better conditions or shelter when the climate gets bad. In theory at least, wingless beetles can allocate more of their energy to other tasks, such as making eggs or foraging. Essentially, flying is an energy-demanding business in mountains, so natural selection simply dictated that walking was better.

So that would seem to clear things up about the ground beetles on sky islands. They just followed the classic evolutionary pathway of island species: after a winged common ancestor colonized the páramo, the loss of dispersal ability dramatically

---

[20] Darwin, Origin, pp. 135–136. It is noteworthy that Darwin explained the reduction of wings as due to natural selection occurring over many successive generations, but considered the possibility that the Lamarckian factor of 'disuse' may have played a role. See Beutel et al. (2009). The entomological study in Madeira was realized by Wollaston (1854).

[21] Darwin Correspondence Project, "Letter no. 1643," accessed on 8 May 2022, https://www.darwinproject.ac.uk/letter/?docId=letters/DCP-LETT-1643.xml

[22] Among them René Jeannel (1879–1965) at the French National Museum of Natural History and Philip Darlington (1904–83) at Harvard's Museum of Comparative Zoology. See Jeannel (1925), Darlington (1943).

limited inter-island movement, virtually ensuring that beetle populations on different islands rapidly became genetically isolated. Consequently, adaptive radiation (the diversification of organisms from an ancestral species into new forms filling different ecological **niches**) was promoted by both evolution within islands and dispersal between islands. But surprisingly, Pierre's research suggests a different evolutionary process. Using distribution data and molecular tools, he found that in the most common ground beetle genus, *Dyscolus* (Fig. 2.5c), rather than a single large radiation from one colonizing winged ancestor, there was a whole series of micro-events in which a very localized lineage from the cloud forest 'ascended' to the páramo. Pierre's studies suggest that the loss of wings occurred in the forest, at an altitude of 2000–2500 m, prior to the beetles' conquest of the páramo.[23]

So why would ground beetles lose their wings in the forest, where it is neither windy nor bitterly cold? In fact, for similar reasons: forest beetles too spend most of their time on the ground, in the undergrowth and the litter, with generally little need to travel widely. Plus, cloud forests can get quite chilly at night, requiring saving energy at the expense of flight. So, from a beetle's perspective, the evolutionary processes on sky islands are not necessarily the same as on ocean islands. While páramos and mountain forests are different, they are not as much so as an ocean island and the surrounding sea. And most importantly, they have closely interacted throughout their historical development.

> Seen from above, the peaks of high-altitude ranges distributed along the spine of South America resemble the islands of an archipelago. There, the assemblage of plants, birds and ground beetles can be explained by both local speciation and immigration from adjacent islands and lowlands. But understanding the Andean cordillera as an archipelago of islands in terms of ecology and evolution is not restricted to summits jutting out from mountain ranges. While these peaks dominate the northern part of the Andean cordillera, a different type of habitat island predominates in the southern Altiplano (or 'high plain'). There, the dry landscape is scattered with blue-green island oases that are some of the most important ecosystems for plants, animals and people living in the arid Andes.

2.3   Island Oases

My first encounter with an oasis was not a mirage. On a windy and cold day of October in the Altiplano, I was driving with **Rosa Isela Meneses**, the head of the National Herbarium of Bolivia, on the road that connects Lake Titicaca to El Alto,

---

[23] Moret (1990), Murienne et al. (2022).

Fig. 2.6 Oasis in the Altiplano. A bofedal in the dry Condoriri valley (4600 m), Bolivia. Bofedales provide local people with valuable resources, including grazing pastures for alpacas and llamas. Canals are constructed to redirect meltwater from nearby glaciers to feed the bofedales (see foreground). A small settlement nearby illustrates the close relationship between the people and the wetlands here. Photo: Olivier Dangles

the largest highest city in the world, where 1 million people live at 4000 m. Rosa has spent a lot of time in the field, traveling to the most remote corners of the Bolivian highlands. She is **Aymara** at heart and feels a vital connection with ***Pachamama*** ('Earth Mother') a goddess that represents a wider belief system in Andean cosmology. This conception is one in which humans, non-humans, and the entire cosmos are intertwined in a network of reciprocal relations. In the **Quechua** language, *Pacha* means 'world,' but it also conveys a temporal and cyclical aspect: a sort of space–time model. For Rosa, this is expressed in the deep bond she has with her fellow earth inhabitants: plants. She knows their names, their ecology, their conservation status, and their traditional uses by local communities. She literally carries a plant book in her head; you learn more from 1 day in the field with her than spending a month reading articles on tropical alpine plants and **bofedales**.

After stopping in a roadside café to warm up with *sopa de mani* (peanut soup) and *té de coca* (coca leaf tea), we turned onto a dirt road to enter the Condoriri Valley ('valley of the condor's head'), one of many carved into the Cordillera Real, a range of dozens of peaks well over 4500 m. We drove for an hour through the moonlike landscape of the Altiplano foothills with an unobstructed view of the glacier-capped cordillera in the background, before reaching a vista with a bird's-eye view of the oasis (Fig. 2.6). A bright green wetland dotted with alpacas, llamas and birds—a haven of life that contrasted clearly with the inhospitable dry and rocky surroundings. "*It's a bofedal,*" Rosa announced happily. "*The treasure of our highlands; a symbol of Pachamama. Like her, the bofedal is life and generates life.*" Here like

Fig. 2.7 Chromatic composition of a bofedal. Beyond their vital importance for local people and their scientific interest, bofedales are aesthetic. This palette of green shades appeared in a single bofedal in Hitchu Khota valley in Bolivia. The shades correspond to different ecological conditions (due to plant species and moisture level). Photos by Olivier Dangles

nowhere else the color green displays why it is a symbol of life, regeneration and renewal. Magnetized by this emerald in the desert, I looked more closely through my binoculars and spotted geese and ducks foraging in the water and mountain viscachas (grey-furred rabbit-like rodents, but with a tail like a squirrel) peacefully nibbling grass alongside herds of alpacas and llamas. The bofedal itself was a chromatic composition of infinite shades of green: seaweed, crocodile, emerald, pickle, lime (Fig. 2.7). Rosa explained that this artistic palette is due to differences in moisture across the bofedal combined with the presence of dense vegetation mats belonging to two different plant genera (*Distichia* and *Oxychloe*). At one end of the spectrum, wet *Distichia* had the densest green; at the other, dry *Oxychloe* patches were yellowish. In the background of this natural 'terrain art,' I spotted a small settlement, the presence of humans. Here, water brings life to the dry landscape and also allows the livelihoods of the local Aymara. Culture and nature are intimately linked.

Rosa stressed that the bofedal landscape cannot be understood without its people: *"The Aymara's interaction with bofedales is based on reciprocity, a respectful partnership. Farmers take care of the bofedal, and the bofedal, by providing food for the llamas and alpacas, takes care of them in turn."* The Aymara have designed ingenious water distribution systems that have sustained the bofedales over the centuries, creating shallow channels with stones that direct water from springs and glacier rivulets (Fig. 2.6). The rushes in the bofedal trap water like a sponge, forming a niche that supports a unique assemblage of plants, invertebrates, birds and mammals. Like kelp, coral, beavers or trees, they are what ecologists call 'foundation species'—species that play a strong role in structuring a community. As the water stagnates, rush mats develop and allow organic matter to build up in the soil, creating hospitable conditions for a whole web of life. As the bofedal expands over decades and centuries, it protects against soil erosion, buffers water flow and boosts grass productivity. In such a water-deprived landscape, which has an 8-month dry season and less than 40 cm of rainfall per year, natural vegetation is scarce, so bofedales represent important (sometimes the only) sources of water and forage for animal herds. In Bolivia, native Aymara communities are directly dependent on these wetlands in regions where conditions are so extreme that they almost preclude human habitation. Millions of people in the central Andes depend on bofedales for their livelihoods. Yet in the context of accelerating climate change and its impacts on precipitation patterns and glacier melting, the future of these oases is uncertain. With Rosa and a group of mountain ecologists, climatologists and glaciologists, we needed to better understand the dynamics of bofedales in the face of climate change. And for this, we would need to scrutinize the colors of these oases from space.

* * *

As the green wetlands of bofedales stand out in a dry brown setting, they are easily recognizable from space. Satellite images allow variation in their size to be monitored. But as bofedales dry out and turn yellow, it is trickier to visually discern them. For this you need satellite images in near infrared (NIR), which shows vegetation density. In this way NIR can stand in for the different shades of green. In an attempt at terrain art à la Jay Hart (see Sect. 1.4), the red in an NIR satellite image indicates the green patches of the bofedales (Fig. 2.8). And because water absorbs NIR, these images are useful for discerning other things that are not obvious in visible light: for example, land–water boundaries and soil water content. These indicate the quality of the bofedale.

By using the image archive from LANDSAT, the longest continuous space-based record of Earth's land in existence, we were able to examine changes in the surface area of almost 1700 bofedales in the Cordillera Real since 1980. We learned that over the last 30 years, bofedales have considerably varied in size, mainly due to variability in rainfall intensity.[24] As páramos contract and enlarge over millennia depending on climatic fluctuations, bofedales do the same over decades, depending on water availability. While it is not surprising that periods with more precipitation lead to bigger, greener bofedales, the data also revealed that the bofedales most influenced by glaciers were less prone to shrinking than others. This is important because it shows that glaciers are not only valuable water reservoirs for major cities, but their sustained flow is vital for the resilience of key ecosystems. This led us to another question: do the changes over time in the size of a bofedal have an effect on the number of wild species it can host? To answer this question, we applied the

---

[24] Dangles et al. (2017).

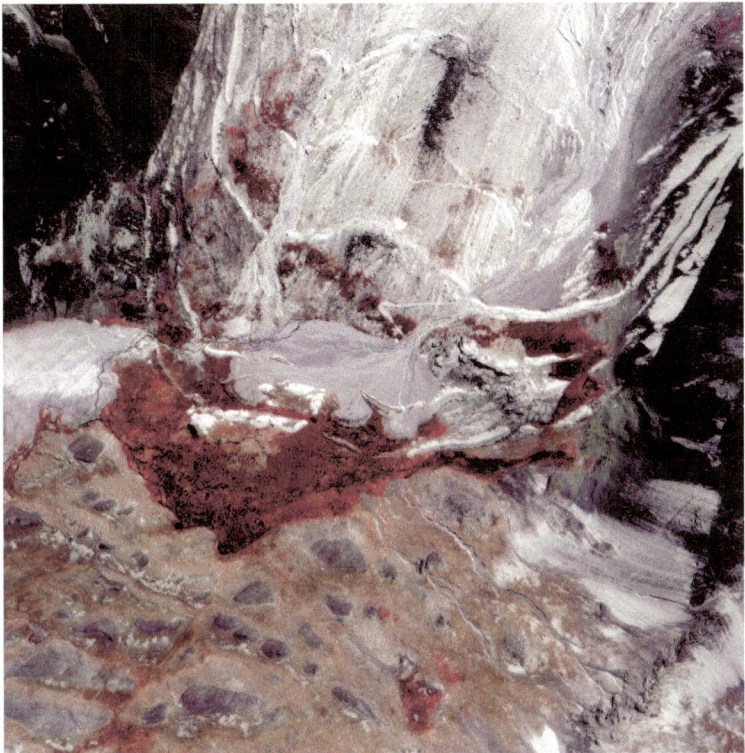

Fig. 2.8 Bofedales as seen in infrared. A near infrared Pleiades satellite image of the Charquini valley in the Cordillera Real, Bolivia. The wetter the area, the redder the shade (water pools appear in black). Photo: CNES/SPOT-Image ISIS program, contract #FC18473

theory of island biogeography to various biological groups—plants, invertebrates and birds—and found the most striking results with birds.

∗ ∗ ∗

Birds have fascinated many naturalists from all fields who have traveled in the South American Andes: the botanist Joseph de Jussieu was fond of woodpeckers,[25] the paleontologist Alcide d'Orbigny discovered a new kind of ostrich (more on this later), and the geographer James Orton admired hummingbirds, feeding their young nesting in the cliffs of Chimborazo "by bringing them flowers of the myrtle."[26] The region lends itself well to interest in birds. Bird diversity is greater in the tropical

[25] Bibliothèque du Muséum d'Histoire Naturelle de Paris, Archives of 'Voyage au Pérou, par Joseph de Jussieu,' Ms 111, planches, fol. 103, 166, 176.

[26] Orton (1871), p. 191. Born in Seneca Falls, some 60 km north of Ithaca, James Orton (1830–77) headed three scientific expeditions to South America. He crossed the continent from west to east, then reversed direction by ascending the Amazon River and explored the Altiplano. He died crossing Lake Titicaca, where he is memorialized by a column of granite on Isla Esteves, facing

Andes than in any other biodiversity hotspot on Earth: 2800 species, of which about one-third are endemic, with new species continually being found through the exploration of remote areas and improvement in genetic techniques. But one of the most remarkable anecdotes concerning early naturalists' interest in birds in the tropical Andes in fact comes from Humboldt. While collecting plants on the slopes of different volcanoes in Ecuador, Humboldt frequently observed groups of condors flying overhead and was fascinated by their ability to live at very high altitudes, *"beyond the limit of perpetual snow."*[27] True to his obsession for measuring everything, Humboldt provided a complete biometric identity card of two individuals, a female and a male, captured on the slopes of Mt Pichincha and Mt Chimborazo. He provided 17 different measurements of the two birds, including eyeball diameter, the thickness of the crest, and the length of each of the four toes and nails (Fig. 2.9). Not content with this remarkable tour de force, Humboldt lamented his oversight in not having described the abundant lice he observed on the condors' bodies—information that may well have interested some entomologists!

I share these naturalists' enthusiasm for birds, and personally consider them the creatures on Earth best able to simultaneously stimulate the two sides of the human brain. Unlike most insects, it is relatively easy to observe birds' beauty, their colors and shapes, their songs and flight, which trigger our emotions and aesthetic sensibilities. Capturing the mating dance of the red-winged blackbird in the evening light of Cayuga Lake[28] or the hypnotic gaze of a pair of crested owls hidden in the branches of the Ecuadorian jungle[29] are some of my best memories as a wildlife photographer. But birds also provoke our curiosity and rational side—to count, list and observe them—which proves very useful, as they represent excellent models to understand how nature works and gauge the state of ecosystems. Many naturalists have been drawn to analyze their behavior and ecological preferences and document their role in ecosystems. Yet bird assemblages in the high Andes are little known, much less how they are influenced by bofedal size.

It is indisputable that bofedales are magnets for birds. They shelter both year-round residents such as the Andean goose,[30] at least a pair of which can always be found on a bofedale, and migrants such as the lesser yellowlegs,[31] which breeds in wetlands in Alaska and northern Canada and winters in South America, using bofedal oases and páramo wetlands as stepping stones on their migratory routes. Some species even nest exclusively in high-altitude wetlands, such as the magnificent diademed plover, a bird currently classified as near threatened. I was convinced that uncovering how changes in a bofedal's size may affect bird diversity was an

---

Puno. The numerous specimens of plants, animals, fossils, rocks and anthropological artifacts he collected made him the best authority on the region since Humboldt. See Miller (1982).

[27] Humboldt (1811–1833), vol. 1, p. 27.

[28] https://www.inaturalist.org/observations/29503405

[29] https://www.inaturalist.org/observations/29443326

[30] https://www.inaturalist.org/observations/65269250

[31] https://www.inaturalist.org/observations/29693432

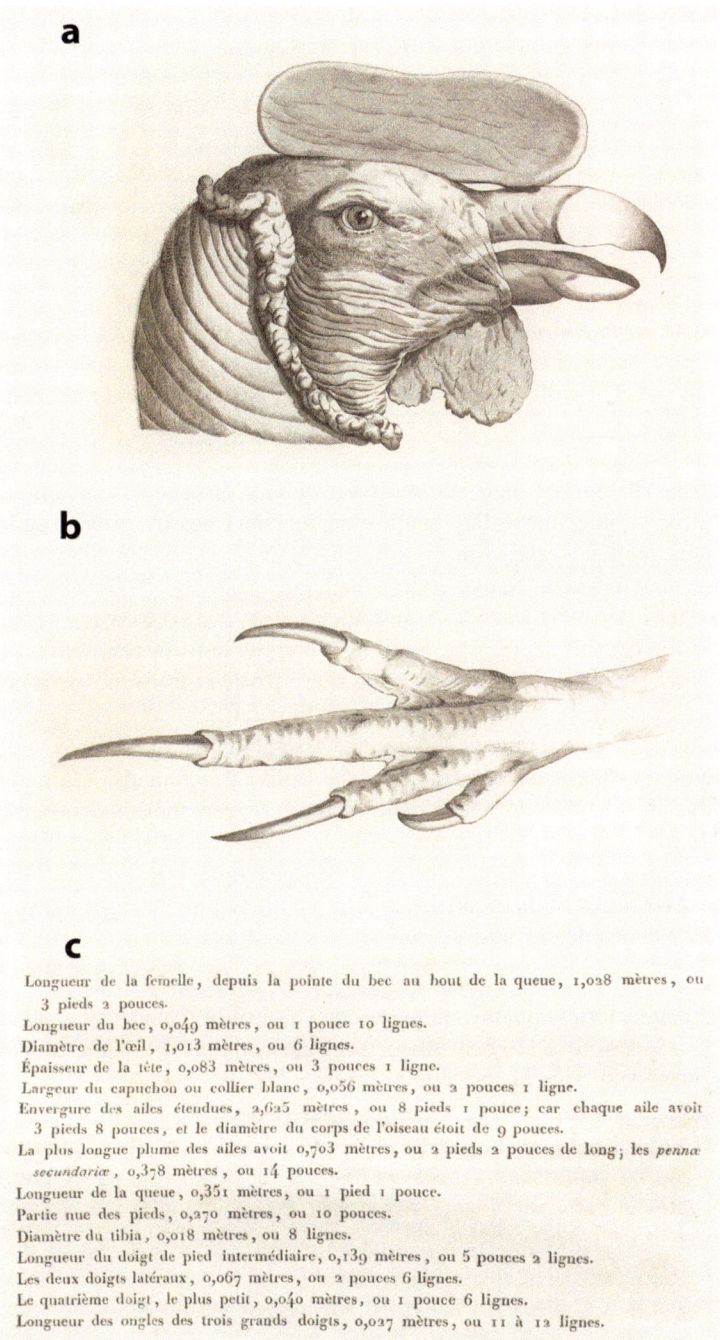

c

Longueur de la femelle, depuis la pointe du bec au bout de la queue, 1,028 mètres, ou
   3 pieds 2 pouces.
Longueur du bec, 0,049 mètres, ou 1 pouce 10 lignes.
Diamètre de l'œil, 1,013 mètres, ou 6 lignes.
Épaisseur de la tête, 0,083 mètres, ou 3 pouces 1 ligne.
Largeur du capuchon ou collier blanc, 0,056 mètres, ou 2 pouces 1 ligne.
Envergure des ailes étendues, 2,625 mètres, ou 8 pieds 1 pouce; car chaque aile avoit
   3 pieds 8 pouces, et le diamètre du corps de l'oiseau étoit de 9 pouces.
La plus longue plume des ailes avoit 0,703 mètres, ou 2 pieds 2 pouces de long; les *pennæ
   secundariæ*, 0,378 mètres, ou 14 pouces.
Longueur de la queue, 0,351 mètres, ou 1 pied 1 pouce.
Partie nue des pieds, 0,270 mètres, ou 10 pouces.
Diamètre du tibia, 0,018 mètres, ou 8 lignes.
Longueur du doigt de pied intermédiaire, 0,139 mètres, ou 5 pouces 2 lignes.
Les deux doigts latéraux, 0,067 mètres, ou 2 pouces 6 lignes.
Le quatrième doigt, le plus petit, 0,040 mètres, ou 1 pouce 6 lignes.
Longueur des ongles des trois grands doigts, 0,027 mètres, ou 11 à 12 lignes.

Fig. 2.9 Humboldt as ornithologist. Humboldt was fascinated by condors and their ability to fly
at very high altitude. He captured a pair of birds, which were illustrated and measured. (**a**) The head
and (**b**) the foot of a condor; (**c**) Humboldt's list of measurements of a female condor. From
Humboldt (1811–33)

important scientific and conservation question. But while I was a decent birder and photographer, I was completely inexperienced in the formal scientific methods required to study them. For this, I needed to partner with an ornithologist.

Kazuya Naoki visited Bolivia for the first time in 2000 to investigate Andean birds in the cloud forest, as part of his PhD studies at Louisiana State University. To compare communities of fruit-eating birds in cloud forests, Naoki had visited several countries—Costa Rica, Ecuador, Peru and Bolivia—but fell in love with Bolivia for its nature, culture and people and moved to La Paz to teach biology at the Universidad Mayor de San Andrés. Naoki enthusiastically accepted my invitation to join us during our field campaigns with Rosa. One particularly cold day in October, we embarked on a trip to survey bird diversity in various bofedales lining the bottom of Condoriri valley. When we reached 4700 m, we stopped by a stunning bofedal still covered with the night's frost. Naoki and his students jumped out and started their survey, scanning the horizon and identifying birds at three equidistant observation points along a **transect**.

During the 30 min of their survey, I sat quietly on a rock, contemplating the romantic desolation of the valley and the encroaching nearby glacier, taking notes and scanning the landscape with my binoculars in search of any signs of wildlife. Once they had finished, Naoki approached me excitedly: *"There's a diademed plover just twenty meters away!"* A new **lifer** for my list! It took me some time to spot the bird in a small depression, and I had to crawl in its direction to try to take a picture, somehow managing to capture the sharp frosted tips of the grass contrasted with the warm softness of the bird's feathers (Fig. 2.10).

Naoki found that an abundant and diverse assemblage of birds, like the diademed plover, could face the harsh conditions of the Bolivian highlands. After conducting 720 surveys in 40 high-altitude bofedales over 1 year, they had observed 2858 individuals from 41 species. It turns out that Andean oases are bird havens. With such a large dataset, it would now be possible to examine how the size of a bofedal affects the diversity of birds found there. This would require the right statistical tools. But first let's take a detour into the past.

* * *

The Arrhenius family features at several key points in the history of science, and some of their ideas proved to be well ahead of their time. In 1957, Gustaf Arrhenius, who was interested in the history contained in ocean sediments, developed a project to reach the planet's mantle by coring to sufficient depth. While this project did not achieve its immediate goal, it later gave rise to a large international ocean-drilling collaboration that continues to this day. Well before that, in a paper published in 1896, Gustaf's grandfather Svante Arrhenius, the recipient of the Nobel Prize in Chemistry in 1903, calculated that *"the temperature in the arctic regions would rise about 8° to 9°C, if the carbonic acid increased to 2.5 or 3 times its present value."*[32] Today we recognize the importance of this discovery in understanding global

---

[32] Arrhenius (1896), p. 268.

Fig. 2.10   A bofedal visitor. The beautiful diademed plover (*Phegornis mitchellii*), a high-altitude wetland breeder, has a small and declining population, making it a potential victim of human-induced change in the Bolivian Andes. Its beauty could make it an effective ambassador for the conservation of its vital habitat, the bofedal. Photo by Olivier Dangles

warming, but this was not appreciated at the time. Svante's son and Gustaf's father, **Olof Arrhenius**, would make his own mark on scientific history. In 1921, he published the results of an analysis in which he counted the number of plant species occurring in different-sized plots in Britain. He then devised a mathematical formulation that explained these results, for which he continues to be known a century later: $S = cA^z$.

As we've seen, an important theorem in island biogeography is that as the size of an island increases, so does its number of species. While this may seem rather intuitive, the observation was first made late in the eighteenth century and slowly took hold in the nineteenth and early twentieth centuries. One of the first to mention the idea was **Johann Forster**, Georg Forster's father, during his travels in the Pacific Islands. "*Islands only produce a greater or lesser number of species, as their circumference is more or less extensive,*" he wrote in his book *Observations Made During a Voyage Round the World* (1778).[33] The first graphs of species–area relationship appeared in the second half of the nineteenth century, and at the turn of the century, Alfred Russell Wallace assembled numerous observations throughout the world (presented in his wonderful book *The World of Life*, 1911) to conclude not

---

[33] Forster (1778), p. 169. For a brilliant description of insular biogeography history and theory, see Quammen (1997).

only that species and area are related for most kinds of organisms and ecosystems, but also finding that *"many small areas [are] often much richer **in proportion to area** than larger ones of which they form a part"* (my emphasis).[34] In other words, while diversity does rise with surface area, it does so *at a rate that is proportionally higher in smaller areas and slows with the increasing size of an area.* A decade later, Olof Arrhenius would be the first to put forward a mathematical formula in which the increase in species diversity occurs at a declining rate (the slope becomes less steep) with an increase in area. After 140 years of contemplation by different thinkers, the species–area relationship turned out not to be as straightforward as it first seemed.

$S = cA^z$. This power function is still today the model that best describes Wallace's observation: the association of increased habitat area (A) with an increasing number of species (S) at a declining rate (z),[35] with c representing the expected mean number of species per unit area. Larger values of z correspond to the faster accumulation of species as area increases, thus indicating higher rates of **species turnover** (change in species composition). So knowing the value of z, we can predict to what extent species richness will increase when the habitat area becomes larger (and vice versa). If, for example, $z = 0.3$ (the average situation in nature), we can calculate that the number of species will decrease by half when the habitat area becomes 10 times smaller: $(1/10)^{0.3} = 0.5$. If you didn't follow that, all you need to remember is that true oceanic islands have z slopes between 0.25 and 0.3, while habitat islands, such as mountaintops or forest patches, have slopes below this range (i.e., the gain in species richness is lower because habitat islands are generally less isolated and/or more easily colonized).

Despite its simplicity, this equation is one of the most important analytical methods in the ecologist's toolkit as it is valid over areas both small and large, with animals as well as plants, from tropical forests to the poles. Since Arrhenius, ecologists have built dozens of alternative mathematical formulas, generally with more parameters, allowing the curve to be more flexible to fit the datasets. But in most cases, this extra flexibility is not worth the cost of having extra parameters to estimate. One hundred years later, Arrhenius's equation is still fending off its competitors.

* * *

So did the bofedal birds agree with Arrhenius's math? They did. The larger oases supported more bird species than the smaller ones, and the species–area curve was fairly neat (Fig. 2.11a).[36] As expected for good dispersers living in a habitat island, we found a fairly small $z = 0.18$. An exhausted bird falling into the sea will invariably die, but one landing in a different landscape, even the arid Altiplano, may well be able to rest and refuel. Moreover, transient birds can colonize a small area, which reduces the steepness of the species–area curve. This is good news for

---

[34] Wallace (1911), p. 29.

[35] The power curve description of the species–area relationship, $S = cA^z$, was proposed by Arrhenius (1921) and modified by Gleason (1922). The rate declines because $z < 1$.

[36] See Cárdenas et al. (2022).

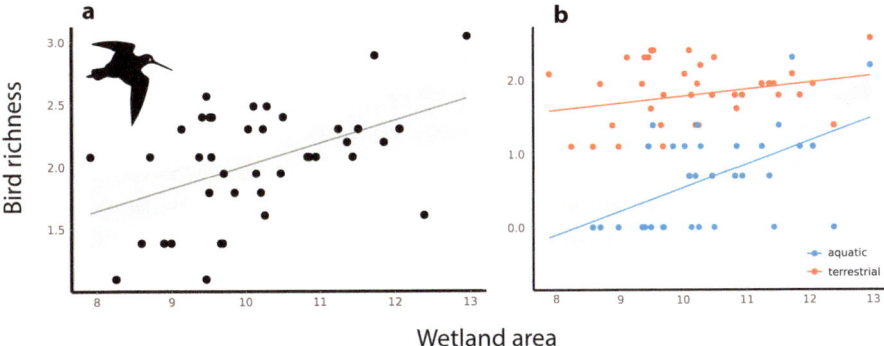

Fig. 2.11 Altiplano birds and the Arrhenius model. Relationships between the number of bird species and the surface area of bofedales in 40 locations of the Cordillera Real, Bolivia. (**a**) All bird species analyzed together, showing the expected increase in number of species with bofedal size (in m^2) following the Arrhenius model. (**b**) Aquatic birds (in blue) and non-aquatic birds (in orange) analyzed separately, showing that while the number of non-aquatic species increases only slightly with bofedal size, size has a much greater impact on aquatic species. The data is displayed in logarithmic space so that Arrhenius's power function becomes linear. Illustration by Tatiana Cárdenas

bofedal birds, as a low z value means there is low sensitivity for an island area, so a lower risk of losing species as a bofedal shrinks when glacier runoff becomes scarce.

However, this analysis only works when the total number of bird species is considered. If we analyze the data separately for aquatic and terrestrial birds living in the bofedal, the results change dramatically (Fig. 2.11b): the slope for the diademed plover and its aquatic kin rose to $z = 0.32$ (i.e. a significant gain in aquatic bird species with an increase in the size of the area), while that for finches, hawks and terrestrial birds dropped to 0.09 (i.e. virtually all terrestrial species are present even if the area is small; there is little gain if it grows in size). While landscapes surrounding bofedales may shelter a number of terrestrial species that are also found in the bofedales, they are not hospitable for aquatic birds. We also found that some waterbirds, such as the yellow-billed pintail and the black-crowned night heron, were found only in large bofedales.

But a problem remained that Arrhenius did not explain. Even when the land and water birds were separated, there was high variability in our data, and many bofedales clearly deviated from the trend predicted by Arrhenius's z. In fact, some large bofedales had as few bird species as small ones. So something other than surface area must come into play to determine the number of species on an 'island.' What could the underlying processes be that leave some large habitat islands so much species-poorer than small ones? Now that we had a correlational model, our challenge was to find and evaluate the causal explanations.

* * *

"The mere simple contact between heterogeneous substances may be a source of movement and of life in all organized beings." [37] This is the assumption reached by Humboldt in his paper on animal electricity, in which he relates his experiments with electric eels. More proof of the visionary thinking of this genius polymath—and what is true for eels may also be true of ecosystems. Many studies have indeed suggested that the best explanation for the species–area relationship is related to the heterogeneity of habitats and resources in nature. In the case of bofedales, our data confirmed that as their area increases, so does the diversity of physical habitats (rocks, pools, streams, tall grasses) and resources, in turn supporting a larger number of bird species. In other words, the greater the range of shades of green, the greater the diversity of feathered bodies. Irrespective of their size, bofedales with productive habitats (for example, undisturbed areas with thick emerald green mats of *Distichia*) and a diverse network of both stagnant and flowing water were among the best predictors of a rich community of birds.

Another interesting result was that bird diversity peaked in bofedales located in watersheds with consistent glacier cover. Large bofedales with low aquatic bird diversity generally received a lower contribution of meltwater, making them more prone to desiccation during the dry season. What relevance does this have for bird conservation? In Naoki's words, *"If you want to protect high Andean bird fauna, leave aside Arrhenius's math and focus on conserving water pools and streams."* Our findings revealed that bofedal heterogeneity is more significant than sheer area in determining how many bird species it can host. And to preserve this heterogeneity and keep seasonal shrinkage within acceptable limits, glacier water is vital.

Bofedales, unique and vulnerable places imbued with ecological and cultural importance, were another example of bridging right brain and left brain. The physical experience of lying in cushion plants, the aesthetic pleasure of the chromatic composition of the landscape, and the unforgettable encounter with a diademed plover, close enough to capture its fragility as a symbol of the challenges facing waterbirds in the high Andes. This merged with mapping by remote-sensing satellites, analysing data to uncover the role of glaciers in the resilience of bofedales, and applying the species–area relationship to better predict future threats to wetland birds. Yet even with this more holistic approach, we had only scratched the surface of how bofedales work. Many other factors influence life in island-like systems such as those in the Andes. An example of this is a micro-island—only a meter in diameter—a plant that contains all the key components of the web of interconnected life forms in the Altiplano.

---

[37] Humboldt (1811–1833), vol. 1, p. 88.

2.4 Micro-islands

I was 13 or 14 the day I first walked into the research library of the National Museum of Natural History in Paris: through its heavy cathedral-like door, over a creaking wooden floor, into a huge, silent, dark room with large forest-green tables illuminated by pools of golden light. The walls were lined with thousands of folio-size manuscripts, there was a strong smell of vanilla and hay, and what seemed to me at the time to be old men leaning over piles of textbooks. I had a singular objective in mind: recording the names (both vernacular and Latin) of all the vertebrate animals living on the planet. Even though I had wisely set aside insects, I still faced an approximate 32,000 species of fish, frogs, reptiles, birds and mammals to list. A daunting task, but I was confident that I could do it since I was starting early in life. I visited the library countless times with my notebook, and though I never completed this giant project, I did become very familiar with the binomial system for naming animals, as well as getting hooked on books and libraries. More than 30 years later I came back to the library to find that the people there did not seem so old; they were mostly my age. And this time I was looking for a single plant, a search which would be served by my basic Latin.

* * *

When in 1735 the Academy of Sciences in Paris sent a mission of astronomers, including Charles-Marie de La Condamine, to present-day Ecuador to measure the terrestrial meridian arc (see Sect. 5.2), they decided that a botanist had to be included on the journey. Tropical regions were famed to host thousands of unknown plants that might be of economic interest. At the time, the brothers Antoine and Bernard de Jussieu were the most famous botanists in France. But the scientists rather took along their brother, the doctor and civil engineer **Joseph de Jussieu**, to study the plants of the New World, particularly cinchona, a shrub whose bark was known to produce the anti-fever agent quinine, potentially useful in the treatment of malaria.

After almost 10 years, the astronomers had completed the purpose of their voyage and were preparing to return to France, but Joseph could not bring himself to leave the region and, from 1745 onwards, he journeyed to the wildest and most unexplored parts of South America. He traveled throughout the Andes, observing the plants, animals and minerals as well as the customs of the inhabitants, providing people with the health care that his medical training allowed. He taught the virtues of plants, examined mines, rebuilt a bridge, sketched dykes and established paths. Joseph de Jussieu returned to France ill in 1771 and died a few years later without having time to write his memoirs. In spite of the loss of most of his manuscripts, a few remain, preserved in the Natural History Museum's archives, and which I had come to the library to study. After putting on white gloves, I listened to the librarian's instructions on how to handle these fragile documents. Carefully—and not without emotion—I opened them one by one until I found what I was looking for (Fig. 2.12a).

One of Joseph de Jussieu's Peruvian drawings represented a cushion plant that resembled an island-like dome (Fig. 2.12b). Below the drawing was a faded, barely

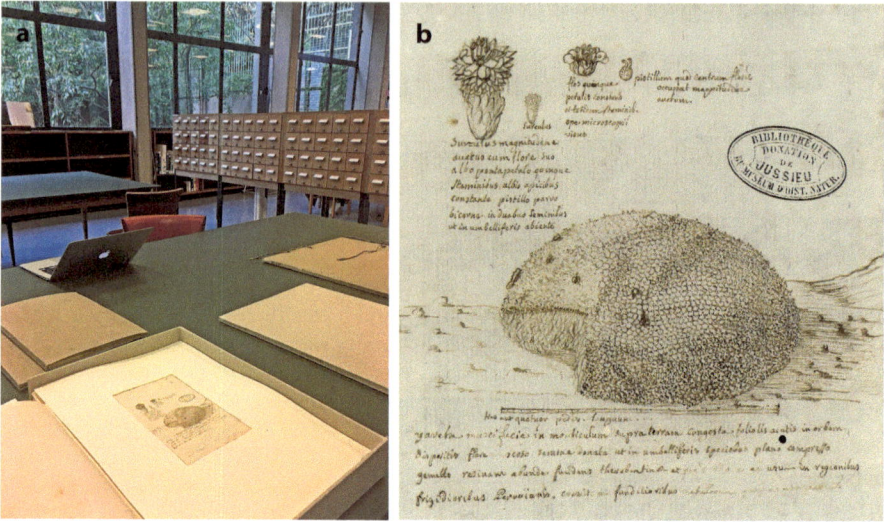

Fig. 2.12 Exploring the archives of an eighteenth-century naturalist. (**a**) Joseph de Jussieu's manuscripts from his '*Voyage au Pérou*' at the research library of the National Museum of Natural History in Paris; (**b**) detail of a drawing of a cushion plant, the **yareta**, and of a branchlet. Photos by Olivier Dangles

decipherable text in Latin written 27 decades ago: "*The Yareta, moss-like, gathered in mounds on the ground, with pointed leaves arranged in a circle, foamy flowers, with a seed as in the umbelliferous species, flat, compressed and twin (?), abundantly producing a resin with the smell of turpentine and pitch that is used in the coldest regions of Peru; it grows in the foundries of the mines (...).*"[38] As de Jussieu had written the description before the Swedish biologist Carl Linnaeus established the binomial system for naming plants and animals, the plant's full name appeared in the antique method, linking a long chain of words into descriptive labels.

As in his drawing, in nature the yareta looks like a moss-covered stone. The plant is actually composed of thousands of tiny scale-like leaf rosettes at the tip of branchlets (Fig. 2.12b), densely packed together and growing at the same rate, so that the surface appears like a very smooth mound—an impenetrable canopy. In fact, the yareta's surface is so dense that you can hardly put a knife into it; not even a Laguiole knife. The plant's scientific name, *Azorella compacta,* reflects this unique characteristic. Perhaps inspired by his medical training, de Jussieu showed a cross-section to reveal the inside. Although this is not very clear in his drawing, the interior

[38] "*Yareta musci facie in monticulum supra terram congesta foliolis acutis in orbem dispositis, flore muscoso, semine donata ut in umbelliferis speciebus plano compresso gemello, resinam abunde fundens therebentinæ et pixis odore [ad] (?) usum in regionibus frigidioribus Peruvianis, crescit in funditionibus metallorum [... ...].*" Bibliothèque du Muséum d'Histoire Naturelle de Paris, Archives of 'Voyage au Pérou, par Joseph de Jussieu,' Ms 111, planches, fol. 55.

of the plant contains a mass of crumbling, damp, dead tissue: the remains of branches crowded out by others.

Following these descriptions of the plant's morphology is information about its uses. Here, de Jussieu refers to a *"resin with the smell of turpentine and pitch,"* a natural product that is still used today in the Altiplano. Placed over skin contusions, yareta resin is a powerful anti-inflammatory. Chewing it is also effective to relieve toothache.[39] In fact, the whole yareta serves as a mobile pharmacy for local people: suffering from liver pain or prostate inflammation? Infuse 5 g of leaves or flowers in one liter of water and take it twice a day for 7 days. Rheumatic pains? 6 g per liter, three times a day for 15 days. Diabetes? 12 g/l, 3 times a day for 1 month. The yareta's roots are said to cure sciatica. What more could you need to stay healthy on the harsh Altiplano? Another important use of yareta lies in Jussieu's mysterious reference, *"it grows in the foundries of the mines."* In the treeless Andes of Chile, Bolivia and Argentina, the mining industry long heavily exploited yareta as a source of slow-burning, almost smokeless, fuel with a high calorific value for the smelting of minerals. During the period from 1915 to 1958, the Chuquicamata copper mine in the north of Chile alone consumed half a million tons of yareta.[40] To a lesser extent, railway companies also used yareta until the mid-1950s to fuel trains connecting the port city of Arica and the farther inland Arequipa to La Paz and Puno, some 3800 m in altitude. At the time de Jussieu was writing, the yareta was an iconic plant of the Altiplano, with a range of recognized medicinal and industrial uses. But almost nothing was known about its remarkable ecology. For this, another scientist—this one interested in mice—would need to enter the story.

* * *

The zoologist and ecologist **Oliver Pearson** set his first trapline in the early 1920s at age eight or ten, capturing a short-tailed shrew. From this revelatory activity as a child, he would later become the director of the Museum of Vertebrate Zoology at UC Berkeley and a pioneer in the modern ecology of the Andes. Pearson was tremendously curious, especially when it came to rodents. His ranging mind had many questions: How many different species of rodents can be found in the Andean Altiplano? Does altitude affect the size of rodent lungs? What are the causes of massive rat drowning episodes observed on the shores of some Andean lakes (up to 325 carcasses in a 100-m transect)? Can a mouse survive without its whiskers or fur? Does the left ovary and uterus of a mountain viscacha function if the right ovary is removed? What is the hourly vole traffic in natural trails? How do volcanic eruptions affect the life of small mammals? What can we learn from 4400-year-old accumulations of mouse droppings in remote Argentinian caves? How fast can Chilean bamboo grow, whose seeds are the favorite meal of a variety of rodents? Within rodentology, Pearson had Humboldt's breadth of curiosity. The research he

---

[39] See De Baldarrago et al. (2012).
[40] Halloy (2002).

conducted to answer these questions has left a strong mark on South American zoology.[41]

Pearson was not only an inquisitive mammalogist, he was also an inspired teacher and accomplished woodcut artist. Referred to by his colleagues as the 'quintessential naturalist,'[42] he published seminal papers not just on rodents, but on the metabolism of high Andean frogs, lizards, vicuñas and birds of particular interest to understand how wildlife responds to temperature variations. Pearson will reappear later in this book, but what was his connection with yareta? In fact, it was his daughter, **Carol Pearson Ralph**.

In 1978, Carol Pearson Ralph published one of the first comprehensive natural history descriptions of yareta. In a style that recalls Humboldt, her paper opens as follows: "*One of the most bizarre sights in the high, dry Andes of southern Peru and neighboring Chile and Bolivia is a stand of the bright green, coral-like heads of Azorella compacta Phil., known as yareta (Fig. 1).*" [43] The corresponding image is quite uncommon in a scientific paper: a picture of her sat atop an exceptionally large, undulating yareta (Fig. 2.13). The tenderness of this picture made me want to learn more about the story behind it and how Carol came to study yareta. I searched her name and found she had mostly published papers on birds and milkweed bugs in the 1970s and 1980s, and that she was now a member of the Northcoast Environmental Center in northern California. I finally tracked down her email address (which promisingly included '@humboldt' in the domain name) and soon received answers to my questions:

> In 1969, I started graduate studies in the Department of Ecology and Systematics at Cornell, planning to be a neotropical ornithologist. After one year of classes, I had the opportunity to go to Universidad del Valle in Cali, Colombia with Dr Richard Root, an ecologist with a bird background who had shifted to insects. Our task, which was to start an ecology program there, was aborted when the students went on strike, so I spent the year studying the dwarf cuckoo and helping Dr Root with his research on the tropical milkweed and its herbivores. By the end of the year, I couldn't pass up a chance to work with a fine professor like Dr Root and asked if I could join his lab, and he accepted me. When I left Colombia, I went straight to Peru to join my family, my head full of milkweeds and herbivores. There I was, practicing looking at plants (when not counting birds and measuring vegetation), suddenly faced with a very unusual plant, so I studied its basic natural history.

I was not surprised that the daughter of a 'quintessential naturalist' had ended up following Richard Root's advice to "let nature tell you what's interesting." What about the story behind the photo? "*That high Andes landscape sits deep inside me, I think because of the year my family (parents, brother and I) lived in a small school bus my dad set up like a camper (which had not been invented yet) in the Altiplano while he did field work. I was 4 years old, and I have a very poor memory, but that place feels like a home.*" It is also not surprising that a strong connection with nature goes hand in hand with good natural history observations.

---

[41] For a list of the papers answering these questions, see Kelt et al. (2007).

[42] Ibid.

[43] Ralph (1978), p. 62.

Fig. 2.13 Atop a millennial cushion plant. Budding naturalist Carol Ralph Pearson, age four, atop an exceptionally large yareta, calculated to be about 3000 years old, in Peru (elevation 4630 m). Photo in Ralph Pearson (1978), used with permission of John Wiley & Sons, from Observations on *Azorella compacta* (Umbelliferae), a Tropical Andean Cushion Plant, Biotropica, 62(10); permission conveyed through Copyright Clearance Center, Inc.

Carol's paper provides a wealth of data concerning the biology and ecology of the yareta, including its reproduction, seasonal changes (or **phenology** in ecology parlance), distribution and the herbivores associated with it. The yareta has flowers that develop all year, with a development from flower to ripe seed that takes 3–4 months (abundant aborted flowers indicate frequent failure of pollination or seed set). These plants tend to grow at the base of boulders, preferentially on the north-facing side (facing the equator) for maximum exposure to annual solar radiation. Both flowering and plant growth are concentrated on the plant's east and north sides, probably due to greater direct sunlight. A yareta's standing crop (the total dried biomass of the living organisms present in a given ecosystem) can reach up to 11 kg/m^2—a biomass per unit area as great as many forests. However, unfortunately for Carol's entomology studies, yaretas host very few insects, apart from a few red mites and pink scale insects (Coccidae). This is probably due to the plant's sticky resin: what may be medicinal for humans is not necessarily good for insects.

But one of Carol's most exciting observations concerned the yareta's growth rate and estimated age. She decided to measure how fast a yareta grows by driving nails into 40 plants, leaving a bit more than one inch of the nail exposed. Her father would return 4 years later and measure the nails again, leading to the discovery that yareta is extremely slow-growing. After 4 years, most individuals had expanded only a few millimeters, suggesting that a yareta extending over several meters of ground is several hundred years old. In fact, Carol estimated that many yareta she measured were at least 850 years old, and occasional individuals—such as the one in Fig. 2.13—might be up to 3000 years old.[44] That would make most yaretas old enough to have coexisted with ancient civilizations, never mind the first European naturalists to take an interest in them! This places yareta among a select group of the most long-lived plants, and probably older than most non-tree species known to date. But beyond this remarkable attribute, yaretas have another important quality. They not only look like coral, as in Carol's description, they also have a similar function, playing a key ecological role in the high Andes.

* * *

From the highest páramo to the most desertic place in the Altiplano, the bright green mounds of *Azorella,* the genus that includes yareta and its kin, stand out in stark contrast against the mineral landscape where they grow. Like many naturalists before me, I was captivated by their unique appearance, but I would never have studied them if not for Fabien Anthelme. This French plant ecologist is fascinated by cushion plants, a collective name describing many species that crowd together to form tightly bound, cushion-like domes. There are an estimated 1300 species of cushion plants, with South America hosting about 350 species, including almost 30 species of *Azorella.*[45] Cushion plants are found in a variety of habitats, including Arctic and alpine fell-fields, Antarctic islands and the windswept mountains of Patagonia. Some species such as *A. compacta* can live at altitudes of up to 5250 m,[46] making it one of the highest occurring woody plant species in the world. These habitats share certain stressful conditions: low temperatures, strong winds and seasonal droughts. By reducing the plant's surface-to-volume ratio, the compact cushion provides a minimum exposure of the plant's surface to external conditions, reducing water vapor loss and conserving heat. Perhaps no region in the world possesses cushion plants as such a dominant component of vegetation cover as the high Andes. This brought Fabien to the region, where he was particularly interested in studying one attribute of *Azorella*'s ecology that had not been described by Carol.

Walking up the flank of the ice-capped **Nevado** Sajama, an extinct volcano and the highest peak in Bolivia (6542 m), Fabien and I observed some large cushions of

---

[44] Later studies would show that yareta growth on the Altiplano is seasonal: it stops growing during the austral winter (or even contracts, a bit like a cactus, with the green body remaining intact) and mainly grows in January–February when two-thirds of the annual precipitation falls. See Halloy (2002).

[45] From the current distribution of *Azorella* species, it seems evident that subantarctic Argentina and Chile is the ancestral area of this group. During cool periods, the Andean cordillera has provided a colonization corridor—a climatic highway—for species that live in cold conditions but at much lower altitudes. Some of these species, like *Azorella aretioides*, spread to the north, as far as Ecuador and Colombia.

[46] Halloy (2002).

Fig. 2.14 A plant that hosts other plants. Ecologist Fabien Anthelme atop a large yareta, probably hundreds of years old, in Sajama National Park, Bolivia (elevation 4630 m). Other plant species (here grasses) can grow on yaretas. In the background, Mt. Sajama (6542 m). Photo by Olivier Dangles

Azorella compacta, impressive enough to deserve a photo (Fig. 2.14). True, when compared to Carol's photo it looks quite small—still a rookie in the Altiplano at no more than 1000 years old. Yet this specimen was more interesting for Fabien, for one clearly visible reason. This yareta was colonized by many other types of plants. In the photo, only the largest colonizers, in particular a grass called *Festuca*, can be observed. But there are many other tiny plants, hidden to the untrained eye. In a study we conducted in the Sajama National Park, Fabien observed 24 different plant species growing on the surface of the yareta's cushion: valerians, fescue grasses and even small paper trees.[47] This is an impressive number for a plant you can hardly push a knife into. We checked this property using a soil compaction tester and measured exceptional compactness, requiring a pressure of over 3000 KPa to penetrate the cushion—more than six times the pressure in a champagne bottle.

Colonization of *Azorella* domes is not limited to the yareta. Further north, in the Ecuadorian páramos, we found that cushions of another species, *Azorella aretioides*, were also hotspots for biodiversity, hosting more than 30 different species of plants.[48] Most interestingly, we discovered that about one-third of these species preferred to grow on an *Azorella* cushion than on the nearby ground. In the high Andes, many plant species tend to aggregate on cushion plants, even though niches are available on the ground. This reflects a pattern of positive interactions between plants, with certain plants being helpful for others: what ecologists call **facilitation**.

---

[47] Anthelme et al. (2017).

[48] Anthelme et al. (2012).

Azorella provides refuge for other plant species that would perish out in the open, just as the survival of hundreds of species of fish is facilitated by the presence of reefs. Beyond their appearance, *Azorella* are indeed in some way the corals of the high Andes.

But before we can jump to the conclusion that a pattern of spatial association in two plant species indicates a process of facilitation, we need to identify the mechanisms at play. A major mechanism identified by plant ecologists is that alpine cushions help other species by making the harsh climatic conditions more habitable, particularly in terms of extreme temperatures. In the high Andes, ground temperatures can vary dramatically, from −15 to +50 °C, sometimes within a single day: a challenging contrast for many life forms. An *Azorella*'s shape is well adapted to face this challenge, and this can also benefit others.

To confirm this, Fabien and I took hundreds of thermal images of *Azorella* cushions across the Altiplano and Ecuadorian páramos (Fig. 2.15). In the microlandscape surrounding the cushion, the transition in temperatures was revealed by shades of color. We found that *Azorella* cushions strongly buffer extreme ambient temperatures: when the ground is a frigid −8 °C, the surface of *Azorella* is a warmer − 2 °C; when the ground bakes at 45 °C, the cushion's surface is a much more tolerable 20 °C for other plants (and animals such as lizards).[49] This phenomenon is explained by what meteorologists call the 'oasis effect,' in which plants create a local microclimate that is cooler than a surrounding dry area due to transpiration and, sometimes, a higher albedo (reflectivity) than bare ground.[50] We found that fissures in the cushion were particularly warm, suggesting that older *Azorella* with complex shapes, bumps and crannies, make even better homes for other plants than smoother, simple cushions.[51] This buffering is even stronger inside *Azorella* domes, where temperatures are 3–4 °C warmer than on the plant's surface. Beyond the reach of the sun, the interior retains moisture, which also helps during dry periods. The tiny thick leaves, their dense growth and the resin content described by Joseph de Jussieu also contribute to reducing water loss and help resist against freezing. The heat capacity of the moisture stored within the cushion can delay nighttime cooling. But this asset may also be *Azorella*'s Achilles' heel in a context of climate change: compact cushions are extremely vulnerable to overheating, and can only survive at high altitudes or sub-polar latitudes—and as long as global warming stays within acceptable limits.

It turns out that *Azorella* offers not only room but board for other plants. On the slopes of Mt Antisana, we found that the presence of *Azorella aretioides* cushions has a strong positive effect on the availability of many soil nutrients: in particular,

---

[49] Anthelme et al. (2014a).

[50] When water evaporates or a plant undergoes transpiration, heat from the surroundings is used to convert liquid to gas, removing energy (endothermic), which results in cooler local temperatures. Vegetation with a higher albedo than bare soil reflects more sunlight, leading to lower surface temperatures.

[51] Anthelme et al. (2017).

Fig. 2.15 The heat of a plant. Thermal images visualizing the buffering effects of a cushion plant *Azorella aretioides* on ground surface temperature (Ecuador, 4700 m a.s.l.). (**a**) The plant increases minimum temperature at night and (**b**) mitigates maximum temperature during the day. As colorful representations of radiant heat, thermal images allow a sensory—even artistic—perception of global warming in natural objects. From Anthelme et al. (2014a) © Anthelme, Cavieres, Dangles, Frontiers in Plant Science. Creative Commons Attribution License (CC BY)

potassium, calcium and magnesium.[52] The accumulation of patches of soil fertility under cushion canopies was the most significant factor involved in the facilitation process—even more important than the improved microclimate. While we were the first to document this process in the tropics, it was already known that cushion plants could be islands of fertility. The renowned alpine plant ecologist Christian Körner put it this way: *"By accumulating 'compost' within their dense branchwork, long-lived cushion plants partly decouple themselves from the substrate, and run their 'private' nutrient recycling with a rich microflora in an otherwise hostile soil environment."*[53]

An important point here is that the facilitation ability of yareta and other *Azorella* species is related to their size and their age. Large cushions are generally better thermoregulators, provide richer humus and offer better shelter from the wind than small cushions. Unfortunately, large, old, bumpy yareta specimens are becoming increasingly rare today in the Altiplano. While the population has somewhat recovered since the heavy harvesting of the cushions in the first half of the 1900s, this

---

[52] Anthelme et al. (2012).

[53] Alpine Plant Life: Functional Plant Ecology of High Mountain Ecosystems by Christian Körner, Springer © 2003, p. 70, Reproduced with permission from *Springer Nature*.

iconic plant is classified as vulnerable in Peru and Chile and endangered in Bolivia. In the rapidly changing tropical Andes, it is uncertain if they will continue to survive, or to grow large and bumpy enough to provide other organisms with the most hospitable homes.

Islands can be perceived in the tropical Andes at different scales. Páramos, bofedales, cushion plants—all are like fractal elements of a similar form that appears repeatedly at different levels in the landscape. They are pieces in the puzzle of the living Andes, tied together by organic links. The size, isolation and age of each of these islands have shaped the evolution of their inhabitants and continue to exert a strong influence today on how they might respond to global changes: wingless ground beetles risk being trapped in páramos constricted by creeping agriculture, waterbirds will search in vain for bofedales as glaciers vanish and these wetlands dry up, Altiplano plants may no longer find shelter in slow-growing yaretas unable to adapt to the speed of change. Humans may also modify the degree of isolation of these islands by making the surrounding landscape more inhospitable and resistant to dispersing plants and animals. But humans can also make the active choice to conserve páramos, bofedales and yaretas. Islands are by nature fragile and require responsible caretaking. Climate change threatens the plants and animals living on these high Andean islands; to understand how we need to begin by going back in time.

Chapter 3
Time Capsules

*A philosophical study of nature seeks, in considering the
changes of phenomena, to connect the present with the past.*

Alexander von Humboldt (1850, p. 104)

The new main protagonists of this chapter are shown in Fig. 3.1.

3.1 Traveling in Space and Time

The avant-garde French novelist **Jules Verne**, one of the fathers of science fiction,
would not have held it against me that I discovered his work in front of a TV screen
rather than in a library. I was nine and, every Wednesday afternoon, avidly followed
the adventures of Phileas Fogg, who takes up the challenge to travel around the
world in 80 days with his valet Jean Passepartout (in this cartoon version, Phileas
Fogg was a lion and his valet, a cat, doubling my excitement). Sharing these
characters' never-ending adventures, from riding on the backs of elephants through
Indian forests to skimming over the frozen ground of Nebraska on sail-powered
sleds, captivated my imagination. Three decades later, when I watched the cartoon
again with my son Nicolas, it was another facet of the journey that intrigued me: the
relationship between space and time. As reflected in its title, the dialectic of space
and time is at the heart of *Around the World in 80 Days*, as it is in Verne's other
works, which systematically include two fundamental dimensions in initiatory
journeys: geography and history.[1] In at least half of his 40 or so books, Verne
included topographic maps, allowing the reader to follow the routes of the explorers
in physical space. Concerning time, Verne's objective was to summarize all the
geological and physical knowledge that had been amassed by modern science to

---

[1] Jules Verne was also the author of an illustrated geography of France and its colonies (1868) and a
history of great voyages and travelers (1878).

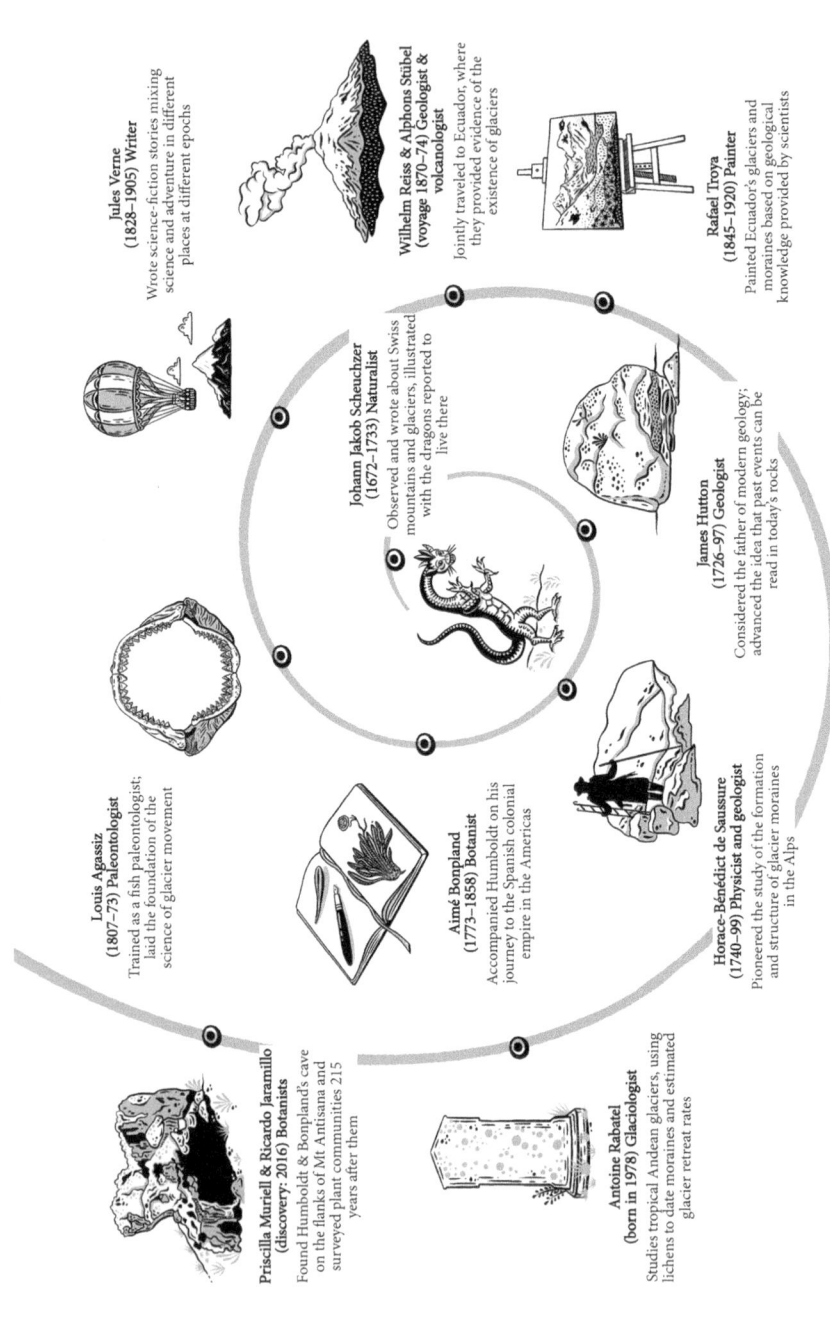

Fig. 3.1 Central figures in this chapter. This spiral timeline shows ten historical or contemporary figures who contributed important ideas mentioned in this section. Drawings by Paula Terán Ospina

reconstruct the history of the Earth. But more importantly, Verne *combined* space and time.[2]

On the 80th day of his voyage, Phileas Fogg arrives back in London 5 minutes past the deadline he had set. But Passepartout realizes that as they had traveled eastward, the days shortened by 4 minutes for each of the 360° of longitude they had crossed. So in total they had gained a day over the whole trip, and Fogg had won his wager. In another of his works, *Journey to the Center of the Earth* (1864), the translation of an ancient runic manuscript requires the main characters, Professor Otto Lidenbrock and his nephew, to go down the conduit of a volcano, which reveals itself a formidable time machine that takes them into the depths of evolutionary time. The further they descend into the bowels of the Earth, the further they travel back in time: at about 120 km underground they reach Mesozoic times and are surrounded by giant dinosaurs. To make this journey to the past credible to readers, Verne describes the different geological layers that follow one another as the explorers go deeper. While he is discreet about the sources of his geological knowledge, his descriptions indicate that he was aware of the uniformitarian ideas first formulated by the Scottish geologist **James Hutton**, the father of modern geology.[3]

Hutton argued that the Earth had formed slowly over a long period of time and by the same physical forces that govern current geological phenomena: among others, erosion, earthquakes and volcanoes. This idea contrasted with the dominant theory at the time, catastrophism, which argued that the Earth had been shaped by epochs of catastrophic actions, interspersed with periods of comparative calm. As a consequence, Hutton stated that it was possible to "read the transactions of time past in the present state of natural bodies." [4] The 'bodies' in question in Hutton's *Theory of the Earth* are rocks, which are difficult to 'read' for ordinary people, but offer a large amount of information to expert eyes, such as those of Professor Lidenbrock during his descent. The trail blazed by Hutton would later inspire many famous nineteenth-century naturalists, including Cuvier, Lyell, d'Orbigny, Sainte-Claire Deville, Élie de Beaumont, and Darwin, pioneers in geophysics and evolutionary biology. Today, an approach that integrates both space and time is still a powerful tool for scientists—including ecologists. One of the things it can help us understand are the consequences of climate change.

* * *

To assess how plant and animal life will change as the climate warms in the tropical Andes and glimpse the possible ecological situation decades from now, like Jules Verne, I decided I needed a time machine to see what the past could reveal. Although there are dozens of Ecuadorian volcanoes at hand—many of them still active—Professor Lidenbrock's method of traveling back to the past was not very tempting. Humboldt himself stated of the crater of Pichincha, the volcano that towers

---

[2] For example, see Dupuys (2000).
[3] For details on the geological information found in Verne's books, see Gèze (1986) and Breyer and Butcher (2003).
[4] Hutton (1795), vol. 1, p. 373.

over Quito, that he did not *"believe that the imagination could conjure up anything as sinister, funereal, and frightening as what we saw below."*[5] As a lesser mortal, I chose two complementary approaches. The first was to travel back in time by unearthing historical data to reexplore sites that had been sampled in the past, in my case during Humboldt's expedition to Latin America. Though this may sound simple, I would discover that the past is indeed *terra incognita* for an ecologist. A trip like this requires a historian, so I brought Pierre Moret (see Sect. 2.2) on board for the adventure.

The second approach was to 'cheat,' by substituting time with space. This line of attack is based on a simple assumption: the factors that underlie plant and animal composition over spatial gradients also drive temporal changes, hence past trajectories of ecological systems can be gleaned from current spatial patterns. Luckily, there are many places in the tropical Andes that can provide spatial gradients, and some host Hutton's 'natural bodies,' namely rocks, that can be 'read' and dated. These two time machines involve no futuristic vessels, and the time we will travel back is only a few centuries, but they will not disappoint. Jules Verne would approve of the journey. In his novel *Twenty Thousand Leagues Under the Sea*, Captain Nemo kept the complete works of Humboldt in the library of his submarine Nautilus, showing Verne's interest in Humboldt[6]—and in stories with twists.

3.2   The Secrets of Humboldt's *Tableau Physique*

In Jules Verne's *In Search of the Castaways* (1868), at the beginning of a trip around the world that will take them to discover the Andes' rocks and glaciers, earthquakes, volcanic eruptions, and condors, the children of Captain Grant and the other passengers of the *Duncan* approach the Pico del Teide on Tenerife. On board, the Secretary of the Geographical Society of Paris[7] laments that there is nothing more to discover on this island since Humboldt had been there before them.

> Oh, ascend it! Ascend it, my dear captain! What would be the good after Humboldt and Bonpland? That Humboldt was a great genius. He made the ascent of this mountain, and has given a description of it, which leaves nothing unsaid. He tells us that it comprises five different zones — the zone of the vines, the zone of the laurels, the zone of the pines, the

---

[5] Humboldt (1803), p. 327.

[6] Jules Verne cited Humboldt in at least six of his books: *Journey to the Center of the Earth* (1864), *The Voyages and Adventures of Captain Hatteras* (1865), *In Search of the Castaways* (1865–67), *Twenty Thousand Leagues Under the Sea* (1869–70), *The Giant Raft: Eight Hundred Leagues on the Amazon* (1881), and *The Mighty Orinoco* (1898). See Clark and Lubrich (2012).

[7] The Geographical Society of Paris, the world's oldest, was founded on 15 December 1821. Humboldt was among its 217 founders and was president of the society in 1845 for 1 year.

zone of the Alpine heaths, and, lastly, the zone of sterility. [...] What could I do, I should like you to tell me, after that great man?[8]

The *Duncan*'s captain replies: "Well, certainly, there isn't much left to glean." In this passage, Verne extols Humboldt's contributions to the study of plants' altitudinal distribution. Yet while famous for his sense of foretelling the future, Verne underestimated how much there would be left to glean for scientists walking in the footsteps of Humboldt and other explorers.

* * *

The past is an indispensable guide for understanding a world in motion. The acceleration of the human footprint since Verne's epoch has made it pressing to document the responses of nature to environmental changes over preceding decades, centuries, or longer. To extend the timeframe of their studies, the holy grail for environmental scientists is to find reliable historical data that allows them to define baseline conditions *before* disturbances caused by widespread human activities, in order to quantify how the planet's state has changed since then.

To achieve this goal, scientists and scholars from other fields have made use of whatever scraps of historical information are available. They have compiled French records of Pinot Noir grape-harvest dates since 1350 in Burgundy and fed them into mathematical models to trace temperature variations over past centuries.[9] They have trawled through hundreds of thousands of restaurant menus collected over the past 150 years in the USA and used the prices of seafood dinners as a proxy for fish catch and scarcity.[10] They have surveyed records of the first flowering dates for over 500 species of wildflowers noted from 1851 to 1858 by writer and naturalist Henry David Thoreau in Concord, Massachusetts to reveal the ecological effects of climate change in this region.[11] They have unearthed samples collected during Robert Falcon Scott's legendary expedition in Antarctica (1902–03), stored for a hundred years in a cupboard in London's Natural History Museum, to provide crucial baseline levels of blue-green algae, free-living bacteria that can be toxic to animals and humans and have risen globally due to climate change.[12] They have studied sixteenth-century epistolary documents from the Spanish court certifying the importation of live white-clawed crayfish from Tuscany by the request of King Philip II to trace the origin of this endangered freshwater species in southern Europe.[13] They have inspected the evidence of insect bitemarks in early 1900 herbarium samples to understand how warming has affected insect appetites and damage to plants over the

---

[8] Verne (1876), p. 76. For a discussion about Verne's references to Humboldt's travel to Mt Teide, see Schifko (2010).

[9] Chuine et al. (2004).

[10] Anderson (2006).

[11] Primack and Miller-Rushing (2012).

[12] Jungblut and Hawes (2017).

[13] Clavero et al. (2016).

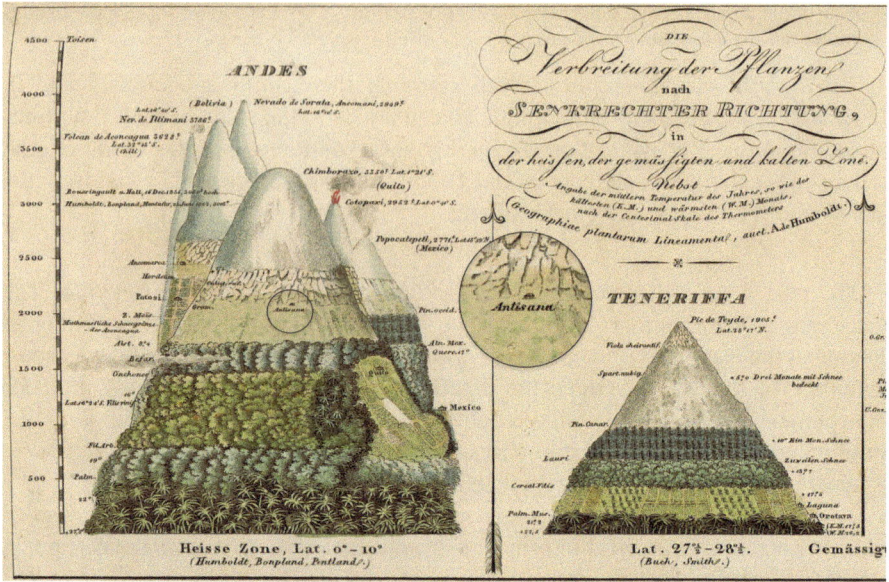

Fig. 3.2 Insightful and beautiful mountains. Alexander Humboldt was a pioneer in combining art and science in the illustrations of his works. Based on information collected by Humboldt (and with his encouragement) German geographer and cartographer Heinrich Karl Wilhelm Berghaus (1797–1884) produced beautifully engraved hand-colored thematic maps such as these plant variation altitudinal profiles for the Andes and Tenerife. Note the presence of the Antisana hacienda that Humboldt mistakenly believed to be the highest inhabited place in the world. Modified from Berghaus (1852), *Physikalischer Atlas*

past century.[14] They have used historical chronicles of ice-thawing dates in northern European harbors to reconstruct winter sea ice severity in the northern Baltic Sea back to the early sixteenth century.[15] They have scrutinized the colors of sunsets in hundreds of paintings by famous artists (including Joseph Mallord William Turner, Caspar David Friedrich and Edgar Degas) to look for clues to estimate pollution levels in past centuries.[16] This creative scholarship bridging science and the humanities is invaluable in helping to understand climate change.

So in view of the potential value of historical materials for investigating long-term environmental change, might there not be "much to glean" from the vegetation belt data obtained by Humboldt and **Aimé Bonpland** while ascending the Pico del Teide more than two centuries ago? Might the now-classic altitudinal vegetation profiles published by Humboldt for Pico del Teide and elsewhere (Fig. 3.2) contain relevant scientific information to set a baseline of the state of nature before industrialization? It was certainly worth a try. Moreover, this might serve not only to

---

[14] Meineke et al. (2019).

[15] Jevrejeva (2001).

[16] Zerefos et al. (2014).

depict changes in vegetation patterns over centuries, but also to test a very topical research hypothesis.

* * *

Ecologists will never tire of finding illustrative names for their concepts and hypotheses. There is, for example, the *portfolio effect*, which puts forward that diversification in natural systems makes them more stable in the face of disturbances. The *Red Queen hypothesis* proposes that species must constantly adapt and evolve in order to survive (from the Red Queen's statement to Alice in Lewis Carroll's *Through the Looking-Glass*: "Now, here, you see, it takes all the running you can do to keep in the same place."). The *green world hypothesis* posits that predators are the key to making our world green, because they keep the number of herbivores from growing out of control and consuming all the plants. Analogous to the role of a keystone in an arch, **keystone species** are those critical to the survival of other organisms, helping hold the ecosystem together. *Landscape amnesia* refers to forgetting how different the surrounding landscape looked decades ago, due to the gradual changes from year to year (more on this in the conclusion). But one concept that can be tested in mountains by means of historical data is the illustrative *escalator effect*.[17] As observed by Humboldt at the Pico del Teide and then in the tropical Andes, mountains compress many climates into a small area: "*on this steep surface climbing from the ocean level to the perpetual snows, various climates follow one another and are superimposed, so to speak.*"[18] Each climate zone corresponds to a specific assemblage of plants and animals adapted to a specific combination of environmental conditions, of which temperature is among the most prominent.

Today, global warming is rapidly reshuffling these mountain climates. Warmer temperatures are pushing mountain-dwelling species ever higher as they try to stay in their comfort zone. The *escalator effect* describes the fact that, in order to survive, species need to ascend the mountain—presumably until there is nowhere left to go. The result is that global warming acts as the escalator to extinction for many species that live near mountaintops.[19] To provide evidence of such an escalator effect, it is necessary to document the vertical migration of species through time; hence the need for historical data. Alarmingly, because the climate is changing so quickly, there is no need to go back that far in the past to observe altitudinal migrations on mountain slopes.

In 1985, the ornithologist John Fitzpatrick of the Cornell Ornithology Lab (one of the research centers the most like paradise I have ever visited[20]) conducted a bird survey in the Cordillera del Pantiacolla, some 130 km northeast of Cuzco in Peru.

---

[17] Marris (2007).

[18] Humboldt and Bonpland, Geography of Plants, p. 78.

[19] Freeman et al. (2018).

[20] The lab is nestled within the Sapsucker Woods, a few kilometers northeast of Ithaca. Visitors to the building are graced with a panoramic view of a pond "with plenty of chairs and spotting scopes available for watching all the activity in the surrounding trees, water, and sky." See https://www.birds.cornell.edu/home/visit/

Resurveying the site 30 years later, he found that birds had indeed responded to recent warming by shifting upslope 40 m on average, with many species experiencing larger shifts of 200–400 m. Mountaintop bird habitats had shrunk by an average of 110 m, with at least four species likely to be locally extinct. In the timespan of just 30 years, warming of less than 0.5 °C had set in motion an escalator to extinction for Andean birds. A sobering observation given that the Earth is expected to warm between 3.3 and 5.7 °C by 2100 in a 'business as usual' scenario.[21] Would older data reveal an even more dramatic effect of warming on mountain dwellers?

* * *

The beautifully illustrated altitudinal profiles published by Humboldt and Bonpland two centuries ago are an irresistible temptation for ecologists to investigate the effect of climate change on plant distribution and to test the escalator hypothesis. Even more so as the vast majority of historical ecological surveys in mountains have been carried out in Europe,[22] and much less is known about centennial vegetation changes elsewhere in the world. Recently, a vegetation resurvey of Pico del Teide was actually carried out,[23] but let's focus on another study using Humboldt's data, which created a lot of buzz when it was published in 2015 in the US Proceedings of the National Academy of Sciences.

In this study, the ecologists Naira Morueta-Holme and colleagues aimed to "follow in Humboldt's footsteps and revisit his pioneering documentation of vegetation elevation ranges."[24] In 1807, a few years after his return to Europe, Humboldt coauthored with Bonpland an *Essay on the Geography of Plants*, in which they presented their amassed data with the innovative *Tableau Physique*. This diagram combines a pictorial view of Mt Chimborazo in Ecuador (and a smoking Cotopaxi volcano in the background) with text denoting the names of plants typical of different elevations in the equatorial Andes (Fig. 3.3a). It is flanked on each side by columns showing the elevation in meters and *toises* (an old French unit of length) and other data such as air temperature, composition and pressure. The concept of the vertical distribution of plants in mountains already existed in the early nineteenth century; as early as 1789, the French geologist and botanist Louis Ramond de Carbonnières had observed the influence of altitude on plants in the French Pyrenees. But the *Tableau Physique* was a breakthrough when it was published, providing for the first time a *unified* view of the physical and ecological implications of mountain verticality, and has since become an iconic milestone—almost a foundation myth—in the history of ecology.[25]

Morueta-Holme's team posited that the vertical zoning model was based on a survey of Chimborazo's southeast slope and could thus be compared with present

---

[21] IPCC (2021).

[22] Steinbauer et al. (2018).

[23] See Renner et al. (2023a).

[24] Morueta-Holme et al. (2015).

[25] For a discussion on the origin of Humboldt's inspiration to draw the *Tableau Physique*, see Päßler (2018); Anthony (2018), pp. 43–44; Renner et al. (2023b).

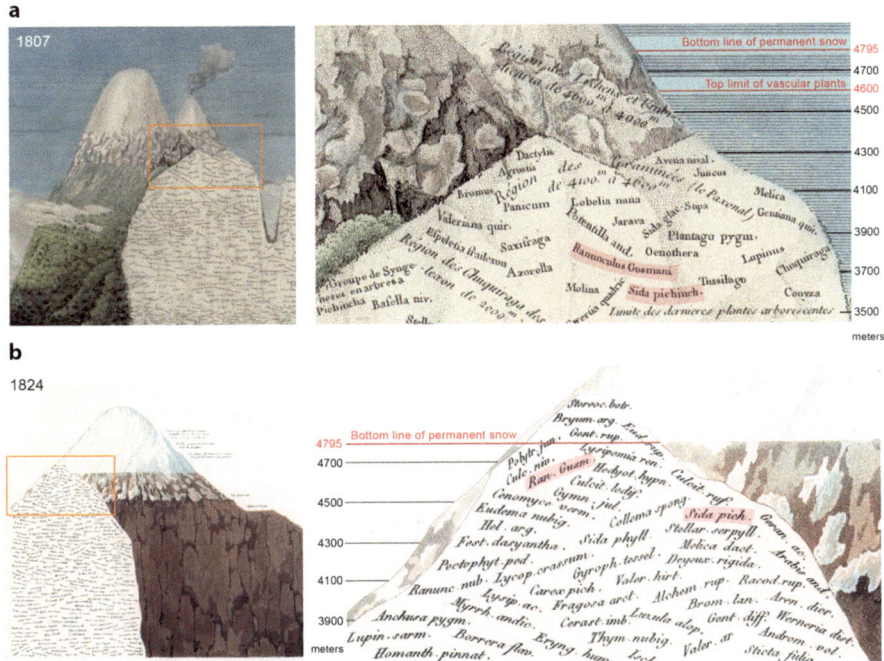

Fig. 3.3 A closer look at the *Tableau Physique*. Details from two of Alexander Humboldt's sketches of the vegetation in the Andes: (**a**) the original *Tableau Physique* in 1807 and (**b**) a later version in 1824. The comparison reveals that part of the data initially published was contradicted in later publications (see, for example, the altitudes of plant species indicated in pale red). The *Tableau Physique* was a dynamic framework used by Humboldt over several decades to refine his unitary view of plant distribution. From Moret et al. (2019). © Moret, Muriel, Jaramillo, Dangles

observations along the same transect. Their findings were groundbreaking. They reported an average upper range limit shift of 675 m for 51 species of vascular plants on Chimborazo since Humboldt's 1802 ascent. That would make the altitudinal migration rate on Chimborazo about 2.5 times faster than those of any plants and animals anywhere else in the world, suggesting that neotropical vegetation may pay a particularly heavy toll concerning climate change. But there were also other disturbing findings: about 86% of the species ascended upslope (as much as 1980 m!), whereas 14% descended (as much as 490 m). These radical findings made me question the methodology, and I contacted Pierre to get a historian's opinion. I suspected that he would also be critical of this study, but I had not anticipated his surprising reaction: "If you want to resurvey plants in the *Tableau Physique*, Chimborazo is certainly not the right mountain to climb!"

This took me aback, as no disciple of Humboldt could resist an ascent of Chimborazo. The influence of the *Tableau Physique* on generations of ecologists, geographers, historians and artists has meant that dozens of textbooks, scientific articles, biographies and documentaries have embraced the legend of the supposed importance of Chimborazo to Humboldt. Shrouded in mystery and romanticism, the myth of Chimborazo is appealing as it feeds the image of a young genius on the slope of a colossal mountain having an 'aha' moment of historic scientific discovery. The historian Andrea Wulf writes, "*As he stood that day on Chimborazo, Humboldt absorbed what lay in front of him while his mind reached back to all the plants, rock formations and measurements he had seen and taken on the slopes of the Alps, the Pyrenees and in Tenerife. Everything that he had ever observed fell into place.*"[26] Though perhaps appealing, this myth is erroneous; this becomes apparent if you read Humboldt's travel diary, a rich and exciting narrative written partly in German and partly in French and which contains the following excerpts:

> On June 23, 1802 (the same day three years ago we had climbed Pico del Teide), we ascended Mount Chimborazo. (...) The bad weather made everything impossible. (...) Our visit of this immense elevation was exceedingly depressing and gloomy. (. . .) Of all the Nevados we visited, Chimborazo possesses the fewest plants. (...) In addition, a vegetation without vigor, not adapted to the beauty of this colossus. [27]

Not exactly the ideal conditions for enlightenment. Worse, because of heavy snowfall the night before their visit and which continued on their way down the mountain, Humboldt could not deploy all the measuring devices he had used on other mountains, and Bonpland could not collect any vascular plants above 3700 m. In fact, the only scientific issues he discusses in his diary concerning his visit to Chimborazo concern volcanism and geodetic measurements, to which I will return in Sect. 5.2. Humboldt probably chose to represent Chimborazo on the *Tableau Physique* for aesthetic and not scientific reasons, as he considered this mountain "the most majestic of all."[28] In his words, "these laborious excursions [to the highest peaks], the narratives of which generally excite the attention of the public, offer but a very small number of results useful to the advance of science."[29] So if Chimborazo is not the right mountain to climb if you want to resurvey plants, is the *Tableau Physique* any help at all for a historical study of the effects of climate change on vegetation and to test the escalator hypothesis? As Jules Verne's scientific heroes like to say when faced with an enigma to be solved: "Let us reason coolly."[30]

* * *

In our case, there are two items of interest in the *Tableau Physique*: plant names and their position on the altitudinal scale. As far as the plant names, we know that the *Tableau Physique*, first published in 1807, was based on a sketch drawn 4 years earlier in Guayaquil, on Ecuador's coast, just before Humboldt and Bonpland left South America. At that time, all the collected samples were packed for the return trip, inaccessible for review, and the identity of many plants was still uncertain at best,

---

[26] The Invention of Nature: Alexander Von Humboldt's New World by Andrea Wulf (2015), Alfred Knopf, p. 87.

[27] Humboldt (2003), pp. 219–222.

[28] Humboldt (1810), p. 106.

[29] Ibid.

[30] Verne (1875), p. 494.

and mostly unknown.[31] With his historian's eye, Pierre dug into Humboldt's trove of publications and found that the German polymath published a revised version of the *Tableau Physique* in 1824 in which almost all the plant species' names had changed, following identifications made by Bonpland and two German plant taxonomists, Carl Ludwig Willdenow and Carl Sigismund Kunth, who collaborated with Humboldt after his return.

It appears the plant names in the famous 1807 *Tableau Physique* cannot be trusted. What about their altitudinal positions in the Chimborazo profile of the *Tableau*? They also changed substantially between the two versions of the *Tableau* (1807 and 1824), with species such as *Ranunculus gusmani* shifting more than 900 m higher in the later publication (Fig. 3.3b).[32] The conclusion of Pierre's historical analysis was that the elevation ranges of Humboldt's plant zones in the *Tableau Physique* should be interpreted as approximations and should not be used as reliable baseline data for quantifying vegetation change over two centuries. His summary: "For the modern scholar, Humboldt's immense work is a veritable jungle, with explanatory models in perpetual update, which tends to hide the primary data registered in the field due to the eagerness so deeply rooted in Humboldt to find universal patterns." Did this finding sound the death knell for our investigation? Certainly not! While Humboldt was a visionary theorist who sometimes made simplifications in search of universal laws, he was also a rigorous observer equipped with the most advanced technology his time could offer. To obtain information based on evidence, we just needed to go back to Humboldt and Bonpland's raw data.

* * *

What is necessary to provide a full description of a species? The ideal triad is a preserved specimen, a detailed description from the field (including information about where the specimen was found) and an illustration (to bring life to museum specimens that are sometimes worse for wear; Fig. 3.4). While the drawings are generally made by artists (or particularly skilled taxonomists), specimen sampling, preservation and field descriptions are the scientist's job. This requires methodical treatment, in particular to associate each specimen (generally preserved in museum collections) with field data (generally written in notebooks). This rigor is sometimes difficult to achieve for a scientist busy with multiple tasks in the field.

During his travels to the Galápagos Islands, Darwin himself unwittingly intermingled bird collections from different islands, as he forgot to note the locality of each specimen he collected. As a result, Darwin's evolutionary understanding of

---

[31] Humboldt himself recognizes this issue in a commentary on the *Tableau Physique*: "After being back in Europe for a few months, I did not dare to add to this tableau a large number of new genera which we will publish but whose names we are not yet sure about." Humboldt and Bonpland (2010), p. 87.

[32] The fact that the raw data used in successive publications was continuously changing between 1807 and 1824 suggests that in the meantime Humboldt was searching for the evidence that would best support his intuition on plant zonation. However, only the first version, the one that contained serious errors, has remained in the collective memory of earth and life sciences as a seminal document.

Fig. 3.4 Three essential elements to describe a plant species. Humboldt and Bonpland developed a rigorous methodology to study the thousands of plant specimens collected during their expedition. (**a**) Bonpland's *Journal Botanique* showing the reference number for an alpine plant species, *Culcitium nivale* (no. 2249); (**b**) specimen of *C. nivale* affixed with small paper strips and labeled with the reference number from the field notebook (no. 2249); (**c**) color copperplate engraving of *C. nivale* by French botanist and illustrator Pierre Jean François Turpin (1775–1840). Sources: (**a**) photo by Olivier Dangles, (**b**) online digital herbarium of the National Museum of Natural History, Paris, Collection: Plantes vasculaires (P), Specimen P00320241, (**c**) Kunth (1820) *Nova genera et species plantarum*

finches relied on the location information from the collections of his travel companions.[33] As the American psychologist Franck Sulloway explains in his fascinating article 'Darwin and his finches: the evolution of a legend':

> Darwin soon realized that the enigma of the finches could largely be explained if they, like the mockingbirds, were confined to separate islands. He therefore began to solicit information from those shipmates on the Beagle who had made their own private ornithological collections and who, unlike himself, had fortunately kept accurate records of the islands from which they had procured their specimens.[34]

Sulloway further explains that Darwin's collecting practices "reflect typological and creationist assumptions he had brought with him to that archipelago. What localities he did record were noted as largely incidental information to remind himself later of scarce species or noteworthy habitats." [35] No original labels in Darwin's hand have ever been found among his finch specimens at the British Museum.

But decades before, Humboldt and Bonpland did not make such an error, and it is precisely their rigorous execution of sampling and measurements and the meticulous documentation of their observations that make their expedition to South America so remarkable for science. The two methodically cooperated in the field. Humboldt measured dozens of variables to describe the physical environment of collection sites, while Bonpland collected, pressed and dried specimens and entered observations in his notebook. Three years after they came back to France, the herbaria samples were sorted, and in 1807, the two scientists offered the Museum of Natural History in Paris 45 boxes of dry plants containing more than 6000 samples, many of which related to previously unknown genera.[36] As it stood, the herbaria formed a collection with the unique characteristic "not to present a plant whose height at which it grows above sea level cannot be indicated."[37] Kudos! Plant names and the altitude where they were collected; this is exactly the data combination we needed for our study. It is not for nothing that Darwin claimed that Humboldt was "the greatest scientific traveler who ever lived."[38]

---

[33] Captain Robert FitzRoy, Harry Fuller (FitzRoy's steward), and Syms Covington (Darwin's servant).

[34] Sulloway (1982), p. 23. Used with permission of Springer Nature BV, from Darwin and his finches: The evolution of a legend, Journal of the History of Biology, 15; permission conveyed through Copyright Clearance Center, Inc.

[35] Ibid., p. 19.

[36] Humboldt and Bonpland collected 5,800 species of which 3,600 had been unknown. This means that in the 4 years spent on the American continent, they collected almost one-tenth of the known plants, while they enriched the world botanical treasury by 5–6%. See Minguet (1969), p. 571.

[37] Hamy (1904), p. 8, translated from French by the author. Humboldt's travel marked the transition from collecting species to understanding the physical variables that explain their distribution: "although new, the discovery of an unknown genus [of plant] seemed to me far less interesting than an observation on the geographical relations of the vegetable world, on the migrations of the social plants, and the limit of the height which their different tribes attain on the flanks of the Cordilleras." Humboldt (1814–1825), vol. 1, p. x. See also Dettelbach (1996).

[38] In a letter to J.D. Hooker, British botanist and explorer, on 6 August 1881, Darwin acknowledged: "I believe you are right in calling Humboldt the greatest scientific traveler who ever lived." In Clark

* * *

We had come a long way, but the task ahead was still daunting. Let's recap. Thanks to Pierre's efforts and incredible ability to orient himself in Humboldt's bibliographic jungle,[39] we managed to assemble the raw material we needed: the digital scans of plant specimens in Bonpland's herbarium, several botanical publications by Humboldt and his botanist companions published between 1805 and 1825, and a few additional tables with precise altitudinal data. But we were still scratching our heads about how to make the links between all this data: in particular the herbarium specimens and the collection locations.

Pierre eventually realized that the key to connecting all the data was a document by Bonpland that had never been published and was accessible only to those who can read Latin and French. After each botanical excursion, Bonpland made a detailed report of collected specimens in his field notebook, the *Journal Botanique*. All entries registered in its seven volumes were numerically ordered (from 1 to 4528) chronologically from the beginning to the end of the expedition. This number corresponded to the number on the handwritten label of the sample in the herbarium. For example, the species *Culcitium nivale* (literally, 'snow mattress' in reference to the dense fuzz of hairs that cover the plant and were once used to make beds for mountain travelers[40]) was given the reference number 2249, which was recorded in both Bonpland's journal and on the herbarium sheet (Fig. 3.4a, b). This pioneering system was completely new for that period and would prove extremely useful in correspondence between Humboldt, Bonpland and Kunth.

How the *Journal Botanique* is still with us today is a story in itself.[41] Just a step away from being packed in Bonpland's bags as he was leaving France for Buenos

---

and Lubrich (2012), p. 206. Darwin also wrote in his diary (entry for 28 February 1832, Bahia, Brazil) that Humboldt, "like another Sun, illumines everything I behold," Keynes (1988), p. 42. For an in-depth analysis of Humboldt's influence on Darwin, see Leask (2003) and Lansley (2018).

[39] Alexander von Humboldt independently wrote some 23–27 books in 49–52 volumes. During his lifetime he published in journals, contributed to the books of others, and wrote 750–1000 studies, articles and essays. Together with their reprints and translations, these were published more than 3500 times between 1789 and 1859. He also wrote thousands of letters. In Lubrich (2019).

[40] This explanation is given by Bonpland in Humboldt and Bonpland (1808–1809), vol. 2, p. 3. Bonpland further explains that *"A large quantity of these plants are collected, and, after having separated the seeds and the thistledown from the receptacle to which they are attached, a first very thick layer of stems with their leaves is formed, either on the ground or on the snow; a second layer is formed with thistledown; a new layer of leaves is placed on top of this; then finally a fourth composed of thistledown. With the help of this bed, the traveler rests from his fatigue and is free from the fear of waking up with frozen feet."* Another species of *Culcitium, C. rufescens*, is known to provide easily flammable dry matter, which was used by Humboldt to boil water. See Humboldt (1854), p. 41.

[41] After the publication of the first volume of *Plantae aequinoctiales*, Bonpland's contributions to this project came to a halt. All planned publications of the results of Humboldt and Bonpland's investigations proceeded only slowly, mainly due to the latter's work in the imperial gardens in Paris. Consequently, in 1810 Humboldt decided to ask Carl Ludwig Willdenow (1765–1812) to assist him in examining the botanical collection. Willdenow accepted, traveled to Paris, and worked with the plant collections in the herbarium for some months until the spring of 1811. From 1813

Aires in 1816, the journal was taken to Berlin by Kunth, who made great use of it for his examinations of particular species. After his death in 1850, the book came into the possession of Humboldt, who deposited it 1 year later, on behalf of Bonpland, "within the scientific treasury of manuscripts at the Jardin des Plantes,"[42] a few shelves away from Joseph de Jussieu's documents, where it has remained to this day. The *Journal Botanique* was fundamental to Humboldt. And 200 years later, it would also prove fundamental to us.

The *Journal Botanique* is a field notebook containing descriptions of plants (and a few animals) observed or collected in the field, information about flower color and smell, as well as preliminary observations. Each entry also indicates the location where the specimen was found, complemented by rough sketches, mostly details of flowers and fruits. For example, for *C. nivale*, Bonpland indicates: "This is the plant we have found at the highest elevation above snow level; it grows sheltered from the wind at the foot of stones or among rocks." As all plants were added in chronological order, it was relatively easy to focus on entries referring to plants of interest for our historical study:[43] those collected in the highest parts of the Andes. We found that Humboldt and Bonpland collected 120 species in the páramos, of which more than half on one mountain: Mt Antisana. This makes sense. Humboldt's diary mentions that the pair spent 4 days in March 1802 on Mt Antisana, making one ascent up to the snowline. In fact, Antisana is the only mountain on which they made plant collections at several altitudes. If Humboldt ever experienced any eureka moment about the interconnectedness of the physical and living worlds, it is far more likely to have occurred on Antisana, not Chimborazo. This was the mountain we had to climb to resurvey alpine plants.

---

onwards, Kunth, who came to Humboldt's attention through a family connection (he was the nephew of Christian Kunth, Humboldt's tutor), worked on the collection. Meanwhile Bonpland, who still had his journal, was setting his sights overseas. In Paris, Bonpland warmly welcomed South American patriots staying in Europe in search of support for independence. As he would later write: "Since the journey I made with Humboldt in South America, I have developed a particular liking for the Americans." In Bell (2010), p. 20. In autumn 1816, he decided to leave France from the port of Le Havre to join the circle around Simón Bolívar, who was striving to achieve independence for the Spanish colonies in Latin America. The irreplaceable *Journal Botanique* was to make the journey with him, but Kunth traveled to Le Havre expressly to recover it. This story was related by Humboldt (1858) in a letter to M. Élie de Beaumont (in Lubrich 2019, p. 435) and as a note in the final volume of the original copy of the *Journal Botanique*: "Conclusion of the botanical manuscripts produced by M. Bonpland during our expeditions, manuscripts that he, upon embarking at Le Havre for Buenos Aires, at my request entrusted to Mr Kunth so as to serve in the publication of our Nova Genera et Species Plantarum." See Lack (2009).

[42] Letter to the museum's professors-administrators in Paris, in Lubrich (2019), pp. 397–398.

[43] A major challenge in this whole process lay in the fact that many entries in the *Journal Botanique* have no species name, or even genus name (or have names that have since been changed), and conversely the number is often missing from the leaves of the herbarium; and in Humboldt, Bonpland, Kunth (HBK), Nova Genera, the number is never mentioned. Result: There are 'orphan' numbers, without identification, and species of HBK that cannot be related to a reference in the *Journal Botanique*. Pierre was able to solve some of the unknowns, but not all.

* * *

So now we had the precise mountain, the plant names and their altitudes. But this is still not enough for a rigorous plant resurvey allowing historical comparison. As discussed in the previous chapter, each Ecuadorian volcano is like an island, with high heterogeneity in climate, humidity, wind patterns, and soil composition between two locations close to each other. Some local conditions can favor the growth of a plant species in one place but not in another at the same altitude. For a robust comparison, we needed to know the exact location where Humboldt and Bonpland collected their plants. For this, we had to reconstruct their itinerary. Thanks to the pair's documentation skills, this was not as complicated as it sounds: Humboldt's travel diaries were our guide.

From 14 to 18 March 1802, the pair explored the Antisana volcano with three young aristocrats from Quito and a dozen porters and servants. This expedition was grueling: "The first night we spent there was cruel. [...] We spent nearly 24 hours without food, we found only potatoes, [...] there was no candle, we filled the small rooms with smoke from the straw fire that served for light to prepare our lodgings. (...) The wind blew and howled like on the open sea."[44] The hacienda would serve as a base camp for their stay, and Humboldt later became fascinated by the landscape's surroundings: large plains "covered with the most beautiful turf of alpine plants, [...] with purple and azure flowers nicely contrasting with the dark green of the carpet."[45] This experience marked him to such an extent that the Antisana hacienda, which he mistakenly believed to be "the highest inhabited place in the world,"[46] appears on most of his representations of the tropical Andes (Fig. 3.2).

Humboldt's account in his travel journal allowed us to trace the route they followed with great precision and to locate the three stations where they stopped to collect plants and measure the altitude. The first station, of course, is the base camp hut (no. 1 in Fig. 3.5), where they collected more than half of their plant specimens. They then attempted the ascent of Antisana by first heading north to a saddle between Antisana and Chusalongo, at about 2218 *toises* (4323 m, no. 2 in Fig. 3.5), and then headed east, toward the summit, with a first stop at 2340 *toises* (4560 m, no. 3 in Fig. 3.5). The accuracy of the two measurements most likely indicates that they were taken with a barometer. Humboldt then mentions a fourth site, at the edge of the snowline; a cave where they had to shelter to protect themselves from the strong wind. This is the altitude where Bonpland collected the highest plant, a specimen of the 'snow mattress.' If we could find this cave, it would be possible to compare, at this precise location, the current altitude at which this plant grows. But for this we required more information about the cave.

* * *

---

[44] Humboldt (2003), pp. 179–180.

[45] Ibid. p. 177.

[46] Ibid. p. 179.

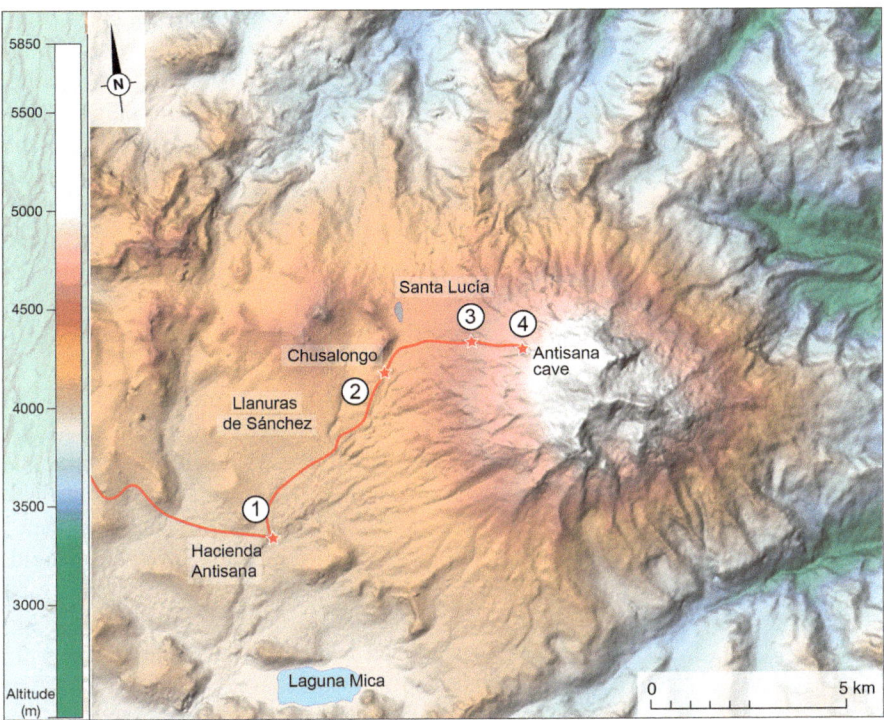

Fig. 3.5   In the footsteps of Humboldt. Map of Mt Antisana (Ecuador) showing the trail taken by Humboldt and Bonpland in 1802 (in red) and the four stations where they made plant collections and altitude measurements. In 2017, 215 years later, a Franco-Ecuadorian team followed this trail and resampled the same locations to assess altitudinal shifts in plant distribution. Map: Pierre and Antoine Moret

If Humboldt is known for a cave, it is not the one on Antisana. Near the village of Caripe, in eastern Venezuela, there is a cave named 'Cueva del Guácharo–Alejandro de Humboldt' that has a statue and a small museum in his memory. Caves are fascinating case studies for geologists and paleontologists as they may reveal secrets hidden by the Earth about its past (in particular, the **Guácharo** cave hides ancient bear remains, see Sect. 5.3). As Jules Verne would later relate in *Journey to the Center of the Earth* (1864), the Guácharo cave "had not given up the secret of its depth to the philosopher [Humboldt] who investigated it to the depth of 760 m."[47]

But in fact, Humboldt had visited the Guácharo cave, in September 1799, less for the geology or paleontology than for ornithology. The cave hosts one of the largest

---

[47] Verne (1867), p.140.

known colonies of guácharos, or oilbirds, a bird so unique that it has its own family (Steatornithidae).[48] Midway between owls and nightjars, these short-legged, chestnut-colored birds speckled with white spots have hook-shaped bills framed by long, whisker-like hairs and roost and nest in large numbers in caves. The oilbird is the only nocturnal flying fruit-eating bird in the world.[49] Throughout their range in the western Amazon, fat young oilbirds are collected by Indigenous people and boiled to extract the oil, which is mainly used in cooking.[50] Also, the crop of the young birds "contains all sorts of hard and dry fruits, which furnish, under the singular name of guácharo seed (*semilla del guácharo*), a very celebrated remedy against intermittent fevers."[51] This strange, unique bird, so well adapted to its environment and embedded in the local culture, captured Humboldt's curiosity and fascination. The Guácharo cave was worth his detour.

The Antisana cave, lying 2000 km southwest as the oilbird flies and 3600 m higher in altitude, also has a connection with ornithology. Humboldt visited this cavity 910 days after the Guácharo cave and found it a good site to estimate the height at which a condor flies. This is an example of Humboldt's obsession with making connections between the physical and the living world, whether plants, insects or birds. If you are wondering how to calculate the altitude of a bird's flight, Humboldt explains the procedure in *Views of Nature*.[52] First you need to know the bird's wingspan: "The largest of the condors found in the Cordilleras near Quito measure 14 feet [4.3 m] across their outstretched wings, and the smaller ones only 8 feet [2.4 m]."[53] Then you need to measure the visual angle (in his case, with a sextant) between the two wingtips when the bird flies just overhead and brush off your trigonometry: height = wingspan/tangent of the visual angle. "A visual angle of four minutes would give a vertical elevation of 1,146 toises [2,234 m]." And finally, you need to know your own altitude: "The cavern (Machay[54]) of Antisana, opposite

---

[48] Humboldt was the first to describe this species previously unknown to science. He wrote in his description: "It is almost the only frugiferous nocturnal bird yet known" (Humboldt 1852–1853, vol. 1, p. 257). The discovery of hundreds of new species since Humboldt's description has not altered his observation of the oilbird's uniqueness.

[49] In the darkness of the cave, the birds navigate using echolocation clicks that are audible to humans. In his description of the cave, Humboldt had an intuition of the use of these clicks: "Their shrill and piercing cries strike upon the vaults of the rocks, and are repeated by the subterranean echoes." (Ibid). Outside of their cave, oilbirds seem to depend on sight (their night vision is aided by retinas packed with one million rods per millimeter, the highest rod density recorded in any vertebrate), and possibly smell, in search of fruit.

[50] These practices led to Humboldt speculating on the conservation of oilbirds: "The race of the guacharos would have been long ago been extinct, had not several circumstances contributed to its preservation. The natives, restrained by their superstitious ideas, seldom have courage to penetrate far into the grotto." (Humboldt 1852–1853, vol. 1, p. 258). Thus only the birds nearest the mouth of the cave were hunted.

[51] Humboldt (1852–1853), vol. 1, p. 259.

[52] Humboldt (1808), Tome 2, p 78.

[53] This maximum wingspan is much larger than that generally observed in condors (about 3.3 m).

[54] 'Cave' in Quechua. Erroneously written "Mackay" in the English version.

the mountain of Chussulongo, and where we measured the bird soaring over the chain of the Andes, lies at an elevation of 2,493 toises [4,859 m] above the surface of the Pacific; the absolute height which the condor reached must therefore be 3,639 toises [7,092 m]." While all this may seem like a lot of work for not much information, the interest of Humboldt's measurement is less the result than the approach. As organisms inexorably ascend the escalator to extinction, estimating the upper altitudinal limit at which they can survive is a major challenge for climate change ecologists. And for the purpose of our resurvey of the *Tableau Physique*, Humboldt's condor-flight estimation gave us precious altitude data to search for the cave. To help us in this, **Priscilla Muriel** and **Ricardo Jaramillo**, two botanists from Pontificia Universidad Católica del Ecuador, joined our team.

Priscilla and Ricardo know the west side of Mt Antisana like the back of their hand. As part of an international long-term monitoring program of the ecological impacts of climate change in alpine environments, they lead projects on this site and have tirelessly surveyed Chussulongo and the side of Antisana facing it in search of suitable study sites. This was the area that had been surveyed more than two centuries earlier by Humboldt and Bonpland. When Pierre and I asked the two Ecuadorians about the cave, they had a good idea where to look for it. A few weeks after our discussion, they followed Humboldt's trail from the Antisana hacienda, left their vehicle near Humboldt's third collection station and, after an hour's walk, spotted a hollow at the foot of a rock wall. The semi-circular opening was high and deep enough for a couple of persons to stand in (Fig. 3.6). On that day, there was no condor flying overhead, but the place had the overall appearance of the cavity described by Humboldt, a rock shelter made of "several huge rocks spewed out from the volcano and piled up."[55] An altitude measurement confirmed their well-grounded hunch. Their GPS satisfyingly displayed 4862 m, a microscopic 0.06% difference from Humboldt's own barometric measurement! The cave was the last clue we needed in our historical investigation started over a year ago. Now we could start surveying the flora and begin to discover what may have changed since Humboldt's time.

* * *

On the cloudy morning of 16 March 2017, 215 years to the day when Humboldt and Bonpland started their 4-day trip up Mt Antisana, Priscilla, Ricardo, Pierre and I headed to the mountain on reconnaissance for our four sampling sites. Our first stop was the Antisana hacienda, where, based on Bonpland's *Journal Botanique*, we knew that a majority of the plants had been collected as this had served as their base camp. We then followed their historical route, marking the second and third sampling stations based on our GPS altitude measurements that agreed with those of Humboldt. There was no information in Humboldt's diary about the exact location and time spent by the scientists at these two altitudes, but we supposed it was not much longer than the time needed to take a barometric measurement: an hour or less.

---

[55]Humboldt (2003), p. 180.

Fig. 3.6 Humboldt's cave today. The cave where Humboldt and Bonpland sheltered in March 1802 on the slopes of Antisana. This cave, near the highest station sampled by Humboldt and Bonpland, would be crucial to documenting vegetation changes over the last two centuries. Note in the foreground carpets of *Senecio nivalis*. Insert: Ecuadorian botanists Priscilla Muriel and Ricardo Jaramillo and field assistant Antonella Bernardi the day they found the cave on 17 November 2016. Photos by Pierre Moret and Ricardo Jaramillo (insert)

We finally reached the cave. The space under the rocks was so small that we were exhilaratingly sure we were standing in the exact footsteps of our illustrious predecessors while they sheltered from the bad weather. To try to connect the physical space in the present with the time in the past, I explored inside and around the cave for any signs of them—a mark on a rock, a piece of cloth or equipment—in vain. No condors either. But this was not why we were here. We knew from Humboldt's diary that due to the bad weather, Bonpland had only collected plants in the immediate surroundings of the cave.

Strictly speaking, for a robust comparative study, our botanical sampling effort at the four sites should be proportional to that of Humboldt and Bonpland. But we had little information on this matter. Weighing up this methodological concern, we finally decided that it would be most useful to sample the four stations as exhaustively as possible. This would allow a thorough recording of the altitudinal limits of the plants they collected along their route so we could compare these ranges with their point measurements. An important realization we came to during the ascension of the peak ourselves is that, knowing that Bonpland collected plants on his way up the mountain, it is likely that the elevation associated with a given plant specimen refers to the first time he saw it: i.e., to its bottom limit. This would be vital information when trying to determine potential changes in plant altitudinal ranges.

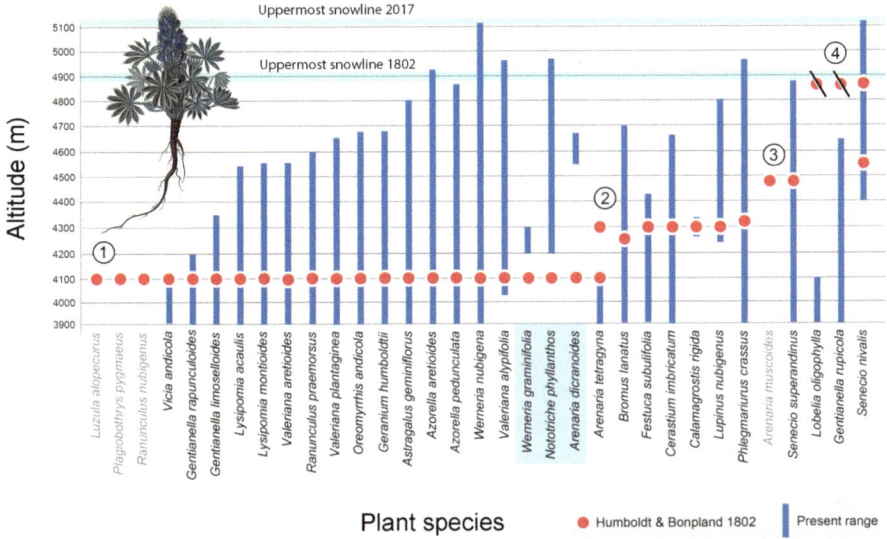

Fig. 3.7 Changing vegetation on a warming mountain. Past (1802, red dots) and current (2017, blue bars) altitudinal ranges of the plant species collected by Humboldt and Bonpland on Mt Antisana. Species in grey indicate plants not found in 2017, which may be locally extinct. Species highlighted in blue suggest an upslope shift from the 1802 bottom limit. From Moret et al. (2019). © Moret, Muriel, Jaramillo, Dangles. Drawing: *Lupinus nubigenus* from Humboldt and Bonpland (1808–09)

After our preliminary visit, Priscilla and Ricardo spent several months resurveying the plants along Humboldt's historical route. To complement their fieldwork, we also compiled all records stored in herbarium databases for the plants collected by Bonpland in this specific zone of the Antisana, collecting 582 additional records. This gave us sufficient reliable historical and present-day data on plant species and altitude to finally answer our initial questions. Was there evidence of an escalator of extinction at Antisana? If so, was its rate faster than in other places in the world? Had it led to extinctions since Humboldt's time?

* * *

Three full years of intense teamwork were summarized in a six-page article we published in 2019 in the US Proceedings of the National Academy of Sciences, as a late rectification to Morueta-Holme et al.'s paper. Our resurvey confirmed that the escalator effect was indeed in evidence at Antisana, though at a less alarming rate than that estimated by those scientists. We found that at least five plant species[56] had extended their present elevation range above the 1802 snowline (Fig. 3.7). One of

---

[56] *Werneria nubigena, Nototriche phyllanthos, Valeriana alypifolia, Phlegmariurus crassus,* and *Senecio nivalis.*

these, *C. nivale* (now called *Senecio nivalis*, Fig. 3.4), the highest plant specimen collected by Humboldt and Bonpland, is now found at altitudes 215–266 m higher than in the early 1800s.[57] This result is remarkably consistent with changes in the freezing altitude on Antisana over the past decades (+10.7 m/decade), as well as with the mean upward shift of 11 m/decade observed at a worldwide scale for both plants and animals. So, the speed of the escalator at Antisana is relatively similar to that recorded in other parts of the world.

What about the bottom range limit of plants in this zone? Has it contracted as plants migrate upward? We found that three plant species[58] now have a bottom altitudinal limit 100–450 m higher than the 1802 record, suggesting a contraction in their lower range, in line with the escalator hypothesis. Has this escalator process led to extinctions? In our intensive resurvey of plants both in the field and in published databases, we were unable to confirm the presence of seven plant species [59] collected in 1802 on Mt Antisana. It is probable that these seven species have become locally extinct, and many other species may follow them up the escalator in the near future.

Our multidisciplinary team was able to decipher Humboldt's *Tableau Physique* and use this 215-year-old data to shed light on the escalator effect in the warming Andes. Yet while historical documents are tempting resources for scientists, they do not give up their secrets easily. Without a historian in the cockpit, this type of time capsule is tricky to navigate. Pierre's disciplines of history and archeology permitted analyses grounded in the material specificity of the place and time. Painstaking attention to historical detail and an exhaustive search for source data, including firsthand accounts, maps, tables and other primary texts in French, German, Spanish and Latin, buttressed our arguments about how to interpret the *Tableau Physique*. A key aspect of this interpretation was to consider Humboldt's *Tableau Physique* as an artistic representation designed to illustrate scientific work in an appealing way and to serve as a supporting visual (in French, *tableau* means both 'data table' and 'painting'). As Pierre put it in one of the dozens of emails exchanged during this study: "The information in Humboldt's work should not be used as if it were objective data, directly processable according to the criteria of current

(continued)

---

[57] While a mean upslope migration rate of about 11 m per decade can be calculated for *S. nivalis* between the two sampling dates (1802 and 2017), it is likely that this displacement was not linear over two centuries. A breaking point in the trend of glacier retreat in the tropical Andes appeared in the late 1970s, with mean annual mass balance (net change in glacier mass) per year almost quadrupling in the period 1976–2010 compared with the period 1964–75.

[58] *Werneria graminifolia*, *Nototriche phyllanthos*, and *Arenaria dicranoides*.

[59] *Arenaria muscoides (Caryophyllaceae), Lupinus sarmentosus (Fabaceae), Luzula alopecurus (Juncaceae), Perezia multiflora (Asteraceae), Ranunculus nubigenus (Ranunculaceae), Plagiobothrys pygmaeus (Boraginaceae), Plagiocheilus peduncularis (Asteraceae)*

science. It requires a critical evaluation that includes taking into account the many contextual parameters from the history of science."

The guidance of historians can assist ecologists in getting the most value out of historical ecological records, helping them to understand their fore-runners who collected data, to interpret this data, and to tease out the discrep-ancies between past and present data, and therefore more accurately identity global ecological responses to climate change. Without a historian, following any past naturalist's footsteps is a risky undertaking. But what if the historical documents themselves are lacking? In that case, how can an ecologist study the effects of time on a place? No fear—there are other ways to investigate space and time in high mountains to assess the effects of climate change. One of these is a substance that Jules Verne admired, but Humboldt was intimidated by: ice.

3.3 Glacier Runes

Ice is a remarkable time capsule for traveling back into environmental archives. In fact, most of our knowledge about historical changes in the Earth's climate and atmosphere comes from ice cores taken at the poles and on mountain glaciers. Drilling down through a glacier reveals snow and ice layers that have captured environmental conditions over a period of time, with the deepest layers dating back from centuries to hundreds of millennia. There are other tools available to look for traces of environmental conditions: tiny **diatoms** and charcoal in lake sediment, growth rings of centennial trees, fossil pollen in muddy bogs, and even piles of mouse droppings in remote caves. But before looking at how ice has been an invaluable tool for climate change scientists, let's take a right brain approach to explore ice as a powerful channel to inspire the imagination, to impress with its beauty and to daunt with its force.

* * *

Ice has long been a source of fascination for humans:[60] the magic of water turning to ice, the lure of immense polar ice caps, the grandeur of snow-covered peaks, the vertigo inspired by crevasses, the quintessential symbol of adventure. The work of Jules Verne exemplifies this, with his heroes often confronting landscapes of ice.[61] His icy environments can be dangerous, "strewn with hummocks, broken obelisks, shattered blocks, overturned pyramids" resembling "a tempest-tossed sea or a ruined

---

[60] See Wilson (2003), McFarlane (2003).

[61] See Rémy (2019a).

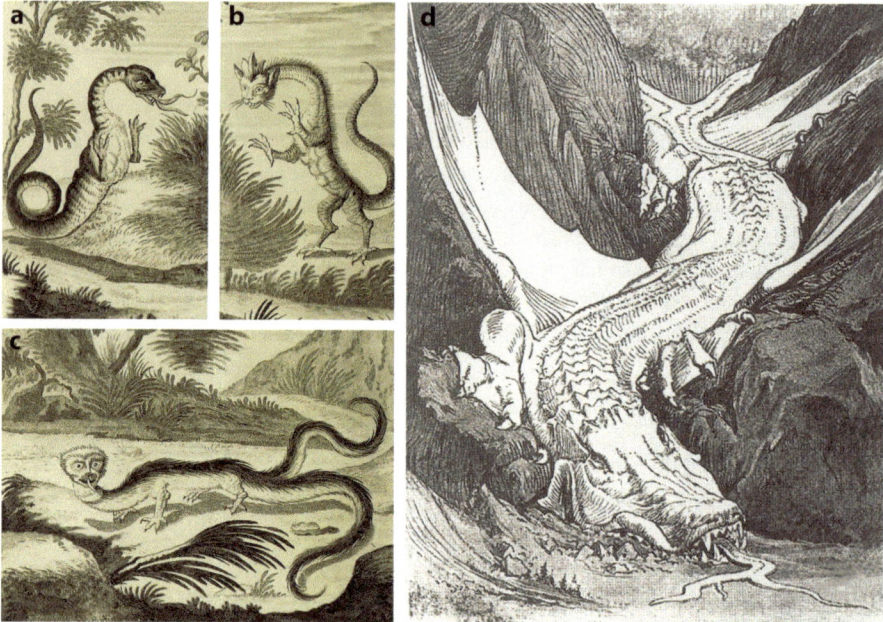

Fig. 3.8 Mountain dragons. (**a–c**) Three different kinds of dragons described by Swiss naturalist Johann Jakob Scheuchzer (1723). (**d**) An imaginative engraving entitled '*Wilderwurm-Gletscher*' portraying the Mer de Glace glacier (on the north face of Mont Blanc in the French Alps) as a dragon. Drawn in 1892 by English mountaineer and artist Henry George Willink (1851–1938)

town."[62] They also harbor extraordinary phenomena explained by science, from frozen giant rhinoceroses buried near the poles to seeming "streams of blood ... flowing beneath the travelers' feet."[63] And they are described as dazzling landscapes "illumined by the rays of the setting sun" with "glaciers flash[ing] back its golden beams with dazzling radiance."[64]

Alongside this, ice was perceived as a romantic symbol of a unique manifestation of nature. Icy environments, including mountain peaks, were considered fearsome places: legendary homes to gods and spirits, witches and fantastic creatures. Of these, dragons. In the early 1700, Swiss naturalist **Johann Jakob Scheuchzer** was so convinced of their existence that he undertook field research in the Alps and published a book enumerating the different kinds of dragons to be found there (Fig. 3.8a–c): some were hairy with two tails and two tongues; others scaly with the head of a cat and a long serpent-like tail; others had a more conventional dragon

---

[62] Verne (1874), p. 275.

[63] Verne (1876), p. 386. This is a reference to 'blood ice', a phenomenon due to the presence of red microscopic algae. This was observed in 1760 by the Swiss naturalist Horace-Bénédict de Saussure in the French Alps and later in 1834 by Charles Darwin while climbing Mt Piuquenes in Argentina. See Rémy (2019b).

[64] Verne (1876), p.136.

appearance, large lizards with bat-like wings and breathing fire.[65] In the nineteenth century, glaciers in the Alps were portrayed as open-jawed dragons clinging to mountain cliffs, slithering down through narrow valleys and threatening to engulf pastures, forests, crops and villages (Fig. 3.8d). Locals also called glacier-fed rivers dragons, describing their fury at full flow, capable of carrying large stones and trees and anything else in their path: *"the dragon was unchained."*[66] The meltwaters were considered dangerous to drink, believed to silt up kidneys and cause "goiters to flourish beneath [farmers'] chins."[67] Dragons may have been fanciful, but the fear inspired by mountains and glaciers was real. As Robert Macfarlane comments in his fascinating *Mountains of the Mind*: *"It was the spread of rationalism in Europe which routed the imaginary dragons from the mountains."*[68] Two contemporaries of Humboldt were brave enough to face these ice dragons: one his mentor, and the other his mentee.

∗ ∗ ∗

The advancing ice at the end of the Little Ice Age (ca. 1300–1850) led scientists to take a close interest in the functioning of glaciers. Among them, **Horace-Bénédict de Saussure**: in the eyes of Humboldt, "the most knowledgeable and intrepid of travelers,"[69] who was both enchanted by glaciers and had a sharp scientific mind. From his appointment as a philosophy professor at the Geneva Academy at the age of 22 until his death at 59, de Saussure traveled the Alps from end to end; his ascent of Mont Blanc (4807 m) in 1787 is considered a major feat of scientific exploration. In his field studies, he collected a wealth of information on which virtually all subsequent theories on glaciers were based.

A key discovery of de Saussure was that glaciers push piles of debris, boulders and sand in front of them, resulting in easily identified buttes marking the limit of the glacier. Savoyards from the Chamonix valley called these buttes *morêna*.[70] A **moraine** is deposited by the glacier and abandoned as it retreats, leaving a rampart of sediment and marking the limit of the ice at a given time. De Saussure was the first to intensively study the formation and structure of moraines and to propose that they could serve to determine the extent of glaciers. He compiled his findings in the monumental four-volume work *Voyages dans les Alpes* (1779), considered by

---

[65] See Scheuchzer (1723), vol. 2, pp. 373–397. See also Fleming (2000).

[66] Bressan, D (2010), Dragons and Geology, http://historyofgeology.fieldofscience.com/2010/10/dragons-and-geology.html, retrieved on 22/01/23.

[67] Mountains of the Mind by Robert McFarlane, Copyright © 2003, GRANTA, p. 105.

[68] Mountains of the Mind by Robert McFarlane, Copyright © 2003, GRANTA, p. 206.

[69] Humboldt (1810), p. 286.

[70] Literally, 'a bulge that forms at the lower edge of a sloping field as a result of descending land:' the scientific term is 'moraine.' See https://www.cnrtl.fr/etymologie/moraine

Humboldt a "precious book which, more than any other, has contributed to the advancement of science."[71]

Yet Humboldt's interest was less in the glaciological knowledge amassed by de Saussure than in his ability to combine dry scientific information with picturesque descriptions and sensations. He wrote about de Saussure's *Voyage dans les Alpes:* "This approach [interweaving science and sensitivity] has been successfully followed in the journey of M. de Saussure," which "amidst dry discussions on meteorology, [...] contains many charming descriptions; such as those of the modes of life of the inhabitants of the mountains, the dangers of hunting chamois, and the sensations felt on the summit of the higher Alps." [72] Humboldt also admired de Saussure's remarkable dedication to taking physical measurements in mountains. De Saussure invented, adapted or improved a myriad of instruments to measure temperature, altitude, magnetic field, sky blueness and clarity, wind, and air humidity, many of which were later adopted by Humboldt during his travels. As Humboldt wrote in reference to de Saussure: "It's a pleasure to walk in the footsteps of a great man."[73] His view of the Swiss naturalist as a mentor is apparent in his first thoughts after ascending Chimborazo, comparing this to de Saussure's ascension of Mont Blanc: "The point where we stopped [...] surpassed by 1,100 m the summit of Mont Blanc, where [...] de Saussure had the joy of arriving, fighting against even greater difficulties than those we had to overcome near the summit of Chimborazo."[74] Humboldt always saw himself as standing on the shoulders of giants.[75]

De Saussure's observations revealed that glaciers come and go and that moraines can be used to track their movements. When a glacier recedes, it leaves behind an imprint in the form of debris, which can provide evidence of the extent of past glaciation. However, de Saussure had no convincing clues about why glaciers move. *Voyages dans les Alpes* sparked an intense scientific discussion, with many of the most brilliant minds of the day starting to think about how glaciers transport stones (in particular large erratic boulders), and how this might be connected to the past climate. In his *Theory of the Earth* (1795), Hutton suggested past "immense valleys of ice sliding down in all directions towards the lower country."[76]

---

[71] Humboldt (1852–1853), vol. 1, p. xx. Instead of the "most valuable book" in the English translation, I used the term "precious book" from the original French text (Relation historique du Voyage aux régions équinoxiales du Nouveau Continent, 1814–1831, p. 32)

[72] Humboldt (1852–1853), vol. 1, p. xxi

[73] Letters to Pictet, p. 156.

[74] Humboldt (1810), p. 106.

[75] In 1813, the German philosopher Johann Wolfgang von Goethe (1749–1832) humorously immortalized de Saussure and Humboldt as two giants of alpine science in an artistic view of nature inspired by the *Tableau Physique*. On the far left, the man who conquered Mont Blanc, on the right the minute figure of Humboldt on Chimborazo, with a condor circling above him. See Goethe J.W. von (1813) "Esquisse des principales hauteurs des deux continents, dressée par M. de Göthe,... d'après l'ouvrage de M. de Humboldt, publié en 1807, sous le titre d'Essai sur la géographie des plantes", printed monograph.

[76] Hutton (1795), vol 2, p. 218.

Later, a trio of Swiss geologists[77] made considerable advances in the study of moraines by taking the first measurements of changes in glacier fronts and studying the boulders far from their terminus. To explain moraine displacements, they posited that the Earth's climate must have oscillated, sometimes colder, sometimes warmer, making glaciers come and go. In 1829, the great German naturalist and philosopher Johann Wolfgang von Goethe also speculated: "For so much ice to exist, cold is needed. I suspect that there was an epoch of great cold in Europe."[78] Observations, measurements, hypotheses: all the scientific ingredients were there to grasp an overall understanding of glacier movement. All that was needed was a champion to fuse them into a theory; a champion whose authority would make these ideas triumph.

* * *

At first glance, Swiss naturalist **Louis Agassiz**, a fish paleontologist by training, seems an unlikely glaciologist. One of his main interests was megalodon sharks, one of the largest marine predators that ever lived. But in fact, the skills he developed to study these sharks' enormous teeth, the largest of which are 17-cm long and weigh more than 1 kg—can equally be applied to the study of rock debris. If fish fossils can tell tales about the world as it once was, why not glacier moraines? In 1837, in the opening address of the Helvetic Society of Natural Sciences meeting in Neuchâtel, Agassiz enthusiastically presented his new ideas on glacial extent, surprising an audience expecting a lecture about ancient fish. He asserted that large fields of ice had covered the globe at one or several points in its long history.

Agassiz's famous talk would be the starting point of modern understanding of ice ages. Like all visionary theories, it had to overcome strong resistance, ranging from virulent sarcasm to disconcerted skepticism—including from the world's most famous scientist of this time: Humboldt. "Your ice frightens me," he would write to Agassiz.[79] But Humboldt's criticisms were those of a friend. The two naturalists had met in Paris in 1832, when Agassiz was a fresh-faced 25 and Humboldt 63.[80] Agassiz would fondly remember one evening at a restaurant at the Palais Royal "which passed like a dream," during which he learned from Humboldt "how to work,

---

[77]Franz Joseph Hugi (1791–1855), Ignace Venetz (1788–1859) and Jean de Charpentier (1786–1857).

[78]For Goethe's citation, see Cameron (1965).

[79]Agassiz (1885), p. 269. Later Humboldt looked more favorably on Agassiz's ideas, as shown in a letter of 15 August 1840 (Agassiz 1885, p. 315): "I cannot close this letter without asking your pardon for some expressions, too sharp, perhaps, in my former letters, about your vast geological conceptions. The very exaggeration of my expressions must have shown you how little weight I attached to my objections ... My desire is always to listen and to learn. Taught from my youth to believe that the organization of past times was somewhat tropical in character, and startled therefore at these glacial interruptions, I cried: 'Heresy!' at first. But should we not always listen to a friendly voice like yours?"

[80]Agassiz (1869), p. 44.

what to do, and what to avoid; how to live; how to distribute my time; what methods of study to pursue."[81]

From that day until Humboldt's death, a close relationship formed that was maintained through letters between the man Agassiz loved "like a father," and the man Humboldt considered "so industrious, so talented, and so deserving of love." [82] This may have been in part because they were so much alike. Like Humboldt, Agassiz was an intrepid explorer, climbing mountains and descending into crevasses. Like Humboldt, who sketched his *Tableau Physique* in Guayaquil before looking at the data, Agassiz announced his ice age theory before having thoroughly investigated alpine glaciers. Like Humboldt, Agassiz had an uncanny ability to condense everything he found into a comprehensive view of the world. Like Humboldt, Agassiz's publications were beautifully illustrated with images that, in addition to their scientific value, were truly artistic. And, like Humboldt on Antisana, Agassiz went into the field to make his observations from a cave.

* * *

To better understand the ice of the past, Agassiz decided to study existing glaciers. In July and August 1838, a year after his Neuchâtel speech, Agassiz started his own study of glaciers, gathering the fundamental facts to support his speculations. His first goal was to find geological evidence. Collecting rocks polished and worn down by ice that had accumulated in moraines would be the key to understanding the glacier's past life.

Agassiz chose the most colossal moraine that exists in Switzerland, in the Aar Valley, for his research. "There is not a glacier in the entire chain of the Alps that can rival the lower Aar glacier for the number and variety of moraines," he wrote.[83] Once there, he set up camp at an altitude of about 1930 m right on the medial moraine of the glacier, "to wrest from it the secrets of its formation and its yearly advance."[84] On this moraine stood a large rock slab balanced atop a heap of boulders that Agassiz made into a rudimentary refuge that he humorously christened the 'Hôtel des Neuchâtelois' (Fig. 3.9). Agassiz shared his cave with a scholarly community of distinguished naturalists: a location, like Humboldt's cave, that allowed them to be in the closest possible contact with what they wished to observe and understand. For more than 5 years, Agassiz and his team made extended stays on the glacier, taking physical, meteorological and biological measurements and surveys. Their work advanced glaciological knowledge in all fields, laying the foundations of modern glacier research. In 1840, Agassiz published a dissertation, *Studies on Glaciers*

---

[81] Agassiz (1869), p. 46. In this text, Agassiz also highlighted that Humboldt's "sympathy for all young students of nature was one of the noblest traits of his long life. His sympathy touched not only the work of those in whom he was interested, but extended also to their material wants and embarrassments." p. 44.

[82] In Irmscher (2013), p. 61

[83] Agassiz (1847), p. 107. Agassiz also chose the Aar glacier due to "its size, its easy access and the proximity of the refuge of Grimsel" (Agassiz 1847, p. iii).

[84] Agassiz (1847), p. iii.

Fig. 3.9 Field work on a glacier. Naturalist Louis Agassiz in the 'Hôtel des Neuchâtelois,' the refuge on the moraine where he stayed with his team during his scientific expeditions on the Aar glacier in Switzerland. Note, on the right, the intrusion of glacier ice (*white arrow*). Entitled '*Agassiz dans l'Hôtel des Neuchâtelois*,' this oil painting was made by Agassiz's personal artist, Swiss painter Jacob Bourckhardt, in 1841, oil on canvas, 33.0 × 42.5 cm—Musée d'Art et d'Histoire of Neuchâtel, No. 19/766, in Charles Gos (1928)

(1840), summarizing the findings and accompanied by a superb atlas of 32 illustrations, which set out his theories about glacial change in detail.

"Of all the phenomena of nature, I know of none more worthy of the attention and curiosity of the naturalist than glaciers."[85] This is how Agassiz enthusiastically starts *Studies on Glaciers*. His meticulous observations of polished and striated rocks, moraine remnants and erratic boulders supported his theory of alpine glaciation, which represented a radical change in the perception of the planet, its history and its climate. Agassiz was also a pioneer in laying other foundations of glaciology. He and his collaborators made crucial discoveries about the structure of the ice, the circulation of water at the surface and below the glacier, and glacial movements.[86] He described glacier dynamics as follows:

> When a glacier is in motion, it rubs against and wears down the bottom over which it moves, scraping its surface which becomes smooth, grinding the broken-off material found between the ice and the rock, pulverizing or reducing it to a clay paste, rounding angular blocks that resist its pressure, and polishing those with large surfaces. At the surface of the glacier,

---

[85] Agassiz (1840), p. 1.

[86] We now know that mountain glacier ice moves by three mechanisms: through sliding across its bed, being squeezed under the pressure of its own weight, and the deformation of the rock on which the glacier rests.

different processes occur. Fragments of rocks that break off adjacent walls and fall on the ice remain there or can be transported to the sides; they advance in this way on top of the glacier, without moving or rubbing against each other [...] and arrive at the extremity of the glacier with their angles, sharp edges, and uneven surfaces intact.[87]

In this way, glaciers are like conveyor belts of eroded rock fragments. He goes on to make a point with far-reaching implications:

Now suppose that, for some reason, one of these immense glaciers loaded with rock debris [...] melts: the result will be that all the angular boulders that were on its surface will collapse onto this irregular heap of rounded debris that the ice was covering.[88]

Indeed, as de Saussure predicted, melting ice leaves behind moraines that are visible long after the glacier retreats, serving as timeless thermometers that can indicate past temperature changes on Earth. Agassiz's discoveries would be the first nuts and bolts of a time machine that would be used by generations of glaciologists and ecologists to come. Let's take it for a test trip in the high Andes.

* * *

Of course, to use glacier moraines as a time machine, a precondition is the former existence of glaciers. So, were there glaciers in the Andes at tropical latitudes? Today, with undisputable proof of the past extent of glaciers, this question seems absurd. Yet it was not formally resolved until the end of the nineteenth century. Humboldt himself, on returning from South America, stated categorically that "The Andes Cordillera does not have any glaciers; this beautiful feature is lacking in this part of the tropics." [89] The reason he gives is that: "There is not sufficient snow, which is very rare at the equator, and the temperature is constant."[90]

Humboldt made some measurements to support his conclusions: "At the equator, the lower limit of snow is one of the most constant phenomena of nature. Bouguer says it is at 4,744 m. A large number of measurements gave me an average of 4,795 m."[91] Humboldt suggests that the difference is due to a measurement error, but he does not consider that this limit may have fluctuated over the past 60 years, when Pierre Bouguer, member of La Condamine's geodetic mission (see Sect. 5.2), made his measurement. This is surprising as, before leaving Europe, Humboldt had "reread, word for word, all of [de Saussure's] works"[92] and had traveled the Mont Blanc region with his *Voyages dans les Alpes* in hand. Humboldt was thus well aware of glacier oscillations described by the Savoyard naturalist. But the permanently snow-capped peaks in Ecuador observed by Humboldt looked quite different from glaciers in the Alps, which were irregular, like tongues snaking down valleys.

---

[87] Agassiz (1842), p. 12.

[88] Ibid.

[89] Humboldt and Bonpland (2010), p. 129.

[90] Ibid.

[91] Ibid. Bouguer was a companion of Joseph de Jussieu during the French geodetic mission.

[92] Humboldt (1868), p. 156.

Eternal snow near the equator appears immobile and resists the impression of a downward-flowing mass of ice.[93]

It was not until the expedition of the German geologist **Wilhelm Reiss** and volcanologist **Alphons Stübel** in 1870–74 that the great masses of snow that cover Ecuadorian volcanoes were recognized, without reservation, as glaciers. And they provided visual proof. Since Humboldt, the scientific study of landscapes had become inseparable from artistic reproduction, and painters became true collaborators of explorers and scientists (see Sect. 5.1). Reiss and Stübel hired the Ecuadorian painter **Rafael Troya** as an illustrator to visually reproduce their geological observations. Stübel trained Troya in the geological interpretation of landscapes so that he could represent them in a way that would bring out the essence of the laws of relief and rock formation.[94] And Troya learned fast. In his 1872 painting of Mt Altar, behind the two condors in the foreground, one can clearly see two large lateral moraines that once flanked the ice (Fig. 3.10). But Troya's painting did not only convey crucial information on Andean geology, it also increased viewers' receptivity and power of observation, allowing them to experience natural phenomena more profoundly. Troya's painting leaves an indelible impression of Mt Altar.

Finally, there was visual proof that the tropical Andes once had and still has glaciers—today we know they make up more than 99% of all glaciers in the tropics (the others are a few ice-covered peaks in eastern Africa and Papua). This means that in the tropical Andes, as in the European Alps, scientists can use moraines to track past glacial extent and travel through time. You just have to find the right moraines to study.

* * *

'Staying alive' sang the Bee Gees on the car radio. This was exactly my concern as our four-wheel drive raced down the tight precipice-lined curves of a dirt road in Milluni Valley, Bolivia (Fig. 3.11). **Antoine Rabatel** liked to drive fast while listening to his favorite playlist of old hits at full volume, no doubt satisfying the sensory side of the right hemisphere of his brain. The other hemisphere is dedicated to pure science. Antoine, a physicist, is a top expert on the glacial past of Bolivia's Cordillera Real. Heir to the tradition initiated by Agassiz's long-term investigation of the Aar glacier, Antoine leads a research network to monitor worldwide glaciers in the context of our warming planet.

---

[93] Another reason why Humboldt erroneously believed there were no glaciers in the Andes is that he was much more focused on tectonic studies. He wrote: "If it be the duty of the men of science who visit the Alps of Switzerland [. . .] to extend our knowledge respecting the glaciers [. . .], it may be expected that a traveler who has journeyed through Spanish America should have chiefly fixed his attention on volcanoes and earthquakes." In Humboldt (1814–1825), vol. 1, pp. 164–165.

[94] Later, the German geologist Hans Meyer (1858–1929), explorer of Kilimanjaro, spent the summer of 1903 exploring the Cordilleras of Ecuador. He was the first to demonstrate that equatorial glaciers are formed, as in the Alps, of a fairly vast upper part consisting of hard surface snow and a lower part formed of ice, sometimes partly covered by rocky debris that makes it easy to confuse with nearby moraines. See Meyer (1908).

Fig. 3.10 Proof in art of the existence of Andean glaciers. The western side of Cerro Altar (Ecuador) showing two clear lateral moraines (*black arrows*), visual evidence of glacier dynamics. The view is from the Collanes Valley at about 4160 m. Note the two condors in the foreground. Painting by Ecuadorian artist Rafael Troya in 1872 (reproduced from Meyer 1908)

Fig. 3.11 Traveling back through time in the mighty Andes. Our vehicle winding along a serpentine road in the Milluni Valley at 4800 m in the Cordillera Real, Bolivia, with Mt Huayna Potosi (6088 m) in the background. Photo by Olivier Dangles

In the same way many ecologists get hooked on wildlife when they are kids, Antoine's passion was sparked by a walk on Glacier Blanc in the French Alps when he was six. By the age of 11, he had already scaled the Dome des Écrins (4014 m). When he discovered that one could make a living exploring glaciers to discover their secrets, this would be the objective of the rest of his studies. In the early 2000s, Antoine obtained a PhD position under the supervision of Bernard Francou, at that time the guru of tropical glacier research. Antoine's thesis was to document the timeline of glacier fluctuations in the Bolivian Andes since the Little Ice Age, some four centuries back. To do this, Antoine spent 2 years crisscrossing the Cordillera Real until he knew every mountain, every glacier, and every moraine intimately. This certainly explained his self-confidence behind the wheel, which went some way to easing my mind. Antoine knew the perfect place to travel back in time, and he was in a hurry to take me there. After all, I was impatient too. I resigned myself to our velocity and watched the spectacular landscape stream by the window in the late afternoon sun: smooth ochre sand dunes, the lush green of bofedales, the dramatic white crags in the distance. The song on the radio once again seemed apt: 'Nothing else matters.'

In these singular circumstances I received my first lesson in lichenometry. "Moraines do not come with a date written on them," explained Antoine. So, glaciologists had to invent ways to reveal their age in order to rebuild a chronological sequence (or **chronosequence**) of glacial retreat. Antoine listed some of the ingenious methods scientists had developed to date rocks. The most common is **radiocarbon dating**—analyzing carbon-14 (C14) in soil and tree debris scraped along by the advancing glacier and found in the moraine. In the high Andes, however, vegetation is scarce (except in bofedales), so moraines do not contain much organic material. Moreover, C14 is not very reliable in dating material less than 300 years old.

Another method reflects our close connection to every part of the universe, as Humboldt marveled in *Cosmos*. Everywhere on Earth is constantly bombarded by cosmic rays originating from high-energy stellar explosions in space. When cosmic rays collide with surface rocks, they fragment the atoms of certain minerals such as chlorine or beryllium. Measuring the amount of these cosmogenic elements gives excellent age estimates for a moraine exposed after glacial retreat. However, it is quite expensive at approximately 2000 dollars per sample, which is unaffordable for large-scale studies.

As a naturalist, I was gratified to find out that Antoine's preferred dating technique is to use lichens. They have a delicate beauty, adding color to grey rocks, testament to the slow time of nature. And while they often go unnoticed, lichens are abundant at high altitudes, and their varied shapes and shades have long been of interest to ecologists. Of course, they did not escape the eye of Humboldt, who reports a "region of lichens and umbilicarias" (a type of lichen with disk-like lobes) in the highest plant zone in his *Tableau Physique*. Among them, Humboldt described the species *Lichen geographicus* (now known as *Rhizocarpon geographicum*), studied by Antoine two centuries later. Humboldt was fascinated by the remarkable ability of lichen to colonize harsh environments, such as the light-deprived environment of mines.[95] Their unexpected value in dating rocks would perhaps have fascinated him even more.

Lichens are tough. They can survive in some of the most inhospitable environments on Earth, from barely cooled lava flows to sunbaked desert stones to boulders scoured by polar winds. This remarkable ability is due to their structure (they lack vascular tissue to pump fluids) and to the symbiotic association that is actually two organisms functioning as a single unit: a fungus and microscopic algal cells. While the fungus provides protection, mineral nutrients and moisture, mainly absorbed from the air and rainwater, the algae offer nutrients from chlorophyll photosynthesis. Recent studies suggest that lichens are even more complex: in many lichens, in addition to the dual support system, a third partner—a yeast—is necessary for the association to be long-lasting. Through this remarkable partnership, lichens have eked out an existence in places across the six continents where no other organisms could possibly live. Yet Antoine's scientific interest is not lichen ecology, it is its longevity, which outranks even the yareta. Some grow at rates of less than 0.1 mm/year and may live over 5000 years. This slow growth and long life make lichens valuable chronometers of glacial retreat. Lichens are visual indicators of the passage of time.

The principle is simple: if you know the growth rate of the lichen, by measuring the diameter (or another attribute related to size) of the largest lichen on the moraine at a site, you can estimate when it started to grow. This allows an approximation of the latest possible date when the glacier was present, as there may be a lag time for lichen colonization after the glacier retreated.[96] But to be reliable, the lichen growth rate has to be calibrated by comparing its growth on supports that can be dated with fairly good precision. Gravestones—with their very precise dates—are one of earth scientists' favorite calibration tools. So, if you see someone wandering around a cemetery with a hammer, he/she may be a glaciologist.[97] The gravestone method being tricky in the Bolivian Cordillera,[98] Antoine calibrated the lichens growing on glacial moraines with the many water channels and *socavones* (mine entrances)

---

[95] One of Humboldt's first books described the growth of plants in dark underground caves; see Humboldt (1793). Lichens were of theoretical interest to Humboldt to support his view of global interconnectedness and considering nature in all its verticality: "The rocky and icy peaks above the clouds, barely discernible to the eye, are covered only with mosses and lichenous plants. Similar cryptogams, sometimes pale, sometimes colorful, branch out on the roofs of mines and underground caves. Thus the opposite limits of plant life produce beings with a similar structure and a physiology equally unknown to us." Humboldt and Bonpland (2010), p. 64; Humboldt even devoted a section of the *Tableau physique des Andes* to "the region of subterranean plants;" see also Anthony (2018), p. 41.

[96] For further information see, for example, Armstrong (2004).

[97] Austrian botanist Roland Beschel (1928–71) pioneered the use of lichens for dating surface rocks. As part of his doctoral research, completed in 1958 at the University of Innsbruck, he studied the ecology of lichens and measured their diameter on dated tombstones in Austrian cemeteries, determining the growth rates of a number of fast-growing species.

[98] The Milluni Valley does have cemeteries, but the gravestones are made of adobe rather than stone.

Fig. 3.12 Reading the passage of time in rocks. Suspended above a bofedal, the South Charquini glacier in Bolivia's Cordillera Real has well-preserved moraines that have been dated using lichens. The confidence interval of the dates is on the order of 10–30 years depending on the moraine. For more details on Charquini's moraines and vegetation colonization, see Małecki et al. (2018). This image was produced in the framework of research conducted by the Laboratory of Physical Geography: Quaternary and Current Environments (LGP, France). Photo by Vincent Jomelli

scattered in the Cordillera Real—perhaps even the one where Joseph de Jussieu observed his first yareta.

* * *

When Antoine stopped the car, we found ourselves facing the ribs of a retreating ice dragon on the opposite slope (Fig. 3.12). More than two centuries of rational scientific studies had not completely kicked the imaginary dragons out of the mountain. The glacier on the southern face of Mt Charquini (5392 m) has left large moraine scars on the flank of the mountain, giving it the appearance of a scene out of a fantastic tale. This glacier may be one of the best places on Earth to observe exceptionally well-defined moraines, providing a palimpsest of recent climate history. Like geological strata when descending to the center of the Earth, the runes left by a glacier have to be read backwards. Measuring the lichens, Antoine could put a date on each moraine, framing the glacier's story between 1679 and 1860.

The moraines reveal that the glacier reached a maximal extent in the valley around 1630–80. As the climate grew drier, it began to retreat during the eighteenth and nineteenth centuries, with only minor readvances on a few occasions. From the end of the nineteenth century, increasingly warm conditions triggered a marked generalized retreat of the glacier, which now reaches a mere quarter of its maximum extent. This scenario is the same for other glaciers in the Bolivian Andes, confirmed by Antoine's moraine dating of dozens of sites. Interestingly, the situation is a bit different in the more humid Andes of Colombia and Ecuador, where Antoine has also dated moraines.[99] There, glaciers retreated later, and at the beginning of the nineteenth century, several glaciers even advanced for a decade or two, potentially due to much colder conditions (the temperature decreased by over 0.8 °C).

Moraine reading is not always as simple as on the desertic Charquini, especially when oscillating climatic conditions overlap processes of erosion, transportation, deposition and deformation of rocks. Sometimes, a major glacial advance can even erase the previous stages in its path, resetting the counter to zero. Moreover, in the humid Andes, vegetation grows fast, rapidly covering moraines. These two characteristics may explain why Humboldt did not observe clear moraines in Ecuador that might have led him to suspect that the Andes' eternal snows are in fact glaciers. If he had seen with his own eyes the power of ice to shape the geological landscape so apparent on Charquini, he may have been more inclined to understand the relevance of Agassiz's theory. If he had observed the succession of life growing on moraines of different ages, he might have given more consideration to the importance of geological time (and ultimately evolution) in understanding the interrelationships between the living and physical worlds and how this can be revealed by spatial phenomena.

Charquini's dragon ribs are gold for an ecologist. While medieval knights quested for the magical draconite stones set in dragons' heads, alpine ecologists search for well-preserved moraines to understand changing ecosystem dynamics in a warming world. A moraine chronosequence has the advantage of providing a set of sites, formed from the same parent material, which differ only in the time they were formed. The distance from the terminus of a glacier is therefore a proxy of the age of the site, allowing an understanding of how its ecology has developed over time. Ecologists call this trick of studying ecological dynamics through spatial datasets 'space-for-time substitution.' While there can be risks in assuming that space and time variations are equivalent, the accurate determination of the history of a site through careful dating validates the use of this approach.

Once a chronosequence has been dated, an ecological study can begin. The most visible change along a chronosequence is plant colonization. As Agassiz observed, "although a moraine is the model of sterility, there are still some plants that grow between the blocks in the sand and gravel."[100] On the South Charquini glacier forefield, we sampled and identified the chronosequence of plant assemblages and found that their mean diversity (about 6–10 species per m^2) was quite homogeneous across all moraines during the nearly 200 years from 1679 to 1860. This level of

---

[99] Jomelli et al. (2009).

[100] Desor (1844), p. 176.

diversity was similar to that recorded in the 11,000-year-old bofedal at the foot of the moraine. In other words, 150 years was enough time for most plants growing in the nearby bofedal to colonize new ground after the glacier retreated.

We also found that these plants significantly enriched the topsoil, particularly with carbon and nitrogen, to a depth of 15 cm. However, when probing deeper, the picture was much different. While the soil in the multi-millennial bofedal was 6.5-m deep, in the 330-year-old moraine it was only about 1-m deep, and in the 150-year-old moraine only about 10-cm deep. It takes time to build an ecosystem. If South Charquini's moraines allowed us to track plant and soil development over the last two centuries, would they also be useful to assess more recent changes and to glimpse the first steps of the formation of alpine environments? A moraine grows larger if the glacier's terminus remains at a given limit for a long period. But when glaciers are in constant retreat, at a steady speed, they no longer leave these stone markers behind. At Charquini, the last moraine dated by Antoine was from the year 1914. From that point, we have to use aerial photographs to see the changes in landscapes, as glaciers are no longer moving at a glacial pace.

* * *

The pace of environmental change on Earth is now so fast that to put the staggering figures into context we need to illustrate them with comparisons. Take two pieces of bad news for nature from 2019. "Worldwide, 37,500 km^2 of tropical forest was destroyed in 2019:" this is the equivalent of destroying one football pitch of rainforest every six seconds. Another example: "267 billion metric tons of Greenland ice is lost every year:" imagine a herd of 2000 elephants charging into the sea every second—that is the mass being added to the sea from the melting ice.

To grasp melting rates in the tropical Andes, consider that at its maximal extent the South Charquini glacier covered about 1.3 km^2, or 736 hockey rinks. Today, 330 years later, it covers the equivalent of only 174 rinks. Even more dramatic is the recent acceleration of the melting rate. Between 1660 and 1840, the glacier lost an equivalent of one hockey rink per year. But since 1970, the glacier has lost five rinks every year (Fig. 3.13). The same scenario is repeated for other glaciers in the tropical Andes—in fact, South Charquini is not the worst off. In Ecuador, Mt Carihuairazo has only one remaining glacier on its southwest face. In ice-hockey-rink equivalents, its surface area has shrunk (for a given year) as follows: from $790_{(1956)}$, $554_{(1991)}$, $324_{(1998)}$, $115_{(2005)}$, $8_{(2019)}$ to $1_{(2021)}$. Further north, in Venezuela, Pico Humboldt is experiencing the same fate: from $2136_{(1910)}$, $1021_{(1952)}$, $244_{(1998)}$, $103_{(2009)}$, $28_{(2019)}$ to less than $10_{(2023)}$ hockey rinks. Glaciers have thawed into miniature versions of their former selves.

Not only are glaciers retreating, some are disappearing forever. In 2009, presaging the many glacier extinctions to come, Bolivia's Mt Chacaltaya glacier (5421 m), the location of the world's highest lift-served ski resort, completely vanished. The imminent death of these millennia-old masses of ice is profoundly concerning for a

Fig. 3.13 Accelerating melting. The retreat of South Charquini glacier in Bolivia's Cordillera Real over the last 340 years. The colored zones show the surface shrinkage during the time period indicated inside each area. Note the green bofedal at the bottom of the glacier. Illustration by Olivier dangles, based on a Google Earth map accessed on September 2021. Data: Maxar Technologies CNES / Airbus

number of reasons.[101] Yet it is also an opportunity to learn about how high-altitude life colonizes new ground. Over the last 10 years, our team of ecologists has partnered with Antoine to study the dynamics of plants and animals on various glacier chronosequences in Bolivia, Peru and Ecuador. We had been in contact with the organisms themselves in the field, but we also wanted to use them to get a comprehensive view of life patterns and functions after glacial retreat. To achieve this, we analyzed hundreds of studies worldwide that described how natural communities respond to glacier loss. Sticking to the analogy of ice hockey, four rules come to mind to explain our key findings about the succession of life after ice.

---

[101] For a multidisciplinary exploration of the scientific, social and economic dimensions of this glacier loss, see Orlove et al. (2008), Pörtner et al. (2019).

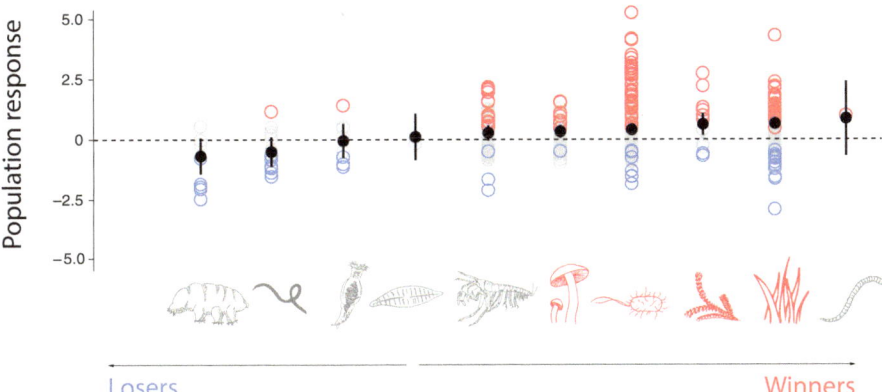

Fig. 3.14 Winners and losers of glacier retreat. Worldwide documented responses of several groups of microorganisms, plants and animals in glacier forefields after ice retreat. From left to right: tardigrades, roundworms, **rotifers**, diatoms, arthropods, fungi, bacteria, moss, vascular plants and segmented worms. Bacteria are clear winners, while the extremophile tardigrades are losers. Modified from Cauvy-Frauné and Dangles (2019). Used with permission of SPRINGER NATURE, from A global synthesis of biodiversity responses to glacier retreat, Nature Ecology & Evolution, 3(12); permission conveyed through Copyright Clearance Center, Inc.

Rule One: There Are Winners and Losers In professional hockey, a game cannot end in a tie. At the end of standard time, teams will go into overtime and, if necessary, a shootout. Someone has to lose. You might think the situation would be different for life in a warmer world since a glacier exposes more land: less ice, new available terrain to be colonized, a more stable substrate, higher temperatures. Such a situation should only lead to winners, right? This is indeed what the broad patterns along a chronosequence show: further from the glacier, soils are enriched by a more diverse and abundant assemblage of microbes, plants, mosses, fungi, insects, spiders and worms. However, this overall increase in diversity is generally to the detriment of cold specialists that have adapted over evolutionary timescales to the harsh environmental conditions of glacier edges. These habitats host a myriad of cold-tolerant life forms, including bacteria, yeasts, algae (cause of the 'blood ice' in Jules Verne), midges, ground beetles, and tardigrades (micro-animals colloquially known as 'water bears' or 'moss piglets' that are almost indestructible and can survive being frozen to almost absolute zero, or –273 °C).

The true breadth of diversity of glacier-adapted life is poorly known, but what is certain is that melting ice is not good news for these unique life forms. This means that glacier retreat does not add new lowland species to the already existing alpine pool, but completely rearranges the whole community. Depending on their ecology and the level of change in local conditions, different life forms will either be favored or disadvantaged after glacier retreat (Fig. 3.14). Most of the time, favored species are generalists that already thrive (or are even invasive) in the lowlands. The losers tend to be specialists with a limited distribution range, sometimes even endemic to a specific mountain. These species disappear because they cannot cope with the new environmental conditions (which become too hot or too dry) or because they are outcompeted by more abundant or successful newcomers. So while the retreat of ice increases local diversity, it results in a more homogenous community across the landscape, lacking the truly unique ice-loving organisms. Not all species are on an equal footing in the face of glacial retreat.

Rule Two: The Smaller the Better   This rule is based on simple physics. In hockey, shorter players have a lower center of gravity that allows them to change direction at high speed without losing their balance. In many situations, smaller players are at an advantage during a hockey game. The situation is similar for the colonization of glacier forelands, the region between the current leading edge of the glacier and the last moraine. Most successful organisms in these environments are small. As low temperatures and rugged terrain make active dispersal such as walking or flying challenging, and upward slopes make dispersal by gravity impossible, not to mention there are very few large animals to hitch a ride on (in fur, feathers or dried mud on an animal's body or feet) or inside (via its intestinal track), most colonizers rely on wind to reach deglaciated ground.

Due to our own weight, humans fail to realize the tremendous influence of the wind on the distribution of organisms. Plant seeds and tiny animals can be swept great heights and distances. And the smaller they are, the further they can disperse. High on the list of these aeronauts are the seeds and spores of algae and lichens. Overall, about nine out of ten plant species that establish recently deglaciated grounds in the tropical Andes are dispersed by wind. Small animals are also effective colonizers of these habitats. On Carihuairazo, DNA analysis revealed that recently deglaciated soils host over 160 taxa, including worms, springtails, tardigrades, spiders and wheel animalcules (rotifers).[102] This bias toward small size is reflected by money spiders, a family (Linyphiidae) of very small spiders that are the fastest to colonize recently deglaciated environments. Their travel secret is ballooning. After climbing to the top of a blade of grass, the spider adopts a 'tiptoe' stance and emits silk into the moving air. When the air generates enough drag to counteract the pull of gravity on the spider, it releases its hold and becomes airborne. It is simple physics.

So successful colonizers of glacier forelands are not a random subset of plants and animals from the lowlands. Dispersal acts like a filter, narrowing the winners down to a handful of small travelers. Of course, the colonization of a new habitat is not simply a matter of getting there. Dispersal is just the first of two crucial steps: to establish a self-sustaining population, plants and animals need food, protection and, depending on their reproduction mode, a mate, all of which demand adaptability, luck and a little help.

Rule Three: Play Like a Team   Hockey is a mesmerizing example of group cohesion. Back-and-forth waves of defending and attacking players are like a murmuration of starlings shape-shifting collectively as they move across the sky. Hockey is an expression of group dynamics and cooperation between players. In the

---

[102]Rosero et al. (2021).

daunting conditions of glacier forelands, many colonizers need help getting established. The exposed, bare ground may overheat in the sun, the ground may freeze during clear nights, the substrate may be unstable, or food resources may be lacking. As we have seen with the cushion plants *Azorella*, some species can ameliorate environmental conditions, facilitating the survival of other species in harsh places. However, *Azorella* may need centuries to grow, while the rate of glacial retreat today is on the order of a few decades or even years. In the forelands of the rapidly melting glaciers of the tropical Andes, nurse plants like *Azorella*, which serve as protection for other organisms, are rare or immature, so colonizers need to find other sources of help. Rocks that provide refuge against wind and cold or small depressions that accumulate water and nutrients are critical homes for newcomers. Many live tightly packed in a few safe sites scattered in an inhospitable matrix of sand and rocks.

But two inhabitants of the high Andes may also be potential facilitators. In the dry Andes of Peru and Bolivia (as well as on the slopes of Ecuador's Mt Carihuairazo and Mt Chimborazo), vicuñas travel in large groups and defecate in communal dung piles that can persist for decades in the rocky moraine landscape. After a few years, these nutrient-rich vicuña latrines may become amazing islands of life, allowing organisms to get a foothold in the barren landscape.[103] Another unexpected helper is lichen. Lichens improve local conditions by buffering temperature extremes of surface soil, facilitating the colonization of pioneer plants and probably animals. Lichens are ecosystem pioneers that fix carbon and nitrogen (sometimes from the air with the help of symbiotic partners) and bind the soil, allowing life to develop in dry, nutrient-starved environments while preventing erosion. They help to form a biological soil crust in glacier forelands, joining a living mantle of microbes that use photosynthesis for energy. In this way, not only do lichens serve as vital clues in reconstructing the deglaciation timeline, they are also likely to be crucial, if poorly studied, protagonists in the recolonization of life after ice.[104]

Rule Four: Don't Discount Luck   In his book *The Success Equation*, US business strategist Michael Mauboussin uses sports analytics tools to place team sports on a continuum where the outcomes range from pure skill to pure luck. He found that hockey is the sport closest to random, with luck contributing more than half (53%) to the season standings. That is not to say that hockey players are any less skilled, but that certain characteristics of this sport (e.g., fewer games in a season, little sustained control on scoring opportunities, more players in action) increase the role of luck in the results. Randomness (or **stochasticity** in ecological parlance) is also a key factor at the glacier forefront. Although different organisms follow individual trajectories during the initial phase of colonization after glacier retreat, their overall diversity is greatly influenced by stochastic drivers: the force and the direction of winds or the incidence of extreme glacial floods. Over the course of succession, the importance of

---

[103] See Reider and Schmidt (2021).

[104] In the tropical Andes, see, for example, Llambi et al. (2021).

randomness is replaced by the skills of organisms to cope with local conditions over the long term. Thriving after glacier loss means navigating along the luck–skill continuum.

"The past is a foreign country, but I also believe in certain continuities— qualities or ideas or struggles or visions that connect the past and present," wrote Aaron Sachs in his book on Humboldt's legacy.[105] Over the last centuries, scientific progress has been a powerful demonstration of the transgenerational transmission of ideas in understanding how nature works. In the Romantic period, de Saussure, Humboldt and Agassiz stood on each other's shoulders to pass on their ideas and perpetuate their legacy as global thinkers. Later, on a parallel literary footpath, Jules Verne used the scientific knowledge of his time to imagine a vision of the future. Today's scientists have invented a myriad of ingenious tools and methods to identify lost continuities between the past and present. Climate change science is perhaps the most prominent example of the use of historical methods in science: it has a particular interest in retracing the history of the Earth to try to anticipate future changes. Across the centuries, the scientific understanding of glaciers has been fed by the human imagination: sometimes feared, sometimes venerated, glaciers have never ceased to thrill us.

The journeys to the past described here provide evidence not only that the rate of warming in the tropical Andes is accelerating, but that this has profound consequences for both the physical and the living worlds. Many mountain plants and animals are either on the escalator to extinction or have not yet migrated due to their slow dispersal ability, a dangerous lag that portends declining diversity of life. On a broader geological timescale, warming has already disrupted the glacial cycles that have been so important in shaping species evolution in the biodiverse tropical Andes. The Earth has now warmed so much that scientists warn that, once totally melted, the ice will not return for a very long time. The period of great melting is on the march. This means that the key to predicting the future of the tropical Andes lies in another critical element: water.

---

[105] Sachs (2006), p. 32.

Chapter 4
Underwater Flies

*When MM Bonpland, Carlos Montufar, and myself on the
23rd of June, 1802, ascended the eastern declivity of Mount
Chimborazo, to a height of 3,016 toises or 5,879 m, and
where the barometer had fallen to 37.69 cm, we found winged
insects buzzing around us. We recognized them to be Diptera,
resembling flies, but it was impossible to catch these insects
standing on the rocky ledges (**cuchilla**), often less than a foot
in breadth, and between masses of snow precipitated from
above.*

Alexander von Humboldt (1850, pp. 232–233)

The new main protagonists of this chapter are shown in Fig. 4.1.

4.1 From Agassiz to Hutchinson

The view in the canyon was staggering. The vertical-walled gorges, dramatic rock
escarpments, tumbling waterfalls and multicolored tree foliage reflecting in the water
of the small river were worthy of a nineteenth-century landscape painting. In my
mind, this place held its own against the most scenic vistas of the high Andes. Even
the turkey vultures overhead in the azure sky brought to mind soaring condors. But
this was Cascadilla Creek in downtown Ithaca, and I was on my way to the Cornell
campus. As I had some time before the weekly Jugatae entomology seminar, I
decided to take a closer look at the stream. Since I was a child to today as a scientist,
running water has always fascinated me, and I am drawn to it wherever I am on the
planet. Crouching on the bank, I picked up a flat stone from the water and inspected
the underside: dozens of creatures scuttled away while others darted deeper into their
burrows. Observing them, mostly immature mayflies and midges, the river of
memory resurrected distant connections to the past.

In 1872, a man in a long black coat and rolled-up trousers leaned over this same
stream, surrounded by his students. He was carefully observing the remarkable
behavior of a small fish, which was diligently gathering little pebbles with its

O. Dangles, *Climate Change on Mountains*,
https://doi.org/10.1007/978-3-031-39528-4_4

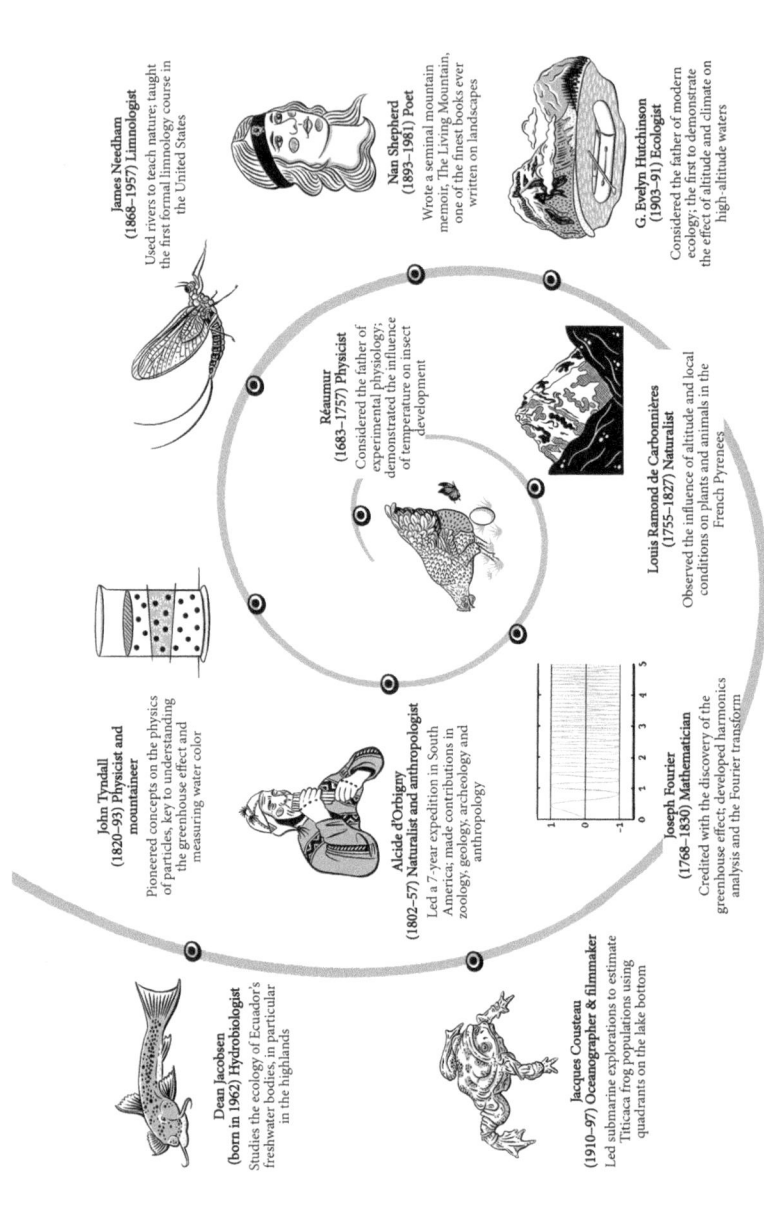

Fig. 4.1 Central figures in this chapter. This spiral timeline shows ten historical or contemporary figures who contributed important ideas mentioned in this section. Drawings by Paula Terán Ospina

mouth and piling them in a small mound. The man explained that this was a male cutlips minnow, trying to build an attractive nest to entice females to spawn with him. He went on to tell the students many other stories about the habits of this and other fish in the region. When food is scarce or competition is high, the cutlips has the gory tendency of plucking out the eyes of congeners or other species as a supplement to its regular diet (mainly minuscule shellfish). The rivers of the region host another underwater nest-builder, the fallfish. It builds a nest that can be up to 1.8 m in diameter and 90 cm high, containing several thousand stones![1]

At the end of his lesson, he declared: *"I was never before in a single locality where there are presented so many branches of natural history as here in this beautiful valley."*[2] For this man who had traveled the world, from the edges of European glaciers to the depths of the Amazon rainforest, observing nature in the act was a source of wonder no matter where he was. This man was Louis Agassiz. Over a century later on the bank of Cascadilla Creek, I could still feel the presence of Humboldt's spiritual son. Cornell was definitely the right place for my sabbatical.

From the moraine of the Aar glacier to Ithaca, what kind of magic had brought Agassiz here? Once again, Humboldt holds the key. In 1846, Agassiz, whose expertise was widely recognized and celebrated at that time, came to the United States on a lecture tour arranged by Humboldt and Lyell. His visit was a marked success, and two years later he was offered a professorship at Harvard University. In 1859, the year Humboldt died, Agassiz founded the Museum of Comparative Zoology there, the first publicly funded science establishment in North America. In New England, Agassiz would not only accomplish his dream of creating a great museum of natural history, he would also discover abundant evidence to support his original ice theory and become an early investigator of the effects of the last Ice Age on North America.

The northeastern United States was a perfect place for this. About 20,000 years ago, the Laurentide ice sheet, an enormous 1800-m-high glacier, flowed across the North American continent, covering what would become New York City, Boston and Ithaca.[3] As the ice retreated some 14,000 years ago, releasing a volume of water equivalent to a 70-m rise in sea level across the world's oceans, it left a profound imprint on the landscape: wide U-shaped valleys, deep glacial lakes and myriad

---

[1] I took the liberty of reconstructing this story based on excerpts from an article by herpetologist and Cornell professor of zoology Albert Hazen Wright (1879–1970) (Wright 1946) and another by James G. Needham (1946). It surely contains some anachronisms, as Louis Agassiz probably did not know about the eye-picking behavior of cutlip minnows or the dimensions of a fallfish nest, which were documented later.

[2] Wright (1946), p. 388.

[3] At the last glacial maximum, some 20,000 years ago, about one-third of the Earth's land area was covered by ice. The largest ice sheet of the last Ice Age was the Laurentide, which covered much of Canada and the northeastern United States. When this enormous ice sheet melted, it formed proglacial lakes. One was the largest lake ever documented on the surface of the Earth (about 349,650 km^2) and was named Lake Agassiz. Its drainage basin was so vast that it is likely that most fish species found today in Canada and the northern US, such as the cutlips, had ancestors that once swam in Lake Agassiz. See Gosnell (2005).

topographical sculptures—moraines, **kettles**, **drumlins**, **esker** and **kames**—all of which reminded Agassiz of similar phenomena in his native country of Switzerland. The region around Ithaca is a true hotspot of these glacial sculptures and graffiti. Cornell itself has various scattered remains left behind as the ice melted away, natural monuments to a dramatic past: a round erratic boulder near Noyes Lodge, a wonderfully striated flat rock shelf across from Boldt Hall Tower, and layers of glacial sediments at the edge of Beebe Lake. And even for people like me who pay more attention to what flies, crawls or climbs than to rocks, there is an unmissable enormous erratic boulder at the southwest corner of McGraw Tower, now a memorial to Ralph Stockman Tarr. This geologist deciphered much of the glacial history of the Finger Lakes region and named the Cornell glacier, a few hundred kilometers southeast of Humboldt's glacier, after his 1896 expedition in Greenland. But despite Cornell's many Ice Age treasures, when Agassiz arrived there in October 1868, he was no longer as interested in glaciers as in nature teaching.

* * *

The concept was quite disruptive for the time. *"No absolute authority imposes appointed textbooks on the student or on any special department of learning. The teacher will come before his class with his own thoughts, with what he brings in his own head rather than in stereotyped print. . ."*[4] With these words pronounced (in a strong French accent) at the inauguration of the university, Louis Agassiz helped lay the foundations of instruction at Cornell.[5] Agassiz's aim was to teach by example and close observation ("'Look, look, look,' was his repeated injunction"[6]), and this

---

[4]Wright (1946), p. 388.

[5]On the occasion of Cornell University's seventeenth anniversary, the memory of Agassiz was commemorated: "Immediately after [his speech, Agassiz] began a course of twenty lectures before very large audiences including nearly all of the student body. They made a deep impression upon all who heard them, and gave a strong impulse to scientific study and research, which has literally remained a power for good in the University from that day to this." From Cornell University Proceedings (1885, p.5).

[6]In a famous sketch reprinted many times ('In the Laboratory with Agassiz,' Every Saturday, (4 April 1874) 16, 369–370), American entomologist and paleontologist Samuel H. Scudder (1837–1911) describes his experience as an assistant in Agassiz's laboratory at the Museum of Comparative Zoology in Harvard (1862–64). Agassiz gave him a scaled-fin grunt specimen (*Haemulon*) preserved in alcohol and asked him to look at it. Scudder reports: "In ten minutes I had seen all that could be seen in that fish, and started in search of the Professor—who had, however, left the Museum; [. . .] Half an hour passes—an hour—another hour; the fish began to look loathsome. I turned it over and around; looked it in the face—ghastly; from behind, beneath, above, sideways, at a three-quarters' view—just as ghastly. I was in despair; at an early hour I concluded that lunch was necessary; [. . .] On my return, I learned that Professor Agassiz had been at the Museum, but had gone, and would not return for several hours. [. . .] Slowly I drew forth that hideous fish, and with a feeling of desperation again looked at it. I might not use a magnifying glass; instruments of all kinds were interdicted. My two hands, my two eyes, and the fish: it seemed a most limited field. I pushed my finger down its throat to feel how sharp the teeth were. I began to count the scales in the different rows, until I was convinced that that was nonsense. At last a happy thought struck me—I would draw the fish; and now with surprise I began to discover new features in the creature. Just then the Professor returned. 'That is right,' said he; 'a pencil is one of the best of eyes.'

is what he put into practice with a large audience of students through lectures and field trips exploring lakes, hills and gorges. Agassiz became a non-resident lecturer at Cornell and, even when absent in person, his spirit exerted a vast influence over the study of natural sciences there. His famous aphorism *"Study nature, not books,"* which hung on his office door, had a lasting influence on many faculty members at the young and innovative university in the late 1890s, decades after his death in 1873.[7] Humboldt shared this conviction, writing to Goethe in 1810: *"Nature must be felt. He who merely sees and abstracts, thinks he describes it, but himself remains forever a stranger to nature."*[8]

Agassiz's inaugural speech thrilled the entire assembly, and in particular one freshman among them, John Henry Comstock. During his whole career at Cornell, he and his wife Anna Botsford Comstock would defend the idea that experience-based education is superior to teaching from textbooks. Anna Botsford Comstock would dedicate her life to leading the nature-study movement that attempted to reconcile scientific investigation with spiritual, personal experiences gained from interaction with the natural world. The goal was not only to acquire scientific knowledge, but to establish sympathy with the 'wonders' of the natural world: whether they be minnows, mayflies or midges. Within this movement, brooks and rivers became an ideal subject for nature study. These are not only a source of childlike pleasures—skipping stones, building rock dams, looking for fish or frogs, swimming, splashing, building mud castles—but a world in itself where one is face to face with nature, from fish in transparent pools to the myriad of creatures hidden under stones. A brook stirs our senses: watching the beauty of water in motion or feeling the power of its flow against one's legs. And one man, walking in the footsteps of Agassiz and Botsford Comstock, would be truly convinced by the potential of running water for science education: **James Needham**.

After obtaining his PhD at Cornell under the supervision of Comstock,[9] James Needham took his passion for fieldwork and freshwater science and spun this into an entirely new field of study: **limnology**, the study of inland aquatic systems. In 1906,

---

[...] With these encouraging words, he added, 'Well, what is it like?' He listened attentively to my brief rehearsal of the structure of parts whose names were still unknown to me: [...] When I finished, he waited as if expecting more, and then, with an air of disappointment, 'You have not looked very carefully; why,' he continued more earnestly, 'you haven't even seen one of the most conspicuous features of the animal, which is as plainly before your eyes as the fish itself; look again, look again!' and he left me to my misery. [...] for three long days he placed that fish before my eyes, forbidding me to look at anything else, or to use any artificial aid. 'Look, look, look,' was his repeated injunction."

[7]Of course, this was not meant to be taken too literally. Agassiz warned about the disconnection between the classroom and the field: *"The pupil studies nature in the classroom, and when he goes out of doors he cannot find her."* See Kohlstedt (2005).

[8]Cited in Nelken (1976, p. 21).

[9]From 1898–1907 J. Needham taught biology at Lake Forest University and then returned to Cornell as assistant professor of limnology. In 1914, when Comstock retired, Needham succeeded him as head of the Department of Entomology at Cornell, a position he held until his retirement in 1935.

Fig. 4.2 Outdoor classroom. Entomologist James Needham (with a hat) and his students during a river ecology class near Ithaca. Needham was head of the Department of Entomology at Cornell University between 1914 and 1935. James G. Needham papers, #21-23-479. Division of Rare and Manuscript Collections, Cornell University Library

at Cornell he taught what would be the first formal limnology course in the United States, and possibly the world.[10] Leading field trips, he taught his students knee-deep in streams and ponds, equipped with nets, buckets and magnifying glasses (Fig. 4.2). In a forested **glen** formed by glaciers and surrounded by steep hills laced with traces of moraines, he set up an outdoor classroom with stones for benches and a speaker's stand. There, as close to his study subject as Agassiz was in his glacier field hut, Needham inspired his students with courses in limnology and insect biology. His verse reflects his teaching style:

> There's hardly a way you can have so much fun
>   As in being outdoors with the brooks as they run[11]

Needham viewed streams as a model to teach nature and stimulate the curiosity of his students, but he was also interested in the taxonomy of particular groups of insects, in particular mayflies and stoneflies, and in understanding how aquatic environments influenced their distribution. He was intrigued the day he received a

---

[10] Hairston and Likens (2009).

[11] Needham (1964), p 28. See also in this volume of the excellent series *Cornell Science Leaflet* some excerpts of Needham's talk 'The common ground of the poet and naturalist' that describes how both approach their work through study that leads to interpretation: "Both begin by observing things; both continue by comparing things; and both end with interpreting things. Each has his message, and although the two messages may be called by names as different as a genetic hypothesis and a sonnet, both may be derived by the contemplation of like materials and may be reached by methods that are not fundamentally different." pp. 5–6.

few glass tubes filled with odd insects originating from lakes above 5200 m (probably winter stoneflies of the family Capniidae[12]), with a letter asking him to take part in a collection effort with an ecological as well as taxonomical aim, and to consider the distribution of these insects at high altitude. The sender was G. Evelyn Hutchinson.[13]

While most ecologists are born naturalists, **G. Evelyn Hutchinson** was a born ecologist. At the age of 5, he was not content just to collect water bugs, but used homemade aquariums to observe the kind of water they preferred. A few years later, he experimented with how different environments affect the emergence of caterpillars and won a school prize, which turned out to be a book on insects by Jean Henri Fabre. At the age of 15, he published his first scientific note on the swimming ability of a grasshopper he observed in a pond in the school woods.[14] From these early adventures in Cambridge, England, and after scientific stopovers in Italy and South Africa, he ended up with a position at Yale University, where he would spend the rest of his life and become the twentieth century's "greatest ecologist."[15] Through his rigorous analytical approach (in particular of water bug interactions[16]), he revolutionized ecology, developing it from a collection of natural history facts to a discipline with a rich theoretical corpus.[17]

Hutchinson's life trajectory and approach to science were strikingly similar to those of the illustrious nineteenth-century naturalists: he acquired diverse knowledge about natural history, geology, archeology and art history, even before entering Cambridge University. He never earned a doctorate in a specific discipline, but bridged fields taught separately such as hydrobiology, physiology, biochemistry, zoology, chemistry and geology, throughout his life retaining an astounding breadth of interests on a dizzying range of topics: nature in illuminated medieval manuscripts, the mathematics of population cycles, prehistoric stone circles, homosexuality, the color of planets, the mating habits of bacteria, etc. He "had a first knowledge of everything that seemed important to the world."[18] He was strikingly modern for his time, embracing new methods (using radioactive tracers to explain the transfer of phosphorus in a pond) and ways of conceptualizing nature, and weaving fascinating and unexpected links between distant regions, from Europe to South Africa, India to the United States. There was something in Hutchinson that

---

[12] J. Needham was the first entomologist to work on Plecoptera in India and was therefore considered by Hutchinson the expert to contact. See Needham (1909).

[13] In Slack (2010), p. 107. Slack erroneously wrote 'John Needham' instead of 'James Needham.'

[14] Hutchinson (1918).

[15] Gould (1989), p. 77.

[16] Water bugs were Hutchinson's favorite study model: "*Hydrobiology satisfies both my physiological interests and naturalist's instinct more than any other branch of zoology.*" In Hutchinson (1979), p. 241.

[17] For a complete biography of E.G. Hutchinson, see Slack (2010).

[18] Quote from the American ecologist Daniel Livingstone (1927–2016) in Slack (2010), p. 392.

made him closer to Humboldt than to any other great naturalist.[19] For Hutchinson, aesthetic considerations were a motive for doing science: *"The values of pure science and the fine arts are identical, the differences being found in the kind of language in which the statement of values are made and in the details of their relationship with the external world."*[20] Hutchinson made his observations with the eye of a scientist and the style and sensitivity of a novelist. Beyond scientific articles, he merged science and literature in fiction that allowed nonscientific readers to experience enjoyment at his descriptions of the world. And as for Humboldt, high summits would be one of his preferred terrains.

* * *

Humboldt's and Hutchinson's ships crossed somewhere in the Atlantic Ocean, although more than 130 years apart. From Europe, Humboldt headed to the Andes; from America, Hutchinson traveled to the Himalayas. In 1932, he was selected as the chief biologist on the Yale North India expedition led by the distinguished geologist and explorer Helmut de Terra, the first modern Humboldt biographer. De Terra may have seen in Hutchinson what he admired in Humboldt: *"a mental prism with facets sparkling on all sides."*[21] Hutchinson would spend five months in Ladakh (sometimes known as 'Little Tibet'), where he conducted pioneering studies on water animals in high-altitude lakes (4300–5500 m). Under strenuous conditions, the expedition carried equipment to the study sites, where, using a small inflatable rubber boat, Hutchinson made vertical profiles of the temperature and oxygen, measured a number of water chemistry parameters, and collected biological samples (mostly crustaceans and insect larvae), providing extraordinary and unique data. Many of the bugs collected by Hutchinson were unknown to science and had to be identified by specialists—hence the parcel sent to Needham containing Himalayan stoneflies (they may have met in 1929 when Hutchinson visited Cornell).[22] Hutchinson then assembled all the data and, in a one-page paper published in *Nature* the year after he returned from Ladakh, showed that fauna assemblages varied according to a lake's chemical composition, in particular its level of oxygen.[23] Hutchinson's trip also triggered his interest in environmental issues. In 1934, in a paper co-authored with De Terra, he hypothesized that the recent rise in lake level observed in Ladakh was associated with large-scale ice movements of *"a similar, though much smaller, order to those which determined glacial and interglacial periods during the Ice Age."*[24] A few years later, he was lecturing on rising greenhouse gases

---

[19] For a comparative analysis of Humboldt's and Hutchinson's view of science, see Kingsland (2010).

[20] The Itinerant Ivory Tower by Everlyn Hutchinson, Copyright © 1953, Yale University Press, p. 228.

[21] de Terra (1955), p. vii.

[22] Hutchinson found Cornell less elitist that Yale with "good workers, good collections and good libraries, rather than endless mock gothic palazzi." Hutchinson (1979), p. 229.

[23] Hutchinson (1933).

[24] De Terra and Hutchinson (1934), p. 320.

and climate change. There is little doubt that if Hutchinson were still around, he would be measuring how climate change is affecting high-altitude aquatic fauna. Using his theories and eclectic methodological approaches, let's do this for him with an appropriate water bug model: flies.

* * *

While lichens may be the champions of survival in the botanical world, they pale in comparison to flies in the animal kingdom. Flies live everywhere: on every continent, on land, underground, in the air, in pools of petroleum, in vicuña dung, and even 1400-m underwater in lakes. They have an unrivaled ability to survive in the most hostile of environments. With over 120,000 species worldwide, **Diptera** represents the third largest order of insects after Coleoptera and Lepidoptera, which have attracted more attention from entomologists. They are found in almost unbelievable abundance—it is estimated that there are about 17 million flies for every human.[25] Most spend the longest part of their lifecycle as soft, pale larvae that dwell inside moist matter or water. More than half of the recorded insects crawling on stream bottoms are fly larvae, which occur in very high densities: sometimes tens of thousands of individuals per square meter.

Fly larvae and adults differ more from each other than a grasshopper from a bee. Most larvae have none of the external characteristics seen in adults—no wings, no segmented legs—and many look like worms or maggots. Yet there is a plethora of dipteran lifeforms: for example, water flies have evolved showy external structures to meet the demands of aquatic life, allowing them to breathe, eat, swim or cling to a fixed support against the force of the current. Invisible to the naked eye and difficult to photograph even through a stereomicroscope, these appendages are best visualized through illustration. Ecuadorian artist Paula Terán Ospina, a former biology student at **La Católica** and now an illustrator, took up the challenge of drawing the amazing physical systems of a few fly larvae from Mt Antisana streams. Such work requires close observation and obsessive attention to detail, artistic creativity and curiosity about the natural world, a connection to living things and excitement about revealing the 'never-before-drawn.' Her "fine lines"[26] were guided by her meticulous observation of each organism to reproduce the exact shape, size and location of dozens of segments, holes, hooks, combs, spurs, brushes and tiny hairs, each fractions of a millimeter in width. "A pencil is the best of eyes" as Louis Agassiz said.[27] Paula's drawings share the strangeness and beauty of what is invisible to most

---

[25] McAlister (2017).

[26] In reference to the remarkable book on Vladimir Nabokov's scientific art by Blackwell and Johnson (2016). Master of prose in multiple languages, Vladimir Nabokov (1899–1977) was also a world expert on butterflies. Nabokov taught at Cornell University from 1948 to 1959 and spent almost every summer traveling west to work on his writing and search for new butterfly specimens. His masterpiece *Lolita* is proof of the cross-fertilization of his efforts: "My pleasures are the most intense known to man: writing and butterfly hunting" [. . .] "Does there not exist a high ridge where the mountainside of 'scientific' knowledge joins the opposite slope of 'artistic' imagination?" he wrote (Nabokov 1973, p. 3 and p. 330).

[27] See note 6.

of us, using "the precision of art" to show "the beauty of science."[28] The six flies in different life stages she illustrated (Fig. 4.3) took over 100 h of effort, longer than the adult life of many of these flies.

The larval stage of flies is typically much longer than the adult stage, and larvae are highly dependent on their environment. This is why flies are particularly relevant study models as climate change indicators. So, from the perspective of an aquatic fly larva in the tropical Andes, what effects might climate change have? Almost 100 years ago in Ladakh, Hutchison explored how aquatic life adapts to high altitudes. But today ecologists also increasingly pose questions about the future. How will high-altitude flies respond to climate change? Will they die, adapt or move? If they die, which species will disappear? If they adapt, what trade-offs will be at play? If they move, will they just shift to a higher altitude on an escalator to extinction? The protagonists we will focus on to explore these questions are five incredible flies: non-biting midges, torrent flies, blackflies, shore flies and sand flies. Each illustrates how flies adapt to different key challenges of a warming climate: changes in water chemistry, flow, oxygen, temperature and predators. The scientific knowledge was amassed with my friend and colleague **Dean Jacobsen**, professor in freshwater ecology at the University of Copenhagen. Without Dean's contagious enthusiasm and generosity, I might never have returned to streams after my PhD on the ecology of streams in the Vosges Mountains, or known the secrets of high Andean waters. Other figures are also key to the story. In the footsteps of Agassiz, Comstock and Needham and their philosophy of using rivers to teach nature to younger generations, high-altitude streams have been magnets for students from all over the world. Two of them, Patricio Andino and Rodrigo Espinosa, tirelessly worked with Dean and me in the Antisana (Fig. 4.4). Along with the 20 or so other students involved in different ways in the project, they have contributed greatly to revealing the secrets of flies.

4.2   The Color of Water

As a child in Copenhagen, Dean Jacobsen was not particularly interested in biology. His dream was to be a pilot. The old beat-up car he ended up in when he started to study the ecology of Ecuador's streams in 1994 was a far cry from a jet. "*I bought a really old Toyota Landcruiser—that was what the project budget could afford—and began driving around like a lunatic, measuring and collecting everything, every-where. Scientifically I was quite unfocused, but it was great fun! It was terribly inefficient because I had to find out everything myself, with a little help from some Danish botanists at La Católica. Constant car problems contributed to this monu-mental waste of time, the least of which was changing flat tires probably fifteen times during the first two-year project.*" Ill-defined research, plans constantly modified by

---

[28] See Sagan (2016), p 244.

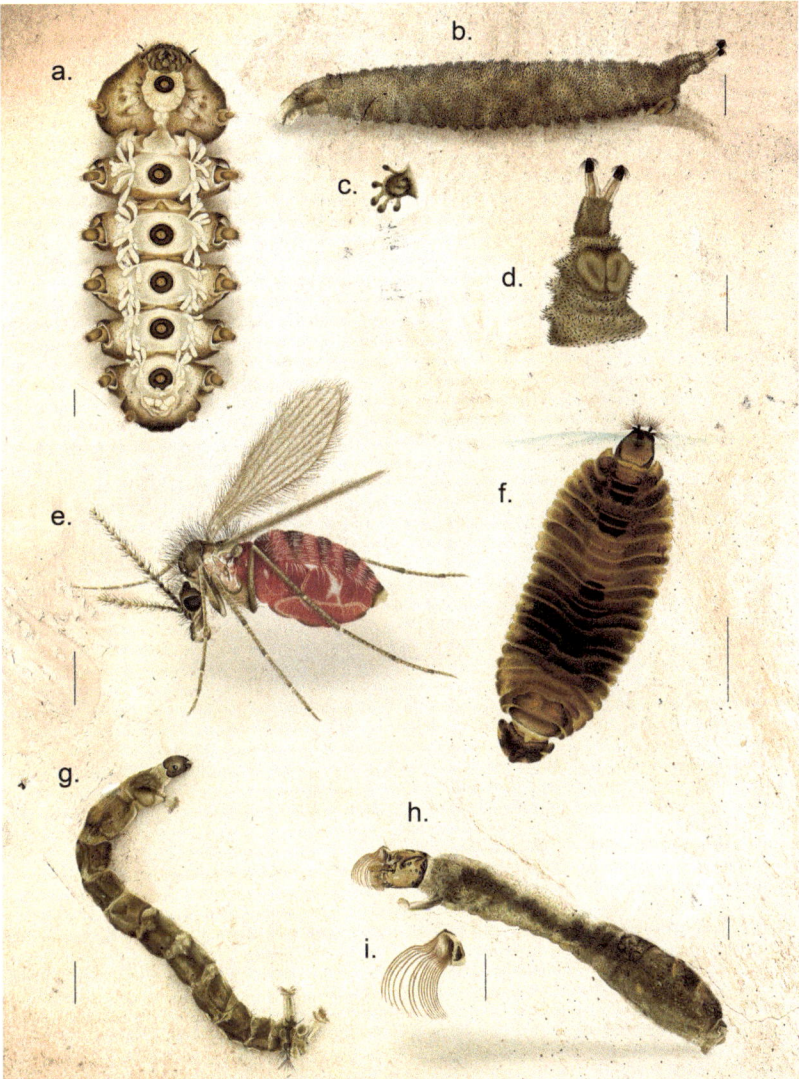

Fig. 4.3 Six fly species. Scientific art showing adaptations of six fly species from Ecuador: (**a**) Ventral view of a torrent midge larva (*Blepharicera*, Blephariceridae) whose six suckers allow it to stay attached to the bottom of raging streams; (**b**) Side view of a shore fly larva (*Dimecoenia*, Ephydridae) and its peculiar respiratory system: (**c**) abdominal spiracles used to take in oxygen during the chrysalis stage between larvae and adult, (**d**) anal spiracle used for breathing during the larval stage; (**e**) side view of a sand fly adult (*Lutzomya*, Psychodidae), vector of Leishmania, after a blood meal; (**f**) dorsal view of a sand fly larva (*Maruina,* Psychodidae) whose anal spiracle allows it to breathe air at the water surface; (**g**) side view of non-biting midge larva (*Podonomus*, Chironomidae) with extended anal spiracles; (**h**) side view of a blackfly larva (*Simulium*, Simuliidae), which attaches to the stream bottom with a silk pad: (**i**) mouth brush to trap food particles. All these species of fly larvae are found in streams on the slopes of Mt Antisana, apart from *Lutzomya*, which is found in the Amazon. Bars = 1 mm. Drawings by Paula Terán Ospina

Fig. 4.4 Under the watchful eye of Humboldt. Ecuadorian undergraduate students Rodrigo Espinosa (left) and Patricio Andino (right) sorting aquatic flies on a folding table outside the hut with the Humboldt plaque (see Fig. 1.2) on the slope of Mt Antisana. Photo by Olivier Dangles

car trouble, and time to think while changing wheels: maybe this was the serendipity that explains why Dean became one of the most influential scholars in tropical freshwater ecology.[29]

A serendipitous path also led me to what would be the starting point of a long and fruitful scientific collaboration with Dean in the study of Andean streams. Soon after arriving at La Católica in May 2006, I knocked on Dean's office door and heard a

---

[29] Dean Jacobsen is professor of freshwater biology at Copenhagen University in Denmark, a department with a long scientific tradition dating back to 1897 when limnologist and entomologist Carl Wesenberg-Lund (1867–1955) established a laboratory outside Copenhagen dedicated to the ecological study of lakes. Influenced by Agassiz's approach to education, Wesenberg-Lund wrote: "It must further be remembered that the study of Nature must always begin with the slightest possible literary ballast. He who has first crammed his head with all that has been written upon a subject will at the moment of observation, when standing face to face with Nature, soon understand that his whole learning is only felt as a burden and restricts his power of observation." Wesenberg-Lund (1920–1921), p. 4.

welcoming *"Adelante!"* The first people I saw on entering were three Ecuadorian students, Patricio Andino (Pucho), Rodrigo Espinosa (Redro) and Verónica Crespo-Pérez (Vero), who were studiously busy sorting and identifying aquatic flies and other bugs under their stereomicroscopes. Then a tall Viking with excellent Spanish appeared from the back of the office. Given my halting Spanish, we settled on English to communicate with each other. After discussing our respective paths that had brought us to Ecuador from Europe, I asked if he could bring me with him next time he went to Antisana for field work. Long before I arrived in Quito, I had spotted the Antisana Ecological Reserve as a potential place to observe and photograph high Andean wildlife. But at that time, the reserve was part of the vast private property of Señor Delgado, an Ecuadorian landlord who grazed his sheep on Antisana, and a permit authorized by him was required to access it. Thanks to Pucho, whose family was friends with the Delgados, Dean had obtained permission to conduct his research in the reserve, so a few weeks later I found myself in his car, unknowingly following in Humboldt's steps as we climbed the twisting road behind the small village of Pintag towards the mythical Mt Antisana. Luckily, by then Dean had bought a new car.

* * *

My first sunny day at Antisana was an unforgettable introduction to the wildlife of the Andean páramo—condor, Andean fox, black-faced ibis, caracara, humming-birds, rabbits, cushion plants, gentians, **queñua, chuquiragua**—all activated my right brain. But my left brain was also engaged, as Dean introduced me to the world of páramo streams and the scientific mysteries he had been trying to elucidate over the years. How do these streams differ from those in temperate mountains? What kind of animals and plants do they host and how are they arranged along altitudinal gradients? Do water bugs suffocate in oxygen-deprived streams above 3500 m? Since the first works in the eighteenth century by the Jesuit historian Juan de Velasco, who proposed a classification of Ecuadorian rivers based on their size, relatively limited knowledge had been gathered on Ecuador's freshwater bodies by scientific explorers until modern times.[30] Dean had found an almost pristine scientific playground in páramo streams. And as he could see I was hanging on his every word given my initial training in hydrobiology, he asked if I would accept co-supervising Pucho, Redro and Vero's undergraduate thesis; his sojourn in Ecuador was coming to an end in a month, and he wanted someone to be officially in charge of his students, though he would carry on supervising their work at a distance. While stream ecology was not exactly in my plans, exploring a mostly unknown world is a major motivation in natural history, so I accepted. Neither of us realized it at the time, but that day on Antisana, Dean set me on a photographic and scientific journey that would define my life over the coming years.

 The Antisana reserve was so close to my home (about an hour by car), and so exclusive to the few who had a permit, that I spent most of my weekends there on

---

[30]Velasco (1841–1844), Tome 1, pp. 13–18; see also Steinitz-Kannan et al. (2020) for a history of hydrobiology in Ecuador.

Fig. 4.5 Fox spirit. Named after the German explorer Wilhelm Reiss (1838–1908), the first European to scale Mt Cotopaxi, the Ecuadorian Andean fox (*Lycalopex culpaeus ssp. reissii*) is endemic to the highlands in Ecuador and southern Colombia. Andean foxes have a strong spiritual significance for Indigenous people in the region, who see it as a symbol of guidance and protection (for a detailed analysis of the place of the fox in the Andean cosmovision, see van Kessel (1994)). We commonly observed the species at our study sites, and I came to regard its generalist and nomadic behavior as a parable for interdisciplinarity (see Sect. 1.1) and serendipitous discovery. Photo by Olivier Dangles

naturalist expeditions. Seized by the beauty of the place, I was initially in more of a contemplative than analytic mode. But eventually the modes overlapped. One day in January 2009 I happened upon a fox that allowed me to take its close-up (Fig. 4.5), I stumbled on a striking side of nature that would keep Dean, myself and our students busy for more than a decade.

* * *

Some of the most important discoveries have been unplanned, from the invention of penicillin at Saint Mary's Hospital in London to the less world-changing but tasty revelation that leaving a piece of cheese in a calcareous cave in Aveyron leads to Roquefort.[31] In his book *Serendipity: An Ecologist's Quest to Understand Nature*, marine biologist James A. Estes considers serendipity as a conceptual high point of

---

[31] Scottish bacteriologist Alexander Fleming (1881–1955) took a vacation from his work in the lab investigating staphylococci bacteria. Upon his return, he discovered that a fungus had contaminated one of his petri dishes, preventing the bacteria around it from growing. He found that the mold produced a chemical in self-defense that could kill bacteria. He named the substance penicillin. The legend behind Roquefort cheese is that a shepherd forgot his meal in a cave one day, and when he

his life as a scientific naturalist. Failures, accidents, inspiring encounters, chance reunions, delays in ship schedules, Estes' major discoveries on the role of marine predators are owed to a myriad of chance events. In a similar way, Humboldt was led to undertake the journey to Latin America as a result of a set of circumstances largely beyond his control:[32] travel and adventure are intrinsically linked to serendipity.

In my own highland voyage, wandering aimlessly through the páramo would ultimately reveal some fascinating secrets of glaciers. As world specialists of serendipity put it: scientists *"must try to take advantage of all their observations, especially accidental observations, because what is called an accident is only an observable effect of a still unknown cause: finding the cause can become the key to a crucial novelty."*[33] I also embraced a more sensory, romantic view of serendipity: *"Often the mountain gives itself most completely when I have no destination, when I reach nowhere in particular, but have gone out merely to be with the mountain as one visits a friend with no intention but to be with him,"* expressed the Scottish poet **Nan Shepherd**.[34] This perfectly describes my experience with Antisana on one particular day.

* * *

Just a few dozen meters from where I had photographed the fox, I was confronted with a remarkable scene. In a flat valley cutting across the western flank of Antisana, two rivers of different colors ran side by side within the same channel, without touching, for several meters (Fig. 4.6). Turbulent eddies eventually mixed the two, gradually diminishing their differences until they merged downstream to become one river. I could see an amazing range of chromatic shades of the light spectrum, from ochres and oranges to blues and greens. I had previously observed each river separately and knew the origin of these colors. On the right was a typical páramo

---

returned a few months later, his cheese had turned blue. The mold (*Penicillium roqueforti*) had transformed his sheep's cheese into Roquefort.

[32] Minguet (1969) reflects on the importance of chance in Humboldt's travels, p. 101: "It was the result of a series of chance events that Alexander found himself embarking for America. [...] What were these fortunate circumstances? First, the Napoleonic campaigns. In 1797, while Humboldt was preparing to visit Italy, he was prevented from doing so by Bonaparte's expedition. A year later, when he was planning to go to Egypt, Bonaparte's sudden decision to attack this region prevented him from entering North Africa. The Directory's project [Humboldt was invited to participate in an expedition to South America, Mexico, California, across the Pacific and then to the South Pole led by French navigator Nicolas Baudin (1754–1803) and funded by the Directory] failed because it was impossible for France to gather the necessary funds due to political events [a war with Austria]. Everything thus conspired to direct Humboldt towards Spain, where he intended to embark for Morocco and Algeria in order to study the Atlas Mountains. Finally, perhaps the happiest of coincidences was the presence of Baron de Forell (1741–1820; mineralogist and a friend of Humboldt's) in Madrid. He was a great friend of the Prime Minister Don Mariano Luis de Urquijo (1769–1817). It is remarkable that Urquijo was there at the right time, since he only served a short time in office, between 17 August 1798 and 13 December 1800."

[33] La Sérendipité. Le Hasard Heureux by Danièle Bourcier & Pek van Andel, Copyright © 2011, Editions Hermann, p. 7.

[34] The Living Mountain by Nan Shepherd, Copyright © 2014, Canongate Books Limited, p. 15.

Fig. 4.6 Where waters meet (1/3). Confluence of a glacier stream (left) and a páramo stream (right) with the cloud-topped Antisana volcano in the background (Antisana Reserve, Ecuador). Photo by Olivier Dangles

river, fed by the water draining from the surrounding peaty soils and vegetation and rich in slowly decomposing organic matter. These absorb most blue-green light, thus the only light available at any depth is orange-red, giving the water its brownish color. The milky water of the river on the left was meltwater flowing from the glacier's snout. The water color is due to suspended fine particles of minerals, usually the size of clay or fine silt, with a pinkish to reddish color depending on the bedrock. The glacier acts as a giant slow-moving mill, reducing mountain slope bedrock to a dust as fine as flour (glaciologists call it 'glacier flour'). In other words, the glacier paints the stream pinkish.

From lake studies, it is known that water color can give high-altitude water inhabitants an advantage, as particles limit light penetration, providing an excellent barrier against harmful UV radiation. But we basically knew nothing about how these visually obvious differences between the two water types and their merging affect the life of stream bugs. Contemplating the scene, watching the convergence of the waters and their chromatic nuances, a myriad of questions began to form. Which stream shelters the highest diversity of bugs or the most unique species? What adaptations are needed to live in these different stream types? How do these streams contribute to aquatic bug diversity on Antisana? How does stream insect distribution vary along the extent of the river? Can insects from one stream colonize the other? What are the ecological consequences of the different colors of the streams and their mixing? How far downstream can this be felt? What would happen if the glacier retreats and the meltwater stream gradually vanishes? This stunning site in front of me with the two streams flowing side by side was a perfect field laboratory, which would allow us to tease apart the effect of water type from confounding factors such as temperature, altitude or vegetation. While I was thinking about hypotheses to test and protocols to set up, my mind suddenly zoomed out to a bird's-eye view of the rainforest, making a short detour to the Amazon.

* * *

Café au lait and black tea. At Manaus in Brazil, the confluence between the two great branches of the Amazon, the muddy waters of Rio Solimões and the dark waters of Rio Negro, creates, as hydrologist Robert Meade once described: *"Six Mississippi Rivers' worth of cafe-au-lait-colored water [. . .] converging here with two Mississippis' worth of black-tea-colored water to produce the greatest hydrologic spectacle on the planet."* [35] This meeting of waters is clearly visible, even from space (Fig. 4.7). This phenomenon occurs due to a difference in the temperature, density and speed of the rivers, the Rio Negro being slower and warmer. In 1853, British naturalist Alfred Russel Wallace was an attentive observer of this sight and the first to propose a classification system for water color in the Amazon. Inspired by *"an earnest desire [...] to see with [his] own eyes all those wonders which [he] had so much delighted to read of in the narratives of travelers,"* including those of Humboldt, Wallace journeyed through the rivers of the Amazon basin between 1848 and 1852.[36] His goal was to draw and describe as much of the flora and fauna as he could, but he was particularly interested in studying fish. Wallace found that Amazonian rivers fall into three main categories (white, black and blue waters) that he believed were *"accounted for by the nature of the country the stream flows through."*[37] He was right. The muddy water, rich in sediment, runs down from thousands of steep mountain valleys of the high Andes, while the nearly sediment-free black waters drain the Colombian hills and forests and are colored by decaying soil and plant matter. In fact, the whole Amazon Basin is a mosaic of different water types connected by the main branch of the Amazon River. Wallace further noticed that most fish in black waters are different from those found in the Amazon. In highlighting that the types of fish living in the Amazon varied according to water color, Wallace was the first to hint at possible ecological reasons for Amazonian fish diversity and the role of water color in this. Was what Wallace posited for fish

---

[35] Earth, Carlowicz M, Friedl L, Ward K, Copyright © 2018, NASA, p. 54.

[36] Citation from Wallace (1853a), p. iii. In his biography, Wallace (1905) reports the influence of Humboldt (vol. 1, p. 232): *"There was in Leicester a very good town library, to which I had access on paying a small subscription, and as I had time for several hours' reading daily, I took full advantage of it. Among the works I read here, which influenced my future, were Humboldt's 'Personal Narrative of Travels in South America,' which was, I think, the first book that gave me a desire to visit the tropics."* He explored the Rio Negro (1849–52) collecting many fish species unknown at that time. Unfortunately, when he was returning to Europe, an accidental fire resulted in the sinking of the ship and all specimens were lost. Only some drawings of the fish remain.

[37] Wallace (1853b), p. 213.

Fig. 4.7 Where waters meet (2/3). Satellite image of the meeting of waters near Manaus in Brazil's Amazonia. Here, the Rio Negro, with its almost black waters, meets the Amazon River (called Rio Solimões in this stretch), with its lighter, mud-colored waters. Map data: Google Landsat/Copernicus May 2022

applicable to insects? To flies? Did Humboldt have anything to say about this? Of course.

* * *

Traveling up the Orinoco River in his search for the Cassiquiare canal that connects to the Amazon basin, Humboldt arrived at the confluence with Rio Arauca in Venezuela. From this point on, Humboldt felt he was entering Dante's *Città dolente*, the gates to hell, so great was the suffering of his company from the millions of bloodthirsty mosquitoes. In his personal narrative, Humboldt dedicates a full section to the misery endured, describing in detail the *"intensity of the plague," "their voracity in certain places,"* the *"effects of the mosquito-sting"* and the *"absence of any remedy."*[38] To escape mosquitoes at night, Humboldt describes dubiously impractical techniques including burying oneself in sand with a handkerchief covering the head, sleeping surrounded by a herd of cows or in the middle of a waterfall, building a treehouse, or filling a small room with smoke and sealing oneself inside. What about using some kind of natural repellent? *"Bonpland and myself tried the expedient of rubbing our hands and arms with the fat of the crocodile, and the oil of turtle-eggs, but we never felt the least relief and were*

---

[38] Humboldt (1852–1853), vol. 2, pp. 272–280.

stung as before."[39] But if Humboldt complained about mosquitoes, it was not so much for the inconvenience of their bite as because they seriously compromised his work.

> Whatever fortitude be exercised to endure pain without complaint, whatever interest may be felt in the objects of scientific research, it is impossible not to be constantly disturbed by the mosquitos [...], that cover the face and hands, pierce the clothes with their long needle-formed suckers, and getting into the mouth and nostrils, occasion coughing and sneezing whenever any attempt is made to speak in the open air.[40]

Incredible as it may seem to anyone who has ever endured swarms of mosquitoes in the rainforest, Humboldt managed to overcome his desperation and turn the mosquito problem to his advantage by making it a new subject of study. Mosquitoes may be a calamity; they are also to be wondered at. Humboldt observed that the distribution of mosquitoes (Culicids) and other Diptera such as black flies (Simulids) and craneflies (Tipulids) presented *"very remarkable phenomena"* [41] related to water color. He wrote:

> If the insects of the genus Simulium abound in the Cassiquiare, which has white waters, the Culex or zancudos are so much the more rare; you scarcely find any there; while on the rivers of black waters, in the Atabapo and the Rio Negro, there are generally some zancudos and no mosquitos.
> [...] tipulid insects, as well as the crocodiles, shun the proximity of the black waters. Possibly these waters, which are a little colder, and chemically different from the white waters, are adverse to the larvae of tipulid insects and gnats.[42]

Thus, 50 years before Wallace, through his own observations as well as local knowledge from native tribes,[43] Humboldt foresaw the effect of water color on the ecology of flies in the Amazon. And six years before Darwin's *On the Origin of Species*, Wallace suggested that the color of Amazon waters influenced the adaptation and diversification of fish assemblages. Reading the accounts of these two learned pioneers would provide some solid clues to the ecological effects of Antisana's meeting of waters. Would the mixing of glacier and páramo streams play a role in generating aquatic diversity, as in the Amazon? If so, would the aquatic diversity of Antisana be endangered by the glacier's retreat and the disappearance of its palette of colors? To answer this question, let's trace one of the Amazon's 1100 tributaries out of Dante's Inferno back to its source in the highlands of Antisana.

---

[39] Ibid, p. 206.

[40] Ibid. p. 273.

[41] Ibid.

[42] Ibid. p. 275.

[43] Humboldt relates an old Indigenous man who assured him that he could tell from the taste of the water where it came from (Ibid. p. 340). "Bring me the waters of three or four great rivers of these countries [...] on tasting each of them I will tell you, without fear of mistake, whence it was taken; whether it comes from a white or black river." Humboldt had great respect for local knowledge. About the Andes Mountains, he said: "The only way to clearly describe the structure of the mountain is to designate the different peaks by the Indian names, which always have a very precise meaning." Humboldt (1854), p. 24.

* * *

Far from the hell of mosquito-ridden lowland jungles, in Humboldt's time, Andean tropical mountains were viewed as a construct of paradise.[44] Antisana is indeed a paradise for a fly ecologist. Here, bugs are of the inoffensive kind: non-biting midges (Chironomidae, Fig. 4.3g) and other two-winged fauna indifferent to humans. Moreover, because of chilly temperatures, adult flies are mostly active only on sunny days or the warmest parts of cloudy days. Otherwise, they are scarcely seen when wandering through the páramo. This is probably why Humboldt, while he had observed streams of different colors on Antisana,[45] failed to notice aquatic flies as he did in the Amazon: *"The elevated plains of Antisana [. . .] contain lakes of fresh water of considerable depth, [. . .] but no fish, and scarcely any aquatic insect livens their solitude. [. . .] All these mountain streams are devoid of organic creatures."*[46]

Humboldt appears not to have upturned stones on the streambeds, where an abundant and diverse assemblage of unusual fly larvae rest. We won't hold it against him, as you need specific sampling equipment to access the world of aquatic flies. While a small sieve may be enough to get a quick overview of the larger animals, professional ecologists have designed various devices to sample the bottom of streams. Just like the Buddhist mandala connects the whole universe through a small circle of sand, aquatic ecologists create their own mandala by throwing a **quadrat** onto a riverbed: the so-called 'Surber.' Invented a century ago by American fish biologist Thaddeus Surber, this device is basically a fine mesh net bag attached to a rectangular aluminum frame. While the frame is held firmly against the substrate, the streambed inside the area is disturbed with the hands to dislodge animals, which are carried into the collecting net by the current. The task is technically easy, but must be meticulously executed to dislodge even the smallest larvae clinging onto rocks. As each Surber sampling takes 1–2 min, when replicated several times in near-freezing glacier meltwater, this can be a bracing task. Once the samples are transferred from the net into plastic vials filled with alcohol, they are taken to the lab for identification under a stereomicroscope.[47] So now that we had our flies, we needed to find a scientific way to measure stream color, beyond my own

---

[44] Historian Jorge Cañizares-Esguerra (2006, pp. 112–128) puts forward an interesting analysis of Andean tropical mountains as a construct of paradise in the early modern period (1500–1800).

[45] "I have observed in the small streams that flow from the volcano and that form the Río de las Tinajillas that some have clear water and others have opaque (lechosa), like water mixed with milk. It is said that the former come from pure snow and the latter from opaque hail with tetrahedral points, which towards the lower limit of the snow clump into masses, similar to sandstone rocks. Is this opacity only due to differences in density, or does it indicate chemical mixtures? Could it be more or less oxygenated snow water?" (Humboldt 2003, p. 178).

[46] Humboldt (1810), p. 291.

[47] Larvae and most soft-bodied adult insects can be kept almost indefinitely in alcohol. But this would have been a challenging task for Humboldt, as he reports (Humboldt 1811–1833, vol. 1, pp. 129–130) *"I must also observe that a long journey on land, very suitable for making a large number of observations, makes it difficult to preserve and transport collections. The jumps of the mules, the*

artistic appreciation of glacier-pink and páramo-brown shades. Yet, as Humboldt warned, *"Anything to do with the color of water is extremely problematic."*[48]

Why problematic? It's not that Humboldt lacked equipment—he took over 40 instruments on his journey and required a train of pack mules to carry them. He cherished these instruments so much that he had them listed in his passport, next to the name of Bonpland, as if they were additional living companions. Humboldt was convinced that the most sensitive instruments were the best way to assemble data for a universal and unified science. Data would indeed soon become the substance of science, and instrumental techniques the means to scientific progress. Humboldt had a complete array of portable barometers, thermometers, hygrometers, electrometers, compasses, dipping needles, magnetic needles, microscopes, timekeepers, eudiometers (for measuring oxygen in the atmosphere), and cyanometers (for measuring sky blueness)—but no water color meters. And there was a good reason for this. This instrument had not been invented yet.

* * *

John Tyndall had not slept all night. The day before, finding himself penniless, he pawned one of his most valuable books for £2 (about $300 today), a deluxe, limited edition that he cherished as it provided information for his scientific reflections and inspiration for his poems, such as:

> The soaring condor plumes his wing
>   on Chimborazo's lofty peak.[49]

The volume was a signed copy of a book by Humboldt, offered to Tyndall by the author himself. Humboldt once attended a public chemistry experiment by the famous German chemist Robert Bunsen (remember the Bunsen burner from your chemistry classes?) and Tyndall, who was his assistant. Tyndall's performance won him a compliment from Humboldt ("When I take up sleight-of-hand work, consider yourself engaged as my first helper."), who offered him one of his books. The night after visiting the pawnbroker, legend has it that Tyndall dreamed that Humboldt came across the volume in a second-hand store, and the very next morning he went to get the book back.[50]

John Tyndall was an intrepid alpinist. He climbed Mont Blanc in the Alps three times, was caught in an avalanche, and miraculously avoided falling into a glacier crevasse, which would have been in his view *"the most beautiful and poetic of all*

---

shock of the boxes against the rocks break delicate and fragile objects. It is almost impossible to preserve jars filled with eau-de-vie."

[48] Humboldt (1852–1853) (French version), vol 1. p. 255. For more discussions about water color, see also Humboldt, Letters to Pictet, p. 198.

[49] Jackson et al. (2020), p. 10.

[50] This paragraph was adapted from Hubbard (1916), pp. 351–352. Many of Tyndall's discoveries were related to observations by Humboldt (e.g. blueness of the skies). See also remarks by Tyndall on Humboldt in Schaumann (2020), pp. 323–324. Note that historian Michael Reidy, a Tyndall specialist, has his doubts about the story of Tyndall selling Humboldt's signed book. "It's not in his correspondence. And, it doesn't sound like Tyndall," he wrote to me.

burials. "[51] Who knows if he would have found his actual death equally poetic—his wife inadvertently gave him a fatal overdose of medicine. But Tyndall's true claim to fame was as one of the most influential scientists of all time. Known for his important contributions to meteorology, magnetism, acoustics and bacteriology, he also studied glacier physics—glacial motion, the fusion of ice—building on earlier theories by Louis Agassiz, among others, on glaciation.[52] In 1859, the year Humboldt died, he formulated the physical basis of the greenhouse effect, an idea that had first been posited three years earlier nearby Ithaca, by US scientist and women's rights activist Eunice Foote.[53] Is that all? No. Tyndall also explained why the sky is blue: its color results from the scattering of the Sun's rays by particles in the atmosphere, a phenomenon now called the Tyndall effect.[54] And it is thanks to Tyndall's discovery that an instrument called a turbidimeter now exists to measure the amount of suspended particles in water, acting as a proxy of water color. It works by passing a light beam through a water sample, whose intensity depends on the concentration, size and color of particles in the water that scatter the light. A light detector set at 90° from the source of the beam measures the reduction in intensity.

* * *

So, on Antisana, when we put a sample of stream water in Dean's portable turbidimeter, what was the result? The turbidimeter's beam went crazy when it passed through glacier stream water. The World Health Organization recommends that the turbidity of drinking water should ideally be below 1 NTU,[55] and urban tap water can achieve levels as low as 0.1 NTU. Our value was 10,000 times higher than that in the most pinkish glacier-fed stream, and generally well above 200 NTU. The brownish páramo streams showed values between 3 and 10 NTU, exceptionally reaching as high as 30 NTU. For both stream types, color intensity correlated with turbidity values, with a higher NTU for more intense colors. Was the instrument up to the task of measuring mixed water? Definitely. The mixed café-au-lait with black-tea waters sat between these two ranges of turbidity values (most were between 50 and 200 NTU). Moreover, the turbidimeter could reveal the proportion of the blend across the many variations of its shades. This magical instrument successfully

---

[51] Hubbard (1916), p. 355.

[52] For a major biography of J. Tyndall, see Jackson (2018).

[53] Eunice Newton Foote (1819–88) demonstrated that certain atmospheric gases, such as carbon dioxide, would absorb solar radiation and generate heat. For further details, see Ortiz and Jackson (2022).

[54] Tyndall was also very interested in the color of snow, ice and meltwater. On the shore of Lake Geneva, he observed the contrast between two waters: "The river was almost white, with the finely divided matter which it held in suspension; while the lake at some distance was of a deep ultramarine," from Tyndall (1860) p. 34. The book also contains a chapter on the "Color of water and ice" describing a method to examine the color of water, pp. 253–256.

[55] Nephelometric Turbidity Units. Nephelometry is another method to measure turbidity, where the intensity of the scattered light is measured (whereas turbidimetry measures the intensity of light transmitted through the sample). Whatever the technique, the same unit (NTU) is used to express turbidity values.

used turbidity to overcome the problematic challenge of measuring the color of the water pouring into Antisana's watershed. Thank you, John Tyndall! Now we would be able to try to find links between water color and the creatures we captured with the Surber sampler.

* * *

"No fear!" This is what the larvae of the non-biting fly *Podonominae* (Fig. 4.3g) seem to say, so at ease are they in the most turbid waters of Antisana's rivers. At an altitude of almost 4900 m, near the glacier front, where turbidity rockets to 800–1000 NTU, the stream bottom is crowded by hundreds of Podonominae larvae per square meter (mainly the genera *Parochlus* and *Podonomus*). They reign as absolute masters, making up between 97 and 100% of the invertebrate life found in the stream, and spending their immature stage in the aquatic equivalent of a London pea souper that penetrates them to their very core. Under the microscope, you can clearly see they have a gut full of glacier rock fragments. How they survive is not yet understood. They have access to plenty of food, either algae or small particles of organic matter blown by the wind, and spiracles to breathe; maybe that's all they need. It's true that living in such an extreme environment has some advantages: there are few predators, and water turbidity is a strong protector against UV, preventing sunburn. Yes, even insects and fish can get sunburned! And it may be that water turbidity doesn't matter to them at all. In the streams we studied, we found that Chironomids were both abundant and diverse. Ladislav Hamerlik, a Chironomid specialist invited by Dean to skim glacier streams, identified 21 different types, with a density approaching 2000 individuals per square meter.[56] Non-biting midges seem to be fond of café au lait.

I have no data for humans, but in the fly world, black tea fans far outweigh those that prefer café au lait. For every one species in the glacial stream, there were three in the páramo stream; and for one individual there were twenty.[57] However, when it comes to non-biting midges, while in páramo streams these are still in the top five of the most abundant organisms (as they are in most of the world's rivers), their diversity is half that found in glacier streams. Several non-biting midge species of the genera *Tanytarsus* (literally, 'stretched-out tarsi'), *Cricotopus* ('ringed leg') and *Metriocnemus* ('ordinary tibia') are absent from páramo rivers, and, from what Ladislav observed in Antisana, these as yet undescribed fly species are strictly endemic to glacier-fed streams. Describing them would require sampling the adults, which has not yet been done. In fact, these midge species may become extinct before being named if the glacier disappears, which is unfortunately the common lot of many species in the tropics.[58] Did we find any flies that have a preference for black tea? Yes—blackflies!

---

[56] Hamerlik et al. (2018).

[57] Instead of 'species,' the right word here is taxa (a group of individuals considered by taxonomists to form a unit), as most invertebrate larvae cannot be identified to species level. For more data, see Espinosa et al. (2020).

[58] See, for example, Coloma et al. (2010).

'Blackfly' can be a misnomer. *"The common name, blackflies, given to the members of this family is not distinctive, for there are many species in other families that are of this color. [. . .] It is like the word blackberry. Some blackberries are white, and not all berries that are black are blackberries,"* wrote the Comstocks.[59] But unlike blackberries, blackflies are not very popular with those who cross their path. With good reason. They are blood feeders, and their massive swarms, clustering over and following their prey, are not only as annoying as mosquitoes, but in extreme cases may cause death through exsanguination. In 1978 in France's Vosges Mountains, 26 cattle were killed by thousands of blackflies. One dead animal had an estimated 60,000 bites.[60] In the páramo though, these flies are more pleasant to observe, as the adults never form the massive swarms that occur in other regions of the world. And even more harmless are the aquatic larvae, wholly devoid of the blood-sucking excesses of adults. Blackfly larvae have cylindrical bodies that are swollen toward the posterior end, which attaches to the supporting surface with a sucking disc (Fig. 4.3h). On the front of the head conveniently near the mouth is a pair of comb-horned 'fans,' whose function is to strain particles of organic matter out of the current as it flows by (Fig. 4.3i). This is why blackflies like black-tea streams—there is plenty of food in the water column there. But wait a minute; didn't Humboldt find that blackflies preferred the white waters of the Cassiquiare canal to the Rio Negro? Did he mix up river names like Darwin did with islands in the Galapagos? Probably not. In fact, the Cassiquiare represents a chemical gradient between clear water at its origin and black water at its mouth, and its many tributaries are a mosaic of different water types. This mixture of different waters may have confounded Humboldt, as it almost confounded us at Antisana.

Surprises happen when you mix café au lait and black tea. But first, let's summarize what we found out about the insects in each specific type of stream color. Pinkish glacier waters shelter some specialized species, especially non-biting midges, but overall, they have low insect diversity and abundance, mostly because of the harsh life bombarded by glacier particles. Blackish páramo streams have more diverse and abundant assemblages, fueled by a plentiful supply of particles of organic matter, which feeds large quantities of bugs, themselves gobbled up by predators. Importantly, along with Chironomidae and blackflies, three other invertebrates complete the top five in non-glacier streams: the mayfly *Andesiops*, the shrimp-like crustacean *Hyallela*, and the caddisfly (take a breath) *Anomalocosmoecus* (what an idea to give it such a name![61]). These are mostly generalist species. Based on the single-color stream patterns, we would have

---

[59]Comstock and Comstock (1895), p. 451.

[60]This was an exceptional blackfly outbreak event caused by contamination of the river by organic waste. It killed many predators and provided food for aquatic larvae. See Noirtin et al. (1981).

[61]The origin of the word *cosmoecus* appears to lie in two Greek roots: the root *kosm-* designates something well ordered or well adorned (cosmetics, for example) and *oecus* (Latinization of *oikos*) is 'house': 'well-adorned' or 'well-built house', referring to the case built by the larva. However, the prefix *Anomalo* refers to 'abnormal,' which adds confusion to the name. See Wiggins and Richardson (1989) for drawings of the larvae and their cases and an alternative etymology.

expected that if glacier and páramo streams mix, just like mixing pink and brown on a painter's palette, this would result in an average of the two in terms of diversity and abundance. This was basically true for abundance, as bug density drastically drops as the waters mix. But the surprise was that, on average, we found consistently *more* species in mixed waters than either glacier or páramo streams. Taking a gradient of colors, from pink to brown, we found a peak in diversity at intermediate levels (Fig. 4.8a). What mechanisms explain this pattern and the coexistence of a greater number of species in merging waters? Here we need a bit of help from G. Evelyn Hutchinson.

* * *

Living Together in Theory and Practice: this is not the title of a self-help guide, but a chapter in Hutchinson's most influential book, *An Introduction to Population Ecology*. One of Hutchinson's lifelong quests was to uncover the mechanisms that allow different species to coexist in the same environment—for example, a lake or a stream. While early ecologists tended to believe that natural systems converge to 'equilibrium,' Hutchinson recognized that non-equilibrium dynamics are crucial in nature. Changes in environmental conditions over time can affect species abundance and competitive hierarchies, and ultimately the number of species that can coexist. Building on these early postulations, in the late 1970s American ecologist Joseph Connell proposed a framework to conceptualize the idea of natural patch dynamics: the intermediate disturbance hypothesis. He pointed out that the richest tropical forests occur where storm disturbance is common, and that coral reefs sustain the highest diversity in areas disturbed by hurricanes, leading to the hypothesis that the most diverse natural systems exist at intermediate levels of disturbance.

Mechanistically, the hypothesis relies on the trade-off between colonization and competition between species. Under very intense disturbance, only a few populations of pioneer species can colonize and become established, leading to minimal diversity. At low levels of disturbance, more competitive species will dominate the ecosystem. At intermediate levels, diversity is maximized, as species that thrive in both systems can coexist. Applied to high Andes streams, glacier streams would play the role of highly disturbed environments where only a few bugs, such as non-biting midges, can survive. Comparatively, páramo streams are low-disturbance systems where the dominant and large mayflies, crustaceans and unpronounceable caddisflies beat many other species in head-to-head competition. Mixed waters are thus intermediate conditions, with turbidity and other stressors keeping down dominant competitors (Fig. 4.8b), reducing competitive exclusion and favoring coexistence. For the moment, this is our best hypothesis to explain the peak of diversity we found in mixed waters. While Humboldt's conception of nature invoked balance and harmony, our study suggests the fundamental role played by disturbance in understanding aquatic life in Antisana, results in line with numerous demonstrations made by ecologists since Hutchinson's original idea. It seems that glaciers do help flies and other organisms live together, not just in theory but in practice.

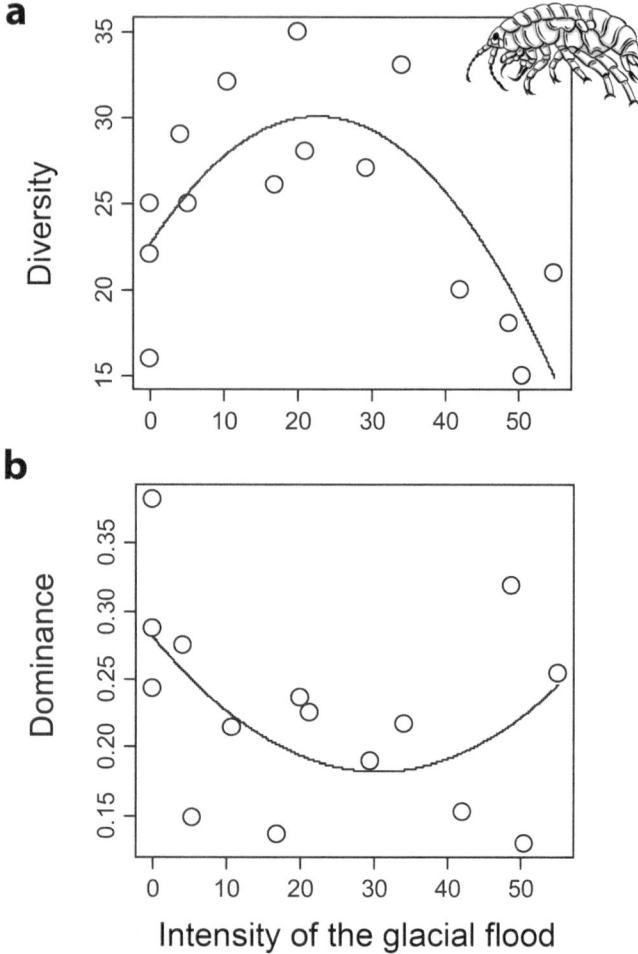

Fig. 4.8 Influence of glacier meltwater on stream insects. Relationships between glacier melt-water intensity and the richness and dominance of insect species in streams in Antisana Reserve, Ecuador. (**a**) Insect diversity peaks at intermediate levels of glacier meltwater intensity, suggesting that glaciers favor the coexistence of many species. (**b**) The concave shape of the curve here suggests that the disturbance provided by meltwater prevents domination by a few species. Modified from Cauvy-Fraunié et al. (2014a). Used with permission of John Wiley & Sons Ltd., from Relationships between stream macroinvertebrate communities and new flood-based indices of glacial influence, Freshwater Biology, 59(9); permission conveyed through Copyright Clearance Center, Inc. Drawing: *Hyallela* by Paula Terán Ospina

Glaciers also have wide-ranging influences. By expanding our sampling to dozens of streams on Antisana, we found that our local observations had basin-scale consequences. Small glacier-fed rivulets flow down the volcano, joining with páramo-filtered water to form larger streams, which in turn discharge their water into wider watercourses. This creates a mosaic of water colors, whose pinkish shades gradually vanish as the water flows downstream, but still provide beneficial pulse disturbances that allow a diverse community to coexist. As Humboldt warned not to be satisfied with *"the mere accumulation of unconnected observations of details, without generalization of ideas,"*[62] we tested whether this observation could be generalized to other places. We joined our dataset with those of colleagues working in Alaska and the French Pyrenees and found that in all three regions, stream invertebrate diversity peaked at intermediate glacier disturbance, estimated at between 5 and 30% of glacial cover in the watershed.[63] As the contribution of glacial meltwater to rivers is declining in many parts of the world, our study revealed the potential scale of this problem: the threat of disappearing glaciers is not just that certain species of café-au-lait-loving bugs could be lost, but that the surprising variety of species in the whole watershed could be reduced. In a world without ice, as the runoff from glaciers dries up, water color in the watershed will become more uniform—to the benefit of the most competitive species, but to the detriment of the 'stretched-out-tarsi,' 'ringed-leg' and 'ordinary-tibia' flies. As on the land left exposed by receding glaciers, there will be winners and losers.

* * *

In the high tropical Andes, glaciers feed streams with rock debris and organic material that give the water a palette of colors and allow diverse species of insects to coexist. The color of water has fascinated generations of naturalists, from Humboldt to Needham to Hutchinson. In Ithaca, James Needham visited and described the McLean bogs a century ago.[64] Following in his footsteps, I looked out over a clear water fen and a black water bog lying side by side, but separated by a ridge left by a glacier. The esker acts as a physical barrier, allowing the differences in water color and acidity between the two bogs to create distinct habitats: the fen has perennial groundwater springs; the peat bog contains acid-producing vegetation. While glaciers here disappeared 10,000 years ago, their effects on natural habitats are still present.

As Hutchinson poetically described of Tibetan lakes in the desert: *"Where the valley of the Indus narrows, the country is sparsely settled, but north and south of the gorge, lying in a dusty land like few remaining stones on a worn-out headdress, many lakes are scattered. They vary in color within a restricted range of blues and greens, just as the turquoise [stones] that they resemble differ one from another. Each shade of blue or green sums up in itself a structure and a history, for each lake is a small world, making its nature known to the larger world of the desert most clearly in its color."*[65]

---

[62] Humboldt (1846–1858), p. 21.

[63] Jacobsen et al. (2012).

[64] Needham (1921).

[65] Hutchinson (1936), p. 99. The 'turquoise' refers to the turquoise jewelry worn by Ladakhi women and to which Hutchinson extensively refers in the first chapters of his book.

4.3   Breathing Glaciers

Glaciers could be said to 'breathe' water. Imagine glacier-fed streams over a thaw–freeze cycle as they alternately swell and shrink. They cycle back and forth: overflow to low flow to overflow to low flow; like the lungs of an ice dragon inflating and emptying (Fig. 4.9). At temperate latitudes such as in the Alps, this is a yearly cycle: streams swell during summer glacial melting and shrink to virtually nothing during winter. Any stream flow in winter often occurs under snow or ice and is barely observable. But glaciers at the equator have a different cycle: as there are no marked seasons, throughout the year the ice follows the daily rhythm of melting by day and freezing by night. This means that to experience a glacier 'breathing' in the tropics, you just need to sleep near one of its streams. Our team set up camp. As Nan Shepherd wrote, *"No one knows the mountain completely who has not slept on it."*[66]

To measure the breathing rate of the glacier, our team monitored the water level in the streams. The lowest flow generally occurs at dawn. Then the glacier-fed stream gradually swells during the day to reach an afternoon peak whose amplitude depends on the weather. If the day is cloudy or foggy, the stream flow will hardly increase because the glacier will not melt much. But during a very hot day, stream discharge at peak flow can be 300 times the volume of low flow in the morning.[67] Why is this of interest? Because these natural rhythms—to the Aymara, expressions of the cyclical *Pachamama*—are likely to be strongly modified as a glacier recedes due to global warming. Hydrologists have drawn a rather clear picture of these changes: at first, as the glacier melts, glacier runoff increases and more water flows downhill. Average discharge conditions resemble those observed at 3 p.m. during a sunny day. Then there will be a turning point (after several years or decades, depending on site-specific features), after which glacier runoff and its contribution to river flow downstream will decline steeply. Everyday flow will be more like 6 a.m. flow conditions. Hydrologists also predict a higher frequency of flow extremes as the glacier shrinks. Finally, the glacier's complete disappearance will sound the death knell of glacier outflows, a rupture in the cyclical nature of the *Pachamama*. Not even a drop of café au lait will remain in the stream, whose existence will exclusively depend on rainfall. In that case, what will become of glacier-stream insects? Like terrestrial plants and animals, the life of stream dwellers is likely to be dramatically affected by changes produced by glacier retreat, yet we are not sure how. Will variations in daily flow affect them? Up to which flow amplitude? Would low flows make them move upstream toward the glacier front (to follow their environmental niche) and high flows flush them downstream? Is it possible to predict what will happen when glaciers pass the turning point and stream levels are drastically

---

[66] The Living Mountain by Nan Shepherd, Copyright © 2014, Canongate Books Limited, p. 90

[67] Jacobsen et al. (2014a). The amplitude of these diel cycles may be variable day to day, depending on air temperature (which influences melting), occasional snowfall (which increases glacier albedo), and wind (which increases sublimation).

| 6:00 a.m. | 11:00 a.m. | 3:00 p.m. | 9:00 p.m. |

Fig. 4.9   A glacier 'breathing.' Daily variations in the flow of a glacier-fed stream on the Antisana volcano, Ecuador (Río Los Crespos, 4480 m). Pictures were taken at four times over a day, showing the early afternoon rise in waters resulting from melting ice. Photos by Olivier Dangles

reduced? To try to understand this, we need to know something about the physics of flow.

* * *

Freshwater ecology field surveys have the advantage that you always camp near water, where you can wash your dishes. You might expect this would be quick work in a raging glacier stream during peak flow in the early evening. Yet disappointingly, even the strongest current is not really effective in removing food debris from your plate. The tiny bugs on the streambed can be compared to this—they do not give in easily against the current. Understanding the non-intuitive physics operating on food debris and stream bugs relies on basic fluid mechanics.

Whenever water flows across a solid surface such as a plate or a stone, the fluid immediately adjacent to it has the same speed as the solid surface: that is, it does not move. This is due to the frictional drag of the solid surface on the water. Farther away from the surface, water has a slightly higher speed, which increases with distance. In other words, there is a velocity gradient adjacent to solid surfaces. The zone between where the water is in contact with the streambed to where the fluid is no longer influenced by the substrate is called the **boundary layer**. Very near to the stream bottom, "*water has properties similar to those that humans would experience when swimming in honey.*" [68] For bottom-dwelling bugs, this practically means living in a different world to the more aquatic environment above (see insert in Fig. 4.10). While living in the boundary layer has some disadvantages, for now, let's focus on the positive: staying on the streambed protects bugs against the buffeting of the water current.

---

[68] Statzner (1988), p. 86. Used with permission of Blackwell Publishing Ltd. from Growth and Reynolds Number of Lotic Macroinvertebrates: A Problem for Adaptation of Shape to Drag, Oikos, 51(1); permission conveyed through Copyright Clearance Center, Inc. Steven Vogel's wonderful book *Life Device* (1988) is the source of much of the information in this paragraph.

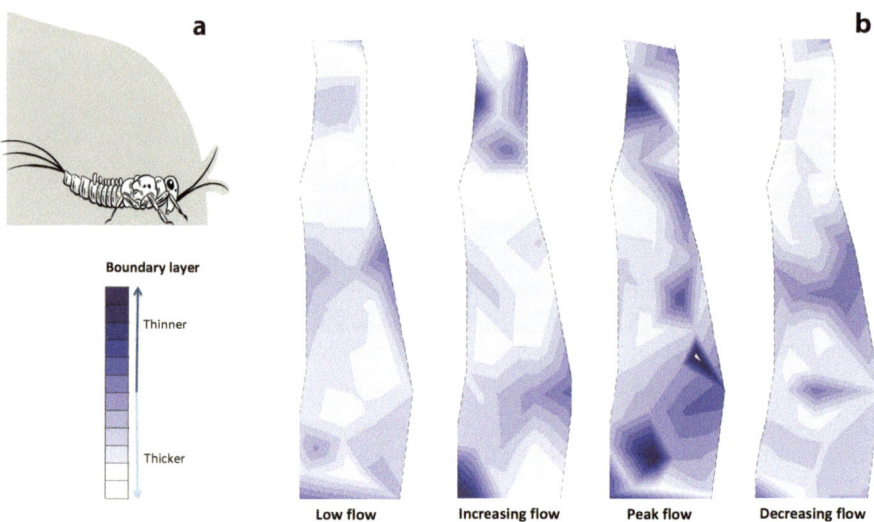

Fig. 4.10 The impact of meltwater flow on a streambed. These maps show the changing thickness of the boundary layer, the narrow but vital space close to the streambed where aquatic bugs live, in a glacier-fed stream on Antisana during a daily meltwater event (the deeper the blue, the thinner the boundary layer, from Cauvy-Fraunié et al. 2014b). Used with permission of John Wiley & Sons Ltd., from Glacial flood pulse effects on benthic fauna in equatorial high-Andean streams, Hydrological Processes, 28(6); permission conveyed through Copyright Clearance Center, Inc. Insert: a mayfly larva inside the boundary layer, drawing by Paula Terán Ospina

Very tiny insects, such as non-biting midges and early stages of many other invertebrates, live within this boundary, shielded from what happens in the water above. But as they grow, insect larvae will leave this protective zone and rise into the same watery world experienced by larger stream dwellers such as fish. To minimize the drag while living within the boundary layer, many aquatic bugs have evolved a flat shape so they are not buoyed up in the turbulent water when they move along the pebbles of the bottom. Others, instead of having a flat surface, have evolved different solutions, such as cylindrical non-biting midge larvae, which have strong hooks on their body that help them cling to the substrate and avoid being swept away. Blackflies have another trick. They can produce silk—which has the unique property of being liquid when stored in their body and sticky in water—to build silk pads that they attach to the substrate. Then they use thousands of microscopic hooks at the tip of their abdomen to anchor firmly to this silk pad. From there they can hang their head downstream, using their mouth-brushes to trap food particles in the water. Fascinatingly, blackflies can also attach a string of silk to the substrate as a safety rope. If they are dislodged by the current or a predator, this lifeline allows the larva to climb back and return to its original position. But despite all these ingenious features, sometimes the mountain current is just too strong. In such situations, flies have a trick that is more similar to octopuses than insects. And this makes them look like some of the strangest creatures on Earth.

* * *

Since the first time I put my hands in a stream to collect bugs with a Surber sampler, I have never worn gloves. It is not that I am particularly resistant to the cold or that I want to impress anyone with my bravado. I just do not want to miss the torrent midge larvae (Blephariceridae, Fig. 4.3a), which are high on my naturalist's must-see list. Scalloped, flat and somewhat limpet-shaped, torrent midge larvae are immediately recognizable by their ventral row of suckers for holding on to a rock in rushing waters. And these are some strong suckers. The maximum velocity they can withstand is unknown, but in one lab experiment some were able to withstand flows of 4.5 m/s. Even more astonishingly, these larvae are able to *move* in strong currents using their suckers, which requires a unique adaptation that allows strong attachment yet rapid detachment. The inner architecture of the sucker is a doughnut shape that is hollow inside, with a piston that is maneuvered up and down by different muscles and a fringe of small hooks. The precise functioning of torrent midges' muscle-controlled suckers has not been fully elucidated, but these are a promising source of bioinspiration for engineers interested in building tiny devices that move in fast flow conditions.[69] The suction disks are so powerful that it takes some effort to pry the larvae off their stones, and wearing gloves is more cumbersome than helpful as they impede 'feeling' the stone . . . and the larvae. With some experience, you can feel the presence of a torrent midge with your naked fingers and then sense the palpable 'pop' when it releases its hold. Finding a torrent midge is both a sensory encounter and an opportunity for remarkable observation under the stereomicroscope. It also means that the stream hosts some healthy fast-flow habitats.

But to understand how these tiny organisms respond to flow variation in the boundary layer, we can't rely on our senses. We need to examine the velocity of the current as close as possible to the streambed, conceptualizing a world that we cannot experience concretely by using maths and physics. Just as thermal imagery can help us detect the temperature on and around cushion plants, equations can reveal the invisible physical world of Lilliputian stream bugs. To crack the code, we would need some mathematics expertise.

* * *

At just 22 years old, French engineering student Sophie Cauvy-Fraunié arrived in Quito with a big smile, anticipating the amazing trekking Ecuador's more than 80 volcanoes would allow her. From her very first encounter with a glacial stream, Sophie grew ever more interested in the rise and fall of the meltwaters. Like generations of students since Needham's field trips at Cornell, she could not resist the call of the water. Her first mission was to measure how boundary layer thickness varies during high meltwater flow, which depends on a number of variables that are heterogeneous over space and time. Fortunately, physicists have developed equations that allow this based on in-stream measurements such as flow velocity and

---

[69] See Frutiger (2002). We know that each suction disk contains a cavity with a piston maneuvered up and down by different muscles. Spine-like tiny hairs cover the suction disks, increasing shear resistance on rough surfaces—similar to the suction pad of the remora fish, which attaches to the rough skin of a shark.

substrate roughness. Sophie performed thousands of such measurements at different times of the daily flow cycle, boiled down the results of her equations and visualized them as a series of maps: visual passports to the physical world of stream bugs (Fig. 4.10). Overall, she found that high meltwater flow decreases the thickness of the boundary layer on the streambed (the deeper the blue on the maps, the thinner the layer)—and therefore the size of the environment of the torrent fly larvae—by up to five times. Imagine climate change reducing the size of Earth's biosphere to one-fifth of its original size: this would certainly require some adjustments! But surprisingly, Sophie found that insect communities did not seem to be affected by changes in boundary layer thickness: they were equally diverse and abundant between low and peak flows. The clue for this apparent mystery was also to be found in Sophie's maps. In a similar way that bofedales are a patchwork of greens (see Sect. 2.3), stream bottoms can be conceptualized as a patchwork of blues. Even during peak flow, many areas of white and pale blues (indicating a thicker boundary layer), persisted—for instance, behind a protruding pebble or boulder—providing refuge to mobile insects.[70] Unlike their temperate kin, which are severely affected by more extreme high meltwater flows during the summer months, water bugs in the tropics had adapted to the regular cycles of their environment. In our Antisana stream, we found that the current of an average daily peak meltwater flow would flush down 150 times fewer aquatic insects than a similar event in a temperate glacier-fed stream (I will come back to how we measured this later).[71] But would these findings hold for the more extreme flows predicted by hydrologists as glaciers shrink? To know this, we would need to extend our analysis though time and space, which required more data.

* * *

As we have seen, Humboldt was an early example of a data freak. He measured everything he encountered, from mountain summits to condor toes, as well as the most minute and unpredictable variations in the environment. He believed that amassing precise, accurate measurements over time could unravel laws of nature that the human senses could not perceive. Modern scientists have a wide range of high-tech equipment available for this—satellites, drones, automatic sensors, DNA barcoding, etc.—so gathering long-term data in nature is no longer a problem. However, it can be challenging to develop analyses that make the most of the massive amount of data generated. Measuring flow amplitude variations in Antisana's glacier streams over long periods was not difficult. Dean had water pressure sensors that could automatically measure the water height every 30 min over several years. The issue was what to do with the high-resolution **hydrographs** we generated with this data (Fig. 4.11a). Could we find a pattern in this messy time series? Which peaks are significant? What difference does it make if a peak flow occurs 1 day, 2 days or 3 days in a row? Fortunately for us, our flow data described

---

[70] Cauvy-Fraunié et al. (2014b).

[71] Jacobsen et al. (2014b).

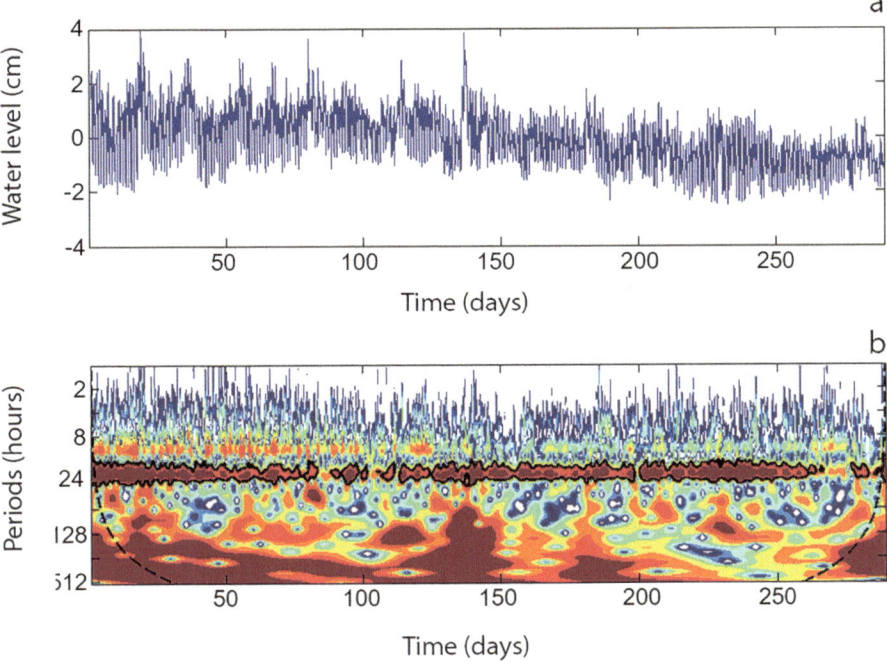

Fig. 4.11 Decoding glacier stream flow harmonics. (**a**) Hygrograph of a glacier-fed stream showing the variations in water level recorded by automatic loggers placed in the water; (**b**) spectrum graph based on a Fourier transform of the hydrograph. The red horizontal band surrounded by a black contour at 24 h indicates a significant flow cycle every 24 h due to the melting of the glacier. Illustration by Sophie Cauvy-Fraunié

repeated fluctuations (swell, shrink, swell, shrink), Sophie was very skilled at math, and the head of a French revolutionary, mathematician and physicist named Fourier was spared from the guillotine.[72]

* * *

Of all the great natural philosophers of the nineteenth century, **Joseph Fourier** was probably one of the most visionary. In 1827, transposing early experiments by de Saussure to the atmosphere, Fourier published an essay in which he recognized that certain atmospheric gases such as carbon dioxide, water vapor and methane contributed to the warming of the Earth.

---

[72] Joseph Fourier was sympathetic to the cause of the French Revolution and joined his local Revolutionary Committee. An incident pitted him against a rival faction in Orléans in 1793. One year later, he was imprisoned during the Reign of Terror, an incident that could have led to the guillotine. But the death of Maximilien de Robespierre (1758–94) brought about political changes in France and Fourier was released. See Herivel (1975).

The establishment and progress of human societies and the action of natural forces can significantly change, over vast areas, the surface of the ground, the distribution of water and the great movements of air. Such effects are likely to vary, over the course of several centuries, the average degree of heat.[73]

Decades before Eunice Foote, Tyndall and Arrhenius, what Fourier described in 1827 would later be called the greenhouse effect. Fourier reasoned as a climatologist, considering the planet as a whole. This globalization of phenomena was by no means self-evident at that time, and Humboldt's lectures may have prompted this perspective, as Fourier mentions him:

Alexander von Humboldt, whose research embraces all the great questions of natural philosophy, considered from a new and very important point of view the observations of the temperatures specific to the various climates.[74]

As a Chinese proverb says, the higher you go, the further you can see.

For all his contributions, Fourier was until recently not widely known. Until the early 1970s, his name did not even appear in the French *Encyclopædia Universalis*. Now, however, Fourier is everywhere, and not just because he is credited with the concept of the greenhouse effect. Fourier became fashionable again for a work published in 1822 that showed how some mathematical functions can be written as an infinite sum of **harmonics**.[75] Like many strokes of genius, the discovery did not excite the scientific community at the time, and it was not until over a century later that Fourier's demonstration would give rise to a now indispensable analytical technique: the Fourier transform. This method allows a signal to be broken down into its constituent frequencies. In other words, it does mathematically what a human brain is incapable of doing with a hydrograph: analyze the messy-looking curves into the simplest of harmonic laws. The Fourier transform is today incredibly valuable for engineers and scientists, used to process cell-phone and Wi-Fi signals, to compress audio, image and video files, to detect gravitational waves, and to represent the underlying patterns of nature—from the crests and troughs of a mosquito population to the flow cycle of a glacier. Fourier—like Humboldt—long disappeared from French collective memory,[76] the mathematician is now taking his revenge in a world of digital bandwidths, waves and natural harmonics.

---

[73] Fourier (1827), p. 592.

[74] Fourier (1822), p. xx.

[75] Fourier completed his memoir in 1807 and won a competition held by the Paris Institute in 1811 on the topic of how heat propagates in solid bodies. But because of the controversy this work generated in the scientific community, Fourier's memoir was not published until 1822. See Grattan-Guinness (1972).

[76] Events have seemed to conspire to erase J. Fourier's name from French collective memory. Fourier is buried in Père-Lachaise in Paris, the most visited cemetery in the world. Unfortunately, the more than 3.5 million annual visitors cannot identify him, as the inscriptions on his grave have become illegible. To make matters worse, the bust on his grave is not his, but comes from a neighboring tomb, that of François Chaussier (1746–1828), a French physician. A large statue of Fourier was erected in his hometown of Auxerre in 1849, but it was requisitioned during the Second World War and melted down to fuel the war effort (as was the case of many statues). And one of the

Using Fourier's method, Sophie could open the hydrograph's black box with a spectrum graph (Fig. 4.11b). Spectrum graphs have a certain beauty, with their lines of vibrant colors in mesmerizing patterns, but they can also be analysed to reveal powerful information. Sophie's showed that there is a significant flow cycle every 24 h. Fourier's method detected even the slightest flow cycle patterns—indicating the slightest pulse of the glacier—in a stream. To her surprise, Sophie discovered that several streams that appeared to have no sign of glacier influence (no milky water or obvious flow fluctuations) nevertheless had a spectrum graph showing a flow cycle every 24 h. The reason for this unexpected result is certainly the complex volcanic subterrain, fissured by past eruptions and favorable to the infiltration of meltwater. As a consequence, a glacier's thawing and freezing can have an effect kilometers away from its terminus. Fourier analyses also provided a series of indicators that precisely described the peak meltwater flows, their intensity, their variability, and their clustering throughout the year. These indicators revealed the extraordinary heterogeneity in these flows and, in the dozens of streams influenced by glaciers on Antisana's slopes, how these shape the lives of stream insects. While boundary layer maps showed that streambed topography helps insects cope with gentle daily flow pulses, Fourier's method revealed that extreme or repeated high flow events over the long term do have consequences on aquatic insect distribution.

By making further analyses at different distances from the glacier, we realized that for stream bugs, irregular pulses of water (either high or low flow) were like playing with the up-and-down switch of the escalator to extinction. In a similar process to that found in birds and plants, low flow periods 'pulled' lowland insects upstream toward the glacier, putting the residents of meltwater streams at risk.[77] But at any moment high flows could flush these aquatic creatures downstream, delaying the extinction process. These findings highlighted the complexity of global warming effects on organisms in glacial streams. But even though Sophie's results were dressed in the elegant garb of a Fourier transform, they had one main caveat: they remained largely observational and correlative. We needed some experimental proof of the effect of flow on stream bugs.

* * *

Since my early days as an ecologist, I have been fascinated by field experiments. Evelyn Hutchinson was probably the first ecologist to bring experimental approaches to a discipline that was largely descriptive at the time. In the mid-1940s, Hutchinson dumped small amounts of radioactive phosphorus-32 in his long-time study site, Linsley Pond, a kettle pond 11 km east of the Yale University campus. His objective was to explain why some lakes experience several plankton blooms a summer. Shouldn't early blooms deplete all the available

---

top ten French universities, created by Fourier in 1811 in Grenoble and until recently named after its founder, was merged with two other nearby universities in 2016, creating Université Grenoble-Alpes. One place you can still see Fourier's name in France is on the Eiffel Tower: he is one of the 72 scientists whose names are inscribed on the first floor (his is the 13th, on the side facing west).

[77] Jacobsen et al. (2014b).

phosphorus in the water, making later blooms impossible? His results indicated that phosphorus moved vertically in the pond, with algae taking up available phosphorus in the water column and then falling to the bottom. This was the first direct demonstration of the phosphorus cycle in an ecosystem[78] and the beginning of experimental ecosystem ecology.

Since Hutchinson, scientists have used their imagination and ingenuity to conduct field experiments on ever-larger scales, allowing major ecological theories to be tested. A few examples of open-air laboratories across the Americas: in Washington state, parts of the shoreline were kept free of starfish, a top predator in that ecosystem, to uncover the causal processes linking species diversity and food web complexity.[79] In Florida Bay, entire islands of mangroves were wrapped in tents and fumigated to remove all the insects to test the island biogeography theory.[80] A few kilometers north of Manaus, a $\sim$1000 km^2 experimental landscape with forest patches of different sizes was created to evaluate the effects of fragmentation on rainforest biodiversity and functioning.[81] In the salty lagoons of the Bolivian Altiplano, an experiment excluded flamingos from the shallow waters to reveal their top-down control on the food web,[82] similar to starfish on the west coast of North America. On the eastern flank of the Pichincha volcano near Quito, our botanist colleagues Priscilla and Ricardo established 2-m-wide hexagonal plastic chambers to assess how experimental warming affects the growth and survival of páramo plant communities.[83]

These controlled experiments on ecosystems of varying sizes are especially useful, as they consider the extraordinary complexity of interactions between environmental conditions and the many species present in an ecosystem. They also allow manipulating environmental disturbance at various levels of severity (e.g. extermination by predators, habitat fragmentation, warming), to help make predictions for the future. Could we apply this approach to glacier-fed streams? Could we manipulate water flow to assess the impact of warming on aquatic insects when ice will have melted to such a point that the glacier expires?

* * *

When Sophie chats about her research with her grandmother, the latter always reminds her: "*As a child, you were always playing in the creek at the bottom of the garden.*" And like many kids, one of Sophie's favorite activities was building dams with rocks. At any age, most dam builders are more interested in what happens

---

[78] Hutchinson and Bowen (1947).

[79] Robert T. Paine (1933–2016) showed that species diversity was much higher when *Pisaster* starfish were present than when they were removed. *Pisaster* maintained balance in the ecosystem by selectively eating barnacles and mussels, preventing their otherwise excessive spread. Paine (1966).

[80] Simberloff and Wilson (1969).

[81] Lovejoy et al. (1986).

[82] Hurlbert and Chang (1983).

[83] Duchicela et al. (2021).

upstream—whether it be the creation of a wading pool or a reservoir—than down-stream. But ecologists are more interested in the downstream consequences: the reduction in water flow.

Building a rock dam as part of a scientific study may seem quite archaic for twenty-first-century researchers, and all of us on our Antisana team were a bit concerned about relying on this 'amateur' way of manipulating the water current, not to mention the potential criticisms from reviewers when publishing the results. However, the remote wildness of the place prevented us from using more elaborate techniques such as building a concrete structure with a notch to calibrate the passing flow.[84] So we improvised more rustic technology and then set up a pilot study to test it over several months. I had spotted a section of a glacier-fed stream where we could stack stones to make a dam that would divert the flow for about 100 m before it re-entered the stream (Fig. 4.12). The pilot study revealed that this rock dam was effective enough to consistently reduce stream flow, allowing us to join the club of ecosystem experimenters. It also permitted, for the first time, a scientific team to experimentally assess the ecosystem-level changes triggered by glacier melting.[85]

While our dam was rudimentary, the experimental design around it was quite sophisticated. Field experiments, unlike more controlled lab experiments, have the advantage of being realistic as they are not isolated from the surrounding complexity and variability of nature. However, this can be double-edged: unexpected natural events unrelated to the disturbance of interest—such as a change in weather or in the population density of a species—can complicate the conclusions. To overcome this issue and generate sufficient relevant data, ecologists must design their experiment at appropriate temporal and spatial scales. To do this, they have developed a useful framework: 'Before–After Control–Impact' design, more commonly known as BACI.

Using the concrete example of our experiment, the *before-after* part of the framework involved a number of steps. Before studying how the flow reduction imposed by our dam would affect the ecosystem, we needed to establish a robust baseline to document how the physics and biota of the stream vary naturally. We equipped our experimental stream section with water level loggers, measured water temperature, oxygen and nutrient contents, and Surber-sampled stream bottom invertebrates and algae repeatedly over one year. The 'after' would require building the rock dam (Fig. 4.12, insert), with the objective of diverting about one-third of the flow from the main channel in order to simulate the kind of reduction in meltwater run-off expected in a few decades. The Crespo glacier, where our experimental stream originated, still covered a surface area equivalent to 1140 ice hockey fields, but it was retreating at an accelerated pace of 10–20 fields per year. After some adjustments of the rocks, we obtained the desired flow reduction and from that point measured stream responses downstream from the dam for another year.

---

[84]But we did use notches in another flow experiment in the páramos. See Rosero-López et al. (2020, 2022).

[85]Cauvy-Fraunié et al. (2016).

Fig. 4.12 Dam-building to mimic glacier flow reduction. View of the experimental dam, with the original stream channel on the left and the diverted water on the right (see *arrows*) in a stream originating from the Crespo glacier on Antisana, Ecuador. In the background, Mt. Cotopaxi. **Insert:** Who says science can't be fun? PhD student Sophie Cauvy-Fraunié and undergraduate student Andrés Morabowen build the rock dam; Photo by Olivier Dangles and Roger Calvez (insert)

In terms of the *control-impact* aspect of the study design, one might think that detecting temporal changes by comparing pre- and post-dam conditions would be sufficient. But field experiments like any other require a control. Thus, we needed to find another stream site similar in all relevant respects to the study site, except that it would not be impacted by the experimental flow reduction. Our control was a stream flowing just a condor's wing flap from the study stream, and we would monitor the water and the invertebrates in both on the same dates.

The BACI design allowed us to elegantly combine time and space to robustly test our conclusions. Post-dam, if the patterns of the water and biota variables in the experimental stream differed from those in the control stream, the differences were unlikely to be due to chance.

So, what were our findings? We expected that the response of flies to reduced flow conditions would be almost immediate, as they are among the quickest creatures to respond to environmental change. Indeed, any novelty in nature, either a small pool of rainwater or a human corpse, are readily detected and colonized by mosquitoes in the first case or carrion flies in the second. But this time flies were beaten by algae. Only two weeks after we had built the dam, algae had already doubled its biomass in the experimental stream, while it remained constant in the control. Flies responded a few weeks later, with, as expected, flow-adapted groups such as blackflies and torrent flies paying the highest price for the flow reduction.

They almost completely disappeared from the experimental stream while they still thrived in the control. But this was not the most dramatic impact of flow reduction. Low-flow conditions also increased the water temperature by 1.4 °C and tripled the concentration of nutrients, which favored algal growth and shifted the whole system toward a more productive state. The biomass of herbivorous bugs, in particular mayflies, rocketed, and the whole food web responded in concert: more predation, more competition, and less space for the delicate flies and other invertebrates adapted to glacier waters. Our simple yet novel experiment had opened our eyes to an issue that had so far been overlooked: food web interactions between species would be central to understanding the effects of climate change on glacier-fed streams. Like west coast shorelines and Altiplano lagoons, the equilibrium of Ecuadorian glacier-fed stream dynamics was fragile in the face of change. And, alarmingly, this fragility was even more profound than we initially thought.

* * *

In streams, it is better to knock down dams once they have served their purpose. Left in place, even a small row of stones can block fish movements and cause flooding during storms. When conscientious recreational dam builders unstack the rocks, they observe an almost instantaneous return to the initial conditions. At least on the surface. When we restored the natural course of our experimental stream, we kept on monitoring the water, insects and algae to see what would happen. We expected it would take some time before the system returned to its original state, but were shocked that in fact it took 16 months until the stream's physical and biological parameters completely recovered. Almost a year and a half before recovering healthy populations of blackflies and torrent flies. This result was worrying. In view of the recurrence interval of low-flow events observed in our hydrographs, glacier-fed streams may have insufficient time to fully recover between each event. Not to mention the predicted increasingly variable flow regimes as glaciers melt, which would increase both the intensity and the duration of drought events.

With our experiment over, Sophie had completed her PhD. But she was left with many questions, one of which was particularly topical. Could we predict low-flow intensity and duration thresholds that would affect glacier-fed stream life? In our experiment, we had set the flow reduction to about one-third of the initial flow during one year, but could an effect already be observed with a weaker or shorter flow? Relentless, Sophie started a new series of experiments, in which we intended to gradually decrease glacier-fed stream flow. But halfway through the experiment, what every ecosystem experimenter dreads occurred: a catastrophic event. The ice dragon violently sneezed. On 8 March 2015 at 4:45 p.m., the sudden collapse of a meltwater pocket below the Crespo glacier completely ruined our experimental set-up. The event was so violent that all our measuring equipment was saturated and we would never know the exact amplitude of this mega-flood. In less than 30 min, the baseline flow of our experimental stream multiplied by more than 3000, and turbidity increased by at least 450%, with thick carpets of sediment completely covering the streambanks. Sophie and I were distraught by the failure of the experiment, but we also learned a great lesson from this. Extreme catastrophic events, seldom observed during a researcher's lifespan and beyond experimental reach, are likely to have crucial yet poorly recognized effects on insects in glacier-fed streams. This is the great value of field experiments—they always keep you tuned to nature.

* * *

The 'breathing' of glaciers is a complex cyclic process of vital importance to stream inhabitants, which would not surprise those with the cosmovision of *Pachamama*. Glacial flow cycles are vital at all scales: from the survival of delicate endemic high-altitude flies to the overall functioning of streams. With glacier retreat, the water cycle will gradually shift from being dominated by ice to being determined by rain. As a consequence, the flow of water courses will be more variable and unpredictable, without the buffering effect of ice storage. As a glacier's breathing becomes more shallow and eventually stops, the flow velocity and turbulence of meltwater streams will be affected, which in turn will impact the life forms that have adapted to different flow habitats. No longer will mountain slopes ring with the raging blast of meltwater torrents at midday, dwindling to the babbling of a gentle brook in the cooler hours. As glacier-fed streams disappear, new freshwater systems will take their place, and a new community of invertebrates will colonize these clearer, calmer waters free of debris. Yet life for an aquatic insect in a post-glacier era will have its own challenges: these will include heat, new predators, and lack of oxygen.

4.4   Craving for Oxygen

Most of the time, we do not realize how much our lives depend on oxygen. We inhale this colorless, odorless gas automatically and are oblivious to the vital role it plays in allowing our cells to burn food molecules and ultimately to provide us with energy. Yet walking at an altitude of 4500 m on the slopes of Antisana quickly brings its essential role to mind. Here is Humboldt on the subject:

> We spent three days in these plains whose cold and temperature are unbearable. [...] More than the cold [is] the scarcity of the air. One always feels uncomfortable in the chest. [...] I measured in the plain the base as well as the height of the volcano as for other geodetic operations. I am accustomed to running without fatigue in full trot from one signal to another, even if they are 800–1,000 toises away [1560–1950 m], without discomfort ... In the plains of Antisana, I could not run 70 toises [135 m] without becoming breathless and feeling strong chest pains.[86]

---

[86] Humboldt (2003), p. 179. One toise equals 1.949 meters.

What in fact happens to the air at high altitude? What does Humboldt mean when he complains about the *"scarcity of the air"*? Is there less oxygen? Curious about this during his travels, Humboldt took and analyzed air samples in countless places, from the base and the summit of Pico del Teide in Tenerife to the banks of the Orinoco River and the altitude he reached on Chimborazo. He reported his findings in the *Tableau Physique,* in a column entitled 'Chemical composition of the atmospheric air' at Chimborazo. And it is unambiguous: *"The quantity of oxygen in the atmosphere appears to be the same in the elevated regions and in the plains."*[87] Yet in another column, titled 'Pressure of the atmospheric air,' Humboldt clearly depicts a decreasing trend as altitude increases: his barometer indicates 0.76 mm of mercury at sea level, while only 0.44 mm at 4500 m. In other words, according to Humboldt's *Tableau,* at higher altitude there is less pressure but the same quantity of oxygen as at sea level. How is that possible? The explanation is that as long as you don't leave Earth's 600-km-thick atmosphere, the *proportion* of oxygen in the air (not its quantity) is constant: a bit less than 21%. But as you go up in altitude, the atmospheric pressure decreases and the same volume of air contains fewer molecules.

To illustrate this through Humboldt's example, at 4500 m above sea level, the atmospheric pressure is reduced by almost half. So even if the air is still made up of about one-fifth oxygen, every time Humboldt took a breath when running on Antisana's slopes, his lungs received half the amount of oxygen that they drew in when he was enduring mosquitoes in the Amazon's lowlands. As a consequence, his heart pumped blood more rapidly and his lungs had to work harder (hence the chest pain) to oxygenate the blood. But like most people spending time at high altitude, Humboldt's body would adapt to these conditions. After a few days spent in the Andean highlands, his blood cells would have almost doubled to reach 7 million per cubic millimeter of blood to counteract oxygen reduction. This adaptation is amazing, but still limited: vicuñas, for example, have 14 million cells per cubic millimeter of blood. Like other animals, aquatic bugs need energy for their daily activities, and oxygen is essential for them. So how do underwater fly larvae get oxygen at high altitude?

* * *

While oxygen is invisible on land, it can be a marvelous spectacle underwater. As Jean-Henri Fabre, who all his life observed *"le délice"* ('the delight') of small ponds, poetically described:

> When the sun is shining [...], it is a sight to see the algae at work. The green carpeted reef is illuminated with an infinite number of glittering dots and takes on the appearance of a fairy-tale ball of velvet into which thousands of diamond-headed pins are stuck. From this exquisite jewelry, beads ceaselessly detach, immediately replaced by others generated on the cushion; softly rising, like globes of light. They ascend every which way. It is a continual fireworks display set off in the water.[88]

---

[87] Humboldt and Bonpland (2010), p. 152.
[88] Fabre (1925), vol. 7, pp. 294–95.

Despite the striking sight of oxygen bubbles produced by organisms in ponds, in fact aquatic plants produce only a tiny fraction of dissolved oxygen in fresh water. The rest comes mainly from the atmosphere. Water contains about 30 times less oxygen than air, so organisms face greater challenges in getting oxygen underwater than on land. But does breathing underwater become harder with altitude, as on land? The general belief is that cold, turbulent mountain streams are particularly oxygen rich, as seemingly evidenced by fizzing air bubbles in the current, as in Fabre's pond. It is true that oxygen is more soluble in water at low temperatures. However, at high altitude, the advantage of cold water is counterbalanced by the low atmospheric pressure: the thin air has more difficulty 'pushing' oxygen into water. The result of these two counteracting processes is that the quantity of dissolved oxygen (O_2) in the water is virtually independent of altitude (Fig. 4.13).

But aquatic organisms are less interested in the *quantity* of oxygen in the water than its *availability*. To understand this, it is necessary to put yourself in their place. Rather than lungs that draw oxygen into the body, most aquatic organisms get dissolved oxygen by diffusion, directly through holes in their skin or through specialized external organs such as gills or spiracles (Fig. 4.3c, d). And oxygen diffusion in the water is closely related to the pressure of oxygen in the water (PO_2). As PO_2 decreases at higher altitudes, oxygen diffusion also decreases (Fig. 4.13). But that's not all. As we've seen, aquatic bugs are subject to the laws of fluid physics and enveloped in a boundary layer of water. Because water movement is low in this boundary, oxygen movement is reduced and hence so is oxygen uptake. The colder the water, the thicker the layer and the more difficult the diffusion.[89] While the boundary layer effect is an ally of aquatic invertebrates in streams to protect them from turbulent flow, it represents an additional challenge for breathing in cold streams.

To sum up, oxygen is more soluble at high altitudes because of lower temperatures, but its diffusivity decreases sharply due to lower oxygen pressure and a thicker boundary layer. The resulting oxygen availability is what physiologists call the oxygen supply index (OSI), which represents the amount of oxygen available for organisms, which steeply decreases as one climbs in altitude. This makes it challenging to breathe on high mountains and in high, cold mountain streams. These extreme environments have sparked nature's creativity to result in **hypoxia**-tolerant species, some of which are among the most bizarre creatures on Earth. Some of the most charismatic of these were discovered by Louis Agassiz's son, Alexander.

* * *

Agassiz Junior buried his father under a glacial boulder. But it was not one of the many boulders he could have found at hand in Massachusetts. It was a 1134-kg boulder taken from Louis Agassiz's native land, from among the moraine debris near the Hôtel des Neuchâtelois, and sent by boat across the Atlantic to Mount Auburn

---

[89] The thickness of the boundary layer varies proportionally with the kinematic viscosity of water, decreasing with increasing temperature, which leads to a reduced thickness of the laminar sublayer surrounding surfaces, facilitating a higher diffusive influx of oxygen. See Verberk et al. (2011).

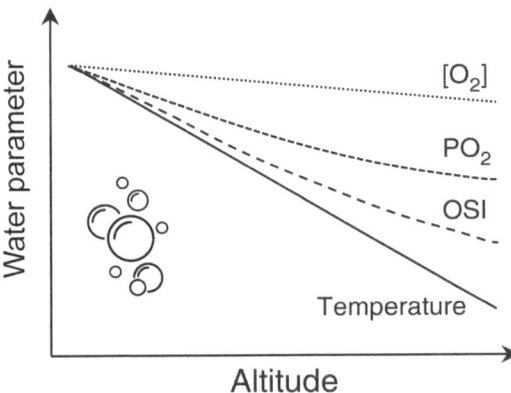

Fig. 4.13 Decreasing oxygen availability in high mountain waters. Relationships between altitude, stream temperature, dissolved oxygen concentration (O_2), partial oxygen pressure in water (PO_2 = oxygen saturation relative to sea level), and the oxygen supply index (OSI). While oxygen concentration (O_2) only weakly decreases in high mountain waters, the oxygen available to aquatic organisms (OSI) steeply decreases due to lower oxygen pressure and a thicker boundary layer. Curves are not to scale. Modified from Madsen et al. (2015). Used with permission of John Wiley & Sons Ltd., from Altitudinal distribution limits of aquatic macroinvertebrates: an experimental test in a tropical alpine stream, Ecological Entomology, 40(5); permission conveyed through Copyright Clearance Center, Inc.

Cemetery in Cambridge. It is probably the most erratic boulder that ever existed—a perfect substrate for lichens to grow on.

Alexander Agassiz was born in Neuchâtel, Switzerland, and immigrated to the United States with his parents. He studied engineering, chemistry and zoology at Harvard, and then worked in the Harvard Museum of Natural History founded by his father. Part scientist, part businessman, he became the president of one of the world's principal copper mines, near Lake Superior, and donated part of his substantial profits to the museum, of which he became curator after his father's death. In November 1874, Agassiz Jr planned a trip to Chile and Peru for the purpose of visiting mines, exploring Lake Titicaca, and collecting Inca antiquities for the Yale Peabody Museum. At an elevation of 3810 m on the border between Peru and Bolivia, Lake Titicaca is a body of water covering some 190 by 80 km, about half the size of Lake Erie. Accompanied by Samuel Walton Garman, a zoologist at the Harvard museum, Agassiz Jr took a large collection of ropes, sounding lines, thermometers to record deep-water temperatures, and all the necessary equipment to preserve whatever they might sample in the lake. Over a period of less than two months, the two scientists carried out, in a Humboldtian spirit, the first coordinated study of this environment, including bathymetric, hydrographic and biological surveys.[90]

---

[90] See Agassiz (1913).

Fig. 4.14 A baggy-skinned frog adapted to low-oxygen waters. Top and side view of the Titicaca water frog (*Telmatobius culeus*), whose baggy skin allows efficient oxygen uptake even in water with low oxygen availability. Drawings made by biologist Samuel Walton Garman from the notes in which he and Alexander Agassiz described and named this species. From Agassiz and Garman (1875)

Among the dozen aquatic creatures found in the lake, one was particularly bizarre: a huge frog that had "*a great baggy skin which hangs in loose folds about it*"[91] (Fig. 4.14). Endemic to Lake Titicaca, the frog was a new species to science and, when Garman named it, he made a tongue-in-cheek reference to the frog's exaggerated skin folds: *Cyclorhamphus culeus* ('round-beaked scrotum' in Latin).[92] The local Aymara, who have long considered this frog a divine creature, gave it a nicer name: **kaira**. Titicaca frogs are revered by the Aymara as animals with special powers. Used as rainmakers during times of drought, a large frog would be carried in a ceramic pot to a hilltop where it was believed the gods would hear the frog's distressed cries and interpret them as calls for rain. Eventually the rain would fall and the pot would overflow, allowing the sacred frog to escape back to the lake.[93] Of course, this required first finding a frog. Agassiz Jr reported that the frog "*remained often for hours perfectly quiet on the bottom, suspended on fronds of myriophyllum* [an aquatic plant], *apparently too lazy to come up to the surface to breathe.*"[94] But what was mistaken for laziness is in fact an ingenious breathing trick.

---

[91] In Agassiz and Garman (1875), p. 277.

[92] The current scientific name of the Titicaca frog is *Telmatobius culeus*.

[93] For further details and bibliographic references, see Jacobsen and Dangles (2017).

[94] Agassiz (1875), p. 285.

* * *

Like other endemic species, the Titicaca water frog is of special interest to any naturalist. But the kaira had a personal appeal for me, as I believe that it gave me my first inspiration to travel to the Andes when I was still a child. In 1973, the kaira was the main protagonist of a show by the French oceanographer **Jacques Cousteau**, whose films on the wonders of the oceans engrossed me. In this show, Cousteau and his team left the sea and traveled to Lake Titicaca. They lifted 20 tons of equipment and two mustard-yellow mini-submarines on a train (hopefully no longer fueled by Yareta cushions) linking the Port of Matarani on the Pacific coast to Puno on the Peruvian shore of the lake. Like Agassiz Jr, whose interest in the lake was in part Inca artefacts as much as biology, Cousteau was in search of submerged treasures. As no light reaches its depths and its low oxygen conditions limit oxidation, the cold waters of Lake Titicaca preserve relics from the past. But instead of treasure, Cousteau found a mysterious giant frog—hundreds of them. Using cords and buoys, Cousteau's team marked off an area of a thousand square meters and picked up all the frogs from the area. From this, they estimated the Lake Titicaca population as about one billion animals. There was hardly a bit of lake bed without a frog.[95] Some weighed over 1 kg and, extended, reached a length over 50 cm. I never forgot seeing this unusual amphibian on Costeau's show, and during my stay in Bolivia, I hoped to encounter this legendary creature. But 40 years after Cousteau, the situation was very different, as overexploitation had decimated its population and it was now critically endangered. In June 2012, during a family weekend at 'Copacabana,' the chilly Bolivian version of the famous Brazilian beach on the southern shore of the lake, it took us a while before we finally found a small kaira hidden among the totora sedges. Nonetheless, observing this individual allowed a fascinating glimpse into aquatic life adaptations for high-altitude breathing.

As the French expression goes, "There aren't 36 solutions." There are basically two ways of coping with low oxygen availability—reducing metabolic demand or increasing oxygen uptake. Living in water that contains only 62.5% of the oxygen pressure found at sea level, the kaira does both. To reduce metabolic demand, the frog stays still. Remaining "*for hours perfectly quiet on the bottom*" as observed by Agassiz Jr, a 50-g individual in 10 °C water will take in just 0.7 mL of oxygen per hour: one of the lowest metabolic rates reported in amphibians (for means of comparison, a 70-kg human at rest consumes about 200 mL of oxygen per minute). To increase oxygen uptake, the frog evolved to absorb oxygen directly from the water through its semipermeable skin, which acts in a similar way to gills. Examining the individual we caught, we could clearly observe the tiny capillaries covering the surface of its body. And its "*great baggy skin*" gives the frog a greater surface-area-to-body-volume ratio, creating even more efficient oxygen uptake. There are several advantages to being a skin-breather. By staying submerged for long periods of time, the kaira avoids sunburns and other hazards of the terrestrial world, from hungry birds to humans. The frog also has other tricks: when the water has too little

---

[95] Cousteau and Diolé (1977), p. 132.

oxygen, it can breathe atmospheric air at the surface using its tiny lungs; plus it has a very high count of blood cells that transport oxygen. No other amphibian possesses this extraordinary combination of adaptations to aquatic life at high altitude. But let's not write off aquatic Diptera: they have also found many solutions to low-oxygen environments.

* * *

Our lives are intimately connected with flies. They invite themselves into our bedrooms and camping tents, get under our shampoo bottles and in our toilet water, make themselves at home in our fruit baskets and flower pots, and have even found their way to the International Space Station (not to mention those that end up in the telepods of mad scientists in science-fiction films[96]). The rudest species intrude right into our bodies—dead or alive—from under our skin to inside our bladders. In my case, my right calf has been branded for life by a fly.

The fly that left me with a tattoo like a cigarette burn belongs to the Psychodidae family, which contains about 3000 species. There are two main clans: sand flies (Phlebotominae, 800 species) and moth flies (Psychodinae, 2000 species). As you might have guessed by the scientific name Phlebotominae, sand flies are the blood-sucking bad guys, at least from a human standpoint. At first sight, these tiny insects have a graceful appearance, with delicate elongated hairy wings (Fig. 4.3e), but their long rostrum for taking blood meals makes them less appealing. In 2010, a sand fly feeding on me left behind something it had picked up in another mammal—a rodent, opossum, or maybe a dog—infected by the protozoan parasite *Leishmania*. This causes Leishmaniasis, which in my case manifested in its cutaneous form: a 2-cm ulcerating pustule. Though I count myself lucky, since another cutaneous form of the disease causes ulcers of the mouth and nose, with disastrous aesthetic consequences locally referred to as 'tapir's nose,' not to mention the deadly visceral form of this parasitic disease. The skin pustule tends to heal on its own, but *Leishmania* are able to persist for many years in their human host by hiding in immune cells. After a treatment of antimony, which has healing properties against the disease, within a few weeks all that was left was the indelible scar from a sand fly.

In contrast to Psychodid sand flies, the family's moth flies are harmless to humans. And to return to our subject, their larvae have developed some of the most ingenious adaptations to overcome low oxygen conditions. In their aquatic larvae stage, instead of absorbing oxygen through water, they breathe via small tubes, or spiracles, located on their posterior, which allow them to obtain oxygen directly from the atmosphere. After all, oxygen in high-altitude air is less rare than in high-altitude water. Shaped by millions of years of evolution, the specific structure of respiratory spiracles—their shape, size, number and location on the body—testify to unique adaptations to different oxygen environments and are useful for taxonomists to differentiate them into groups. The moth fly species found in Antisana belongs to the genus *Maruina* (Fig. 4.3f), one of the first species of which was

---

[96]Telepods are fictional teleportation devices featured in the movie 'The Fly' (1986) by David Cronenberg. See McAlister (2017) for further references on the distribution of flies.

collected by James Needham in Puerto Rico in 1935 and is preserved at Cornell's Department of Entomology. Its respiratory siphons are fringed by a dense array of hairs whose function is unknown. They may serve to retain an air bubble when larvae are submerged, although they usually remain very close to the water line. Larvae of another fly family also occurring in Antisana, shore flies (Ephydridae, genus *Dimecoenia*), have anal spiracles, making them look like Mr Krabs from SpongeBob SquarePants from behind (Fig. 4.3d). Species of the same genus living in salt marshes use aquatic plants like a snorkel, inserting their spine-like spiracles into the roots to breathe!

Flies without tube-shape spiracles breathe through holes that run down the side of their body, sometimes assisted by additional biological adaptations such as small gills or even the presence of hemoglobin in their body fluids. Why are fly larvae so inventive at getting oxygen out of the water? Maintaining energy is probably more vital for aquatic flies than for any other animal. A larva's life is to eat and accumulate enough reserves to provide not only enough energy for them, but for their transformation into an adult (the chrysalis) and for life thereafter. Many adult flies, such as non-biting midges, have reduced mouthparts or effectively none at all, and are generally said not to feed. Still, in their short life of just a few days, they need enough energy to find a mate and reproduce. The males die within a day or so, and the females live just long enough to mature and lay their eggs. This is why fly larvae absolutely require oxygen. And their array of respiratory inventions is key to explaining their success at acquiring this in harsh environments, including high-altitude streams. Yet even these sophisticated inventions may be seriously challenged by climate change.

* * *

Dean Jacobsen retraced much of the same ground as Humboldt, but investigating stream bugs rather than plants. His main quest in Ecuador, beyond finding the best *el hornado* (roast pig, cooked whole, generally served in highland markets and restaurants and accompanied by *llapingachos*, fried potato pancakes, and *mote*, hominy), was to understand how altitude affects the distribution of aquatic insects. The data he amassed during his first surveys at various altitudes across the cordillera indicated that water bug diversity in any given site was best predicted by one variable: oxygen. Dean further explored this pattern, and the results suggested that the upper altitudinal limit of several species was related to oxygen pressure.[97] These preliminary investigations led us to think that, like water chemistry and flow rate, oxygen might be an important variable to predict the response of glacier stream invertebrates to global warming.

To understand how oxygen might be an issue in climate change, we need to remember a key law in animal physiology: the respiration of cold-blooded organisms increases exponentially with temperature as well as scaling to body mass. This is important, since as we have seen, oxygen availability is closely related to

---

[97] Crespo-Pérez et al. (2016).

temperature. Let's imagine the physiological consequences for a stream bug facing climate change in Antisana. Warmer air temperatures in the future mean warmer waters, hence more available oxygen (higher OSI), but also higher metabolic demands, especially for large insects. If they react by moving upstream to follow their preferred temperature, their metabolic needs will not increase, but less oxygen will be available due to the lower air pressure at higher altitude. Dean has called this problem "the dilemma of altitudinal shifts:"[98] with global warming, insects are caught between a rock of high temperatures and a hard place of low oxygen. We needed to know whether oxygen supply and demand are equivalent for insects on Antisana's slopes, which would require a field experiment in which we could manipulate oxygen and temperature simultaneously.

* * *

Experimental physics has been part of natural history research since the French entomologist **Réaumur** became interested in uncovering the secrets of animal behavior. Trained as a physicist and mathematician, Réaumur conceived a new way of studying natural history that relied on experiments and instruments. In his most memorable experiment, Réaumur made hollow glass bowls about the size of a chicken egg with a hole in one end. He then introduced butterfly chrysalises inside, closed the opening with a notched cork so that the insect would not suffocate (insect chrysalises also need to breathe, and they do so through spiracles), and placed the bowls under a hen. He kept track of the temperature with a thermometer he built himself, and demonstrated that the warmth from the hen could speed up the transformation to a butterfly from 14 to 4 days. On the fourth day, Réaumur reported "the first butterfly ever to be born under a chicken, and the first of this kind of butterfly to have spent such a short time as a chrysalis."[99]

Half a century later, Humboldt not only used the scale invented by Réaumur for his temperature measurements, he also deepened the practice of experimental physics in natural history. One particular application concerned the respiration of aquatic animals.[100] *"The respiration of animals that usually live underwater is one of the most interesting problems in physiology"* he wrote in 1808 in the opening sentence of a publication on fish respiration.[101] Over seven months, Humboldt performed dozens of experiments in Paris with the French zoologist Jean-Michel Provençal to assess how fish affect, through their respiration, the quantity of oxygen in water. They placed tench in bell jars filled with water from the Seine and, after a few days, boiled the remaining water to calculate how much oxygen was removed. Humboldt was also one of the first to combine experimental physics and natural history in the

---

[98] Jacobsen (2020).

[99] Réaumur (1734–1742), vol. 2, p. 14; cited and translated in Terrall (2014), p. 53.

[100] Building on works by the chemist Antoine Lavoisier (1743–94), who first demonstrated the role of oxygen for animal respiration in 1780, Humboldt was particularly interested in how animals breathe in seemingly unfavorable environments: at very high altitude, underwater, underground, or even in Carrara marble. See Humboldt (1811–1833), vol. 1, pp. 298–303; Humboldt (1808), p. 81.

[101] Provençal and Humboldt (1808), p. 359.

field—in fact, in a study concerning the breathing of an aquatic animal. And not one of the easiest to handle … a crocodile! He collected about 40 individuals in May 1801 on the banks of Río Magdalena in Colombia during a trip from Cartagena to Quito, saying in the introduction of the publication relating his experiment: *"It is easy to understand that the ferocity, the size and the enormous strength of the sharp-snout crocodile make it impossible to carry out any experiments on an adult animal; I had to content myself with enclosing under bell jars young animals hatched out of the egg since fifteen to twenty days."*[102]

Free from animal welfare considerations and the strict experimental protocols of modern zoologists, Humboldt forced three crocodiles "to keep their snout out of the water and at the same time prevent them from breaking the jars by whipping them with their tail" by bounding them "by the back legs, the forelegs and the extremity of the tail on bamboo crosses."[103] By measuring the quantity of air in the bell jar at different points, Humboldt could define the oxygen threshold below which crocodiles started suffering and finally perished. He repeated the experiment five times and found consistent results "to the nearest hundredth." [104] This was the kind of physiological threshold data we wanted to obtain for aquatic bugs, creatures less tricky to handle, but a no less complicated experiment.

* * *

The American geographer Jared Diamond, an avid birdwatcher, reports having to restrain himself from identifying birds at inappropriate moments. At his garden wedding ceremony, he continued to identify birds, finally chiding himself, "Jared, you're about to get married: pay attention to the experience, not to the birds."[105] I admit to this problem myself during days of fieldwork in the tropical Andes, where you are constantly distracted by furry, feathery or scaly creatures. The Antisana Ecological Reserve is particularly problematic in this respect, and it was frustrating to miss potential observations while focusing on sampling a stream or counting flowers on a cushion plant. In January 2012, Philip Madsen, a Danish student brought to Ecuador by Dean to carry out the physiological threshold experiment (like Humboldt's crocodile experiment)—and also a fanatic birdwatcher—was sorely tested by the birdwatching temptations flying around. But concentration was needed, as the experiment required military-level planning: in the words of Dwight Eisenhower: "Plans are nothing, planning is everything."[106]

Dean's experimental design required the team to collect stream bugs whose altitudinal distribution was limited by oxygen and place them in experimental trays filled with stream water, where the oxygen saturation could be manipulated (Fig. 4.15a). The experiment needed to be conducted in the field in a stream at an

---

[102] Humboldt (1811–1833), vol. 1, pp. 253-259.

[103] Ibid.

[104] Ibid.

[105] The Cornell Lab of Ornithology and Vyn (2015), p 179.

[106] Eisenhower, D.: A speech to the National Defense Executive Reserve Conference in Washington, D.C., 14 November 1957.

altitude of 4180 m, as it would be difficult to recreate natural conditions in the lab. In addition, a protocol had to be established to disentangle oxygen saturation and temperature. As mentioned, the two factors are linked, and both decrease more or less proportionally with altitude. To achieve this, Dean had the idea of using sodium hydrosulfite, a white crystalline powder with a weak volcanic odor (because of the sulfur) and known to reduce the oxygen level in water. This would allow oxygen saturation to be varied in the trays (at 55 or 62%) while keeping the temperature constant. Likewise, by putting water in a cooler, the temperature could be varied (5 or 10 °C), while keeping the oxygen constant. Moving from lower to higher values of oxygen and temperature would approximately mimic a 300-m ascent for the invertebrates (from 4180 to 4480 m).

We worked with five species—four insects (stonefly, mayfly, caddisfly and beetle) and one crustacean—of which we had collected several hundred individuals in the stream. Unfortunately, other types of flies were not included: moth flies and shore flies were not abundant enough in the streams, non-biting midges were too small to handle and impossible to identify in the field, while blackflies required water flow conditions incompatible with our experimental trays. In total, the experiment involved three treatments (oxygen, temperature and a combination of both), two levels per treatment (high and low) and eight replicate individuals for each of the five species. This meant that Philip had $3 \times 2 \times 8 \times 5 = 240$ experimental arenas to oversee! Even with only 15 arenas running at the same time, Philip had to constantly adjust the oxygen and temperature levels in the trays, as they tended to return to equilibrium with the atmosphere. To be successful, planning was everything. There was definitely no time for birdwatching.

But one thing was still missing. As we could not measure an invertebrate's oxygen consumption directly, we needed a proxy of the response to changing water conditions. While Humboldt had set two levels, 'suffering' and 'dead,' to assess the response of young crocodiles to oxygen deprivation, we wanted a more quantitative and objective metric. Like crocodiles in bell jars, invertebrates in Petri dishes do not generally keep still and tend to search for a way to escape. If oxygen conditions get uncomfortable for them, they would be expected to stop moving to save energy. Philip therefore divided the bottom of each Petri dish into four numbered fields by drawing two perpendicular lines through the center (Fig. 4.15b). Every minute for 16 min, Philip registered the field number of each individual in its Petri dish and used the total number of movements between fields as a metric of movement activity. Everything was ready. From Réaumur's glass eggs to Humboldt's bell jars, a leap in time and space brought us to Dean's Petri dishes.

* * *

When all the data was collected, it was time to analyze it. While creatures to distract naturalists abound on the campus of La Católica, from marsupial frogs to booted racket-tail hummingbirds to rhinoceros beetles, students of the entomology department have tiny *cubiculos* that help them to focus on their lab work. As the data results were quite messy at first glance, Philip would need to concentrate (Fig. 4.16). Each of the five species seemed to respond in its own way to changing oxygen

Fig. 4.15 Analysing oxygen deprivation in insects. (**a**) Undergraduate student Philip Madsen adjusting water parameters during the experiment on Antisana at 4180 m; (**b**) bugs placed in Petri dishes in an experimental tray. Small holes in the Petri dishes allowed the exchange of external water so that Philip only had to adjust the temperature and oxygen parameters in the tray rather than each dish. Numbered fields allowed the insects' activity to be monitored throughout the experiment. Photos by Olivier Dangles (**a**) and Dean Jacobsen (**b**)

pressure and temperature conditions: the crustacean and stonefly did not really react to temperature, the others liked it warmer, some significantly, others not. All the species were more active in conditions of higher oxygen, but only the crustacean and beetle in a significant way. Refuting metabolic theory, the stonefly was not significantly less active at both low oxygen and temperature. The lack of statistical differences was partly explained by the high variability within each treatment, as each individual exhibited different ways of coping with oxygen and temperature stress. This was a reminder that field experiments, despite the best planning and realization, rarely allow reproducibility "to the nearest hundredth," as Humboldt proudly reported in his crocodile results. Philip may have had more conclusive results with a higher number of replicates (for example, 40—the number of young crocodiles—rather than 8), but then the experiment would have been unrealistic to conduct in field conditions.

Nonetheless, the results broadly supported our initial hypothesis that low temperature and/or oxygen saturation would suppress invertebrate movement. To confirm this, we conducted a complementary experiment directly in the stream. In net bags, we transplanted the five species from sites where they naturally abound (4180 m: high oxygen, high temperature) to sites where they occur in low numbers or are absent (4480 m: low oxygen, low temperature). We found that the decrease in activity, from a high to a low temperature and oxygen environment, was strongly correlated with mortality. These results suggest that even relatively small differences in temperature and oxygen pressure, like those observed in a 300-m gradient along Antisana, may influence the distribution of stream insects. Philip and Dean's experiment was one of the first to provide clues about this poorly investigated issue.

* * *

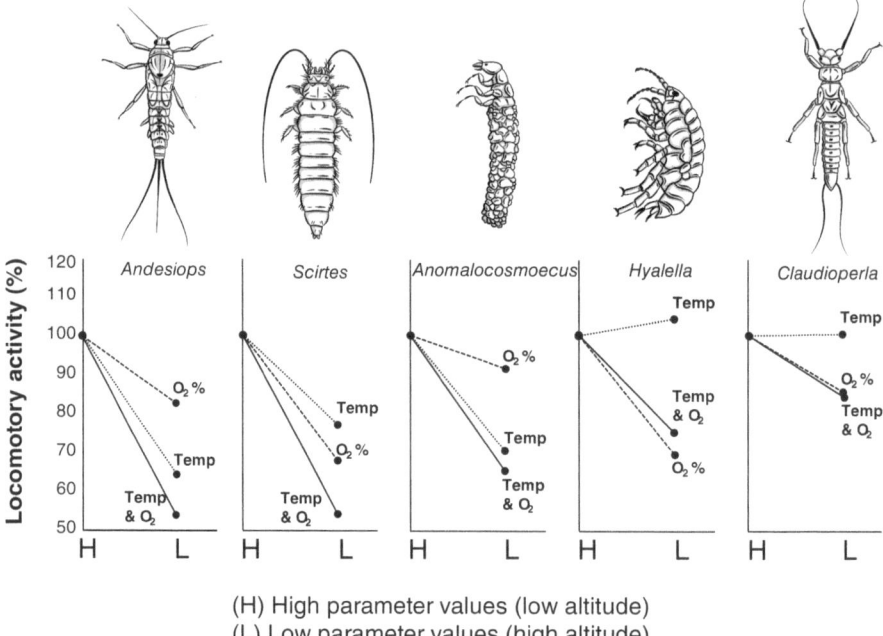

(H) High parameter values (low altitude)
(L) Low parameter values (high altitude)

Fig. 4.16 Insects are less active at low oxygen levels in high altitude streams. Graphs showing the activity of five stream invertebrate species at two levels of oxygen saturation (O_2%): high (H) 62% (low altitude) and low (L) 55% (high altitude). Because bug metabolism is affected by water temperature, we also tested the combined effect of oxygen level and temperature ('Temp & O_2' on the graph) for two temperatures (about 5 and 10 °C). We also measured the separate effect of temperature as a control. Asterisks indicate significant results. Modified from Madsen et al. (2015). Used with permission of John Wiley & Sons Ltd., from Altitudinal distribution limits of aquatic macroinvertebrates: an experimental test in a tropical alpine stream, Ecological Entomology, 40(5); permission conveyed through Copyright Clearance Center, Inc. Drawings by Paula Terán Ospina

Could moth flies, stoneflies or crustaceans ever reach the top of Chimborazo? Would their adaptive breathing capabilities allow them to overcome the dilemma of altitudinal shifts and the rock and a hard place of high temperatures or low oxygen and ride the escalator up from their current habitats? Or will they be blocked by an invisible physiological barrier, like a climber encountering a crevasse? Answering this question is difficult, not only because insects' ability to cope with changing oxygen and temperature conditions is little known, but also because concomitant changes in other parameters across the altitudinal gradient, such as UV radiation, water mineral content or food availability, will complicate predictions. Perhaps one of the best things we can do is to keep learning about organisms' adaptations to extreme altitude

(continued)

conditions. In 2020, on the slope of Llullaillaco volcano in northern Chile, a team of biologists found the highest-dwelling mammal in the world. This small mouse lives at just over 6450 m above sea level, and it is able to shift from a lipid-based to a carbohydrate-based metabolism to yield an extra 15% of energy with the same quantity of oxygen. [107] Such useful adaptations will be key to enable aquatic flies to survive at higher altitudes. Yet they will not only have to adapt to what lies above, but also to the threats that come from below.

4.5 Dangers in the Water

French explorer Germain Paterne could hardly believe his eyes. "There is a sight which has never been seen in the Seine or the Loire, nor even in the Garonne—a spectacle worthy of being seen!"[108] he exclaimed. In front of him, a herd of oxen crossing a river were suddenly struck by convulsions, paralyzed and, turning over on their sides, waved their legs one last time before disappearing under water. Paterne was on the Río Orinoco of Venezuela, and the beasts attacking the buffaloes were electric eels. This naturalist "who would risk his neck just to discover a new sprig of grass"[109] is one of the main characters in Jules Verne's novel *The Mighty Orinoco*.

However, this incredible story of electrocuted oxen did not come from the novelist's imagination, but was inspired by a field observation by Humboldt. At the end of February 1800, Humboldt and Bonpland set out across the llanos of Venezuela, where they had an opportunity to study the powerful current given off by the electric eel, which despite its name and appearance is a knifefish. In Europe, Humboldt had done several experiments on the effect of electricity on nerves and muscles and remained skeptical about "animal electricity." Humboldt "was impatient, from the time of [his] arrival at Cumaná [the port where he landed in Venezuela], to procure electric eels" locally known as *tembladores* because of the trembling that one feels when touching them.[110] However, catching these fish was not an easy task. The local people proposed 'horse fishing:' an extraordinary fishing method for an extraordinary fish. They assembled about 30 horses and mules and forced them to enter a muddy pool. Dislodged from their hiding places by the trampling, the electric eels leapt out of the water and released electric shocks on the horses. This quickly exhausted the fish, allowing fishermen to safely grab five

---

[107] See Storz et al. (2020) and Schippers et al. (2012).

[108] Verne (2002), p. 259.

[109] Ibid. p. 44.

[110] In Steleanu (1959), p. 434.

large specimens with small harpoons and ropes. Humboldt's account of this no doubt inspired Jules Verne's story.

While Humboldt probably had more interest in these eels as living batteries than fish,[111] his insatiable thirst for knowledge also led him to take a great interest in the aquatic fauna of the New World.[112] He collected and described at least 20 fish species and two fish genera new to science, collected information about the altitude and temperature of the waters in which they live, and amassed a myriad of ecological and behavioral information on fish. For example, Humboldt dissected a flying fish that had leapt onto the deck of his ship, observed, years before Wallace, the influence of water color on river fish assemblages in the Amazon,[113] and reported the habits of the piranha: "no other fish has such a thirst of blood [. . .] if a person be only slightly bitten, it is difficult for him to get out of the water without receiving a more severe wound."[114] But in my view, the most extraordinary fish described by Humboldt lives in the freshwaters of the Andes. Even the electrocution of oxen seems more believable than what these fish are capable of.

Local lore claimed there was a fish that was spewed from volcanoes. Although Humboldt did not witness the phenomenon himself during his 7-month stay in the Kingdom of Quito, he reported: "The fish vomited from the volcanoes is a phenomenon so common, and so generally known by all the inhabitants of that country, that there cannot remain the least doubt of its authenticity [. . .] All the inhabitants agree that they are identical with those which are found in the rivulets at the foot of these volcanoes, called *preñadillas.*" Humboldt goes on: "The Cotopaxi and Tungurahua throw out fish, sometimes by the crater which is at the top of these mountains, sometimes by lateral vents." [115] While attacks by electric eels have been confirmed scientifically,[116] proving that a fish can survive the 1000 °C heat of an erupting volcano seems more challenging.

Humboldt went to great pains trying to explain this phenomenon. "Perhaps the clay mud in which these animals are enveloped protect them from the strong heat," [117] he guessed without much conviction. But he finally put forward a more reasonable explanation: the fish may "perhaps come out of a concavity far from those from

---

[111] Humboldt carried out both dissections and physiological experiments on his captured eels, but came to no definitive conclusion on the mode of action or purpose of the electric organs. "The mode of action, the means by which this current is produced, is shrouded in the deepest darkness." Humboldt (1811–1833), vol. 1, p. 88. Some of his experiments involved subjecting himself to painful and dangerous electric shocks from the fish.

[112] See Nico (2001).

[113] "It is far from the case that over this vast area of land [the Amazon and the Orinoco], the river fish are the same. The temperature, depth and speed of the water, its clearness, its chemical properties, the riverbeds, sometimes muddy, sometimes rocky, powerfully influence the animal organization." Humboldt (1811–1833), vol. 2, p 147.

[114] Humboldt (1852–1853), vol. 2, pp. 166–167.

[115] Humboldt (1811–1833), vol. 1; pp. 22–23.

[116] Catania (2016).

[117] Humboldt (1811–1833), vol. 1, p. 25.

Fig. 4.17 Volcano-erupting fish. Illustration by French artist Albin Mesnel (1830–75) of *preñadilla* (*Pimelodus cyclopum*) ejected from an active volcano as described by Humboldt (1811–33). From Pouchet (1872), p. 608

which erupts the volcanic fire."[118] After heavy rain or during eruptions that provoke movements of subterranean waters, the fish may be flushed to the surface from underground caverns.[119] Indeed, the **preñadilla**, a catfish initially described by Humboldt as *Pimelodus cyclopum* (Fig. 4.17), probably in reference to the mythical one-eyed giant living under Mt Etna in Sicily, belongs to a family of cave-dwelling fish. Some populations come and go from the bowels of the Earth to the surface, while others are exclusively subterranean. Living up to altitudes of 3350 m in rivulets of the tropical Andes, the small *preñadillas* (about 6 cm long) are perfectly adapted to strong currents. While torrent midges use their six suckers to anchor themselves on the streambed, the *preñadilla* has another trick: it alternates the contraction of thick ventral fins and the attachment of its sucker mouth to rocks to swim against the current. This sophisticated technique allows them to climb wet vertical surfaces of at least 4.5 m and to explore habitats unreachable to other fish, in particular in the highlands.[120] It is thus understandable that Humboldt posited that

---

[118] Ibid.

[119] There are other examples of cave creatures being expulsed from underground during climatic or geological events. In 1689 the naturalist and polymath Janez Vajkard Valvasor (1641–93) reported that after heavy rains, the olms (*Proteus anguinus*) washes up from underground water and is believed by local people to be a cave dragon's offspring. See Flannery (2019), p. 92.

[120] These fish can easily climb vertical surfaces of 5.5 m. See Johnson (1912).

they were "without doubt the fish that live in the highest regions of our globe,"[121] although this would prove to be wrong.

* * *

The message was intimidating. "If it rains tomorrow, it will be your fault because you have annoyed *lu* [the water spirits] by your fishing!"[122] The message was addressed to a scientist by a distressed Tibetan who discovered him collecting a fish in the lake near his home, at 4940 m. The fish was a bowl fish relative called (take a breath) *Schizopygopsis younghusbandi*. The scientist was Dean Jacobsen. He was following the example of G. Evelyn Hutchinson, who had made extensive ecological observations of high-altitude fish assemblages during his Yale expedition in the Himalaya. In addition to fulfilling his lifelong dream of walking on the roof of the world, Dean wanted to study the ecology of these fish, including their altitudinal limits and feeding habits. Like most scientists, to Dean the fish was a sample, but to the local people it embodied a spirit—an aspect that is generally absent from the inventory lists of ecologists. In addition to learning that Tibetan fish beat the supposed altitude record of *preñadillas* [123] and are fond of eating non-biting midges and blackflies, he discovered the reluctance of Tibetans to fish. In Tibet, it is believed that water is sacred and that fish protect the water. A similar esteem was once given to *preñadillas*, which the Incas used as currency and to offer tribute. But the respect of Tibetans for fish goes farther. Most do not eat fish. As Dean explains: "They have a fear of getting fish bones stuck in their throats. And they believe that fish don't have tongues and therefore cannot gossip: Tibetans detest gossip and reward fish by not eating them." [124] This may be why, unlike people in the Andes, Tibetans have protected their waters from a formidable predator: trout.

* * *

Just like Joseph Fourier, the botanist **Louis Ramond de Carbonnières** narrowly escaped the guillotine. After publishing his monograph on the natural history of the Pyrenees in 1789, the year of the French Revolution, he tried becoming a deputy, but soon realized that collecting plants was less dangerous and probably more fun than politics. In 1792, he left Paris to set up home at the foot of the Pyrenees and dedicated himself to studying their geology, taking temperature and altitude measurements, gauging prevailing winds and cataloging the vegetation (from his training with botanist Antoine-Laurent de Jussieu, Joseph's nephew). At the beginning of the nineteenth century, Ramond de Carbonnières was probably one of the best

---

[121] Humboldt (1811–1833), vol. 1, p. 23.

[122] *Lu* is the holy spirit of water in the Tibetan Buddhist and Bon religions. In Jacobsen et al. (2013), p. 50.

[123] Fu et al. (2004) reported both *Herzensteinia microcephalus* (Cyprinidae) and *Triplophysa stewarti* from 5200 m in the headwaters of the Yangtze River, perhaps the highest fish ever recorded.

[124] Jacobsen et al. (2013), p. 50. Used with permission of John Wiley & Sons Books, from Sacred fish: on beliefs, fieldwork, and freshwater food webs in Tibet, Frontiers in ecology and the environment, 11(1); permission conveyed through Copyright Clearance Center, Inc.

specialists of what we would now call mountain ecology. It is therefore not surprising that when Humboldt published the results of his expedition in 1804, he exchanged ideas with this "savant who combines the varied knowledge of the naturalist with the great views of the physicist."[125] With the *preñadillas* in mind, Humboldt was interested in comparing the upper altitudinal limit of fish in European mountains and in the equatorial Andes. Ramond de Carbonnières informed him: "Trout were the only fish I observed in the high places. [...] We find fish up to the elevation of 1,170 toises [2,280 m] and above there are no more."[126] In some ways, the brown trout was for the Pyrenees what the *preñadillas* were for the Andes and *Schizopygopsis* for the Himalayas. To each its own mountain range.

But since Humboldt's time, humans have brought about a great reshuffling of species distribution through both intentional and accidental introductions of thousands of plants and animals around the world. Among them, two species of trout, the brown trout and the rainbow trout, have been introduced in at least 125 countries, mainly for sport fishing and aquaculture. As reported by Louis Agassiz's wonderful atlas-size book *The Natural History of Freshwater Fish of Central Europe,* the brown trout is native to the old continent (Fig. 4.18).[127] As for the rainbow trout, it comes from the Pacific coast of North America and Kamchatka, where it is locally known as *mykizha* (hence its curious scientific name *Onchorrhynchus mykiss*). The first trout to swim in the waters of the tropical Andes arrived in the 1920s, and they found the cold, clean waters of the páramos particularly to their liking. So much so that now it is the trout that lives in the highest elevations of Ecuador.

Wherever they are, introduced trout are the chief suspects when it comes to the extinction or decline of native aquatic fauna. They are large carnivorous fish, and they can feed on a broad range of aquatic prey, including insects, tadpoles and fish, with little concern whether these are red-listed species or not. They can also outcompete native species or affect them through the spread of fatal diseases or parasites. In Lake Titicaca, generations of trout have now been feeding on kaira tadpoles, contributing to pushing this species further to the brink of extinction. The same situation occurred some 50 years ago with the now extinct Titicaca flat-headed fish (*Orestias cuvieri*), the largest of the around 45 species found in the high desert Andes. Other factors in this decline include overexploitation of native frogs and fish and the degradation of water quality in the lake due to the lack of water sewage

---

[125] Humboldt (1811–1833), vol. 2, p.147.

[126] Ibid. p. 147 and 149.

[127] Agassiz and Vogt (1839). This volume focuses on the Salmonidae family only. Agassiz's overarching project of a complete ichthyological study for Europe was never finished. Agassiz continued his passion for the study of fish in the US, in particular during his 1850 exploration of Lake Superior. See Schlett (2015). In 1865–66, Agassiz would lead the Thayer expedition to Brazil, from which he would bring back to Harvard about 80,000 specimens of Brazilian fauna, mostly fish. He was joined by his lifelong artist partner and friend, Jakob Bourckhardt (1808–67), who had painted him in the Hôtel des Neuchâtelois (see Fig. 3.9). Bourckhardt produced more than 800 watercolors of fish specimens, some of them kept in an aquarium to maintain their coloring. See Britski and Figueiredo (2019).

Fig. 4.18 Mountain water predator. Adult male brown trout (*Salmo trutta*) caught in its native Switzerland and illustrated in Louis Agassiz's *Natural History of Freshwater Fish of Central Europe*. The sketches below the fish show its jaws when the mouth is open and the contours of the fish seen from above; those on the right represent magnified scales from various parts of the body. Agassiz's book blends science and art with its precise technical drawings and beautiful hand-colored lithographic plates by artists Joseph Dinkel and Hercule Nicolet. The fish was painted while swimming in a glass jar in front of the artist. From Agassiz and Vogt (1839)

treatment. What about the trout's effect on *preñadillas*? Although we know little about the historical distribution of *preñadillas*, it is likely that trout have displaced them to lower elevations as both species rarely occur together in the same stream in the páramos. Can they share the same space without coexisting? In the rugged terrain of the tropical Andes, climbing *preñadillas* may be able to find safe refuges upstream from waterfalls, which are impassable for trout. Yet the latter are often helped by sport fishermen to overcome natural barriers. Today, it is hard to find any páramo lake or stream without trout. Boosted by the constant stocking of recreational fishing lakes, trout populations have spread to higher and higher altitudes and are now arriving at the limit of glacier streams. This is not only a danger for *preñadillas*, it is a new threat for aquatic flies; a threat that could affect one of the most essential components of the water bug lifecycle: drift.

* * *

Human attention to the fly–fish relationship probably dates back to the distant origins of fly-fishing, some 20 centuries ago. The Roman writer Claudius Aelianus described Macedonian anglers fishing using a hook and a lure of red wool thread and two cock feathers. In the time of William Shakespeare, probably himself an angler, angling treatises detail how, where and when to prepare fishing flies and let them

Fig. 4.19 Go with the flow. (**a**) A typical stream in the vicinity of Ithaca, NY; (**b**) The first published photograph of a drift net positioned to catch drifting organisms (insects, crustaceans, etc.) from a stream in the region of Ithaca. Drift data are instrumental for assessing the response of stream invertebrates to the presence of predatory fish. Photos (**a**) by Olivier Dangles and (**b**) from Needham (1928) used with permission of John Wiley & Sons from A Net For the Capture of Stream Drift Organisms, Ecology, 3(9); permission conveyed through Copyright Clearance Center, Inc.

drift. It is therefore surprising that the phenomenon of drift, the downstream transport of aquatic organisms in the current, did not attract the interest of ecologists until the mid-1950s. One ecologist though made some pioneering experiments in the 1920s in streams in the vicinity of Ithaca (Fig. 4.19a) to assess the amount of food that drifting organisms provide to fish. This ecologist was Paul Needham, James Needham's son, a limnology professor at Cornell who was passionate about trout. Needham Jr's drift net consisted of a 6-m-long and 1-m-wide fine-mesh copper-wire cloth with a catch basket as a central collecting point, and maintained across the stream using a flexible wire clothesline tied to objects on either bank (Fig. 4.19b). The study served as a proof of concept allowing future quantitative studies and, a few decades later, much had been revealed about the drifting behavior of stream life.

While there may appear to be little movement of organisms in a stream apart from fish, the water is a real invertebrate highway. If you set up a net in a stream, overnight it will collect hundreds of individuals that have been drifting past. But remarkably, water bugs not only drift passively, due to accidental loss of anchoring on the substrate or because they die, but also drift *deliberately* to search for suitable food, escape unsuitable water conditions, or avoid predators. To what extent do introduced trout have an impact on the traffic of native invertebrates in Andean streams?

To find out, we need to turn to Alex Flecker. His knowledge on freshwater ecology is the size of his office library: XL. Heir to Needham, whose legacy he continues to perpetuate in his courses, Alex is a fish ecologist at Cornell and, as host of my sabbatical, bore the burden of quenching my thirst for knowledge on the subject. Like others at Cornell, Alex has a trick to survive Ithaca's long winters. He flies to the tropical Andes, his second home. Back in the 1980s, Alex retraced

Humboldt's path to the Rio Apure basin, one of the largest tributaries of the Rio Orinoco, and then moved up the slopes of the Andes. At about 2000 m, Alex found the perfect place to test a hypothesis he had about drift.

At that time, limnologists in Europe and North America had convincing evidence that the drift activity of aquatic invertebrates is typically greatest at night. This behavior was thought to be an adaptive response to minimize exposure to visual-hunting, drift-feeding fish such as trout, yet few works had experimentally assessed this fish-avoidance hypothesis. The introduction of trout in Andean mountain streams had set up a natural lab for Alex to test this. In a similar way to Hutchinson, who had taken advantage of the natural configuration of Tibetan lakes with and without fish to test the response of plankton to predation, Alex could compare bug drift along a gradient of fish predation. Alex found that drift was generally irregular in fishless streams. In contrast, in naturally fishless streams containing introduced trout, he observed nocturnal peaks in drift, suggesting that aquatic insects had rapidly adapted their behavior in response to the new predators.[128]

Alex's pioneering study was a precursor to our investigations on Antisana to test the effect of introduced trout on stream invertebrates. At the foot of the mountain lies a large reservoir, the Laguna de la Mica, which has been stocked with rainbow trout for sport fishing. We did not know since when or with how many individuals, but we suspected that trout were now present throughout the basin, as we observed eggs and fry in rivulets upstream from the laguna. We also suspected that trout might be an additional threat to high Andes stream invertebrates, finding themselves squeezed between oxygen-deprived heights and new fish predators. To test our hypothesis, we needed to investigate three questions. First, could trout colonize glacier streams? Based on Ramond de Carbonnières' reports from the Pyrenees,[129] turbidity might be disabling for visual hunters such as trout. Second, would trout feeding habits in the páramo threaten some invertebrates more than others? In their native range, trout feed in particular on mayflies, blackflies and crustaceans, but this might not be the case elsewhere. Finally, was bug drift along Antisana's water highways affected by patrolling trout? Perhaps the bugs in Ecuador would adapt to this fish's presence. To find out, first we had to catch some trout.

Fortunately, it is much easier to catch trout than electric eels. Dean donned a pair of rubber boots and enclosed the downstream end of a reach of stream with a net, securing it to the bed with rocks (Fig. 4.20). Then he walked from upstream to downstream, forcing the trout to swim toward the net. The smartest fish escaped, the others were caught in the net and then placed in a bucket. There they were sacrificed for the sake of science, and to the advantage of the local *preñadillas*. Dean suspected

---

[128] Flecker (1992).

[129] In his diaries of his expeditions in the Pyrenees, Ramond de Carbonnières reported about the Lac de Gaube: "There are many fish. Doctor Labbat [with whom de Carbonnières had a relationship because he was a deputy at the French legislative Assembly] tells me that this is so only on the stream side, and he attributes the aversion that the trout show for this part of the lake to the calcareous sediments brought from the Vignemale and deposited at the mouth of the stream." From Ramond de Carbonnières (1931), vol. 2, p. 156.

Fig. 4.20 Sampling trout on Antisana. At 4200 m, freshwater biologist Dean Jacobsen nets off the downstream end of a stretch of stream to survey the trout population in high-altitude waters. Photo by Olivier Dangles

that this technique may have missed some fish in the milky water of glacier-fed streams, so we decided to complement the survey using another technique: electro-fishing. This sounds brutal, but it is an efficient way to collect fish with minimal stress for the animals by delivering voltage to the water, which temporarily stuns the fish and allows them to be safely collected. Instead of the fish giving the electric shock, in this case it was us! (I don't know whether electrofishing works with electric eels.)

This allowed us to answer our first question. We caught 35 rainbow trout in the stream with no glacier influence (compared to 11 with the net technique), and 4 trout in the lowest part of the glacier stream (compared to 0 with a net). Trout seemed to be at ease in the confluence of milky and black waters, but the turbid glacier waters are evidently not optimal environments for them. Nonetheless, our survey demonstrated that these environments are not impermeable barriers that protect glacier stream invertebrates from ascending predators.

What about the trout's food preferences? Are they picky or do they just feed on whatever passes in front of them? In our analysis of their stomach contents, we found that in most stream sites, trout were not selective and fed on the most abundant invertebrates drifting in the current: non-biting midges, mayflies and crustaceans, as observed in other streams around the world. These invertebrates are either quite large (such as mayflies and crustaceans) or very abundant, and therefore are energetically cost-effective for the fish to catch. In the lower part of the glacial stream, however, we found that non-biting midges were a delicacy for trout: while midges represented less than 1% of the drifting organisms, they represented about one-fourth of the trout diet. A similar situation was observed for blackflies, which were proportionally four times more abundant in the trout diet than in the drift. Currently, we have no explanation for this preference, but it indicates that trout may be an additional stress on these flies. And with the predicted decrease in glacier meltwater, their situation could worsen.

Finally, did trout affect bug drift? Here we had to deal with certain sampling constraints. The fine silt particles from the glacier transported in the streams tended to rapidly clog up the mesh. We also needed to collect the drifting organisms at regular intervals, including at night, when it is dark and bitterly cold (Fig. 4.21). Our drift sampling required anchoring two steel bars into the rocky substrate with a sledgehammer, attaching the net and adjusting its height just above the substrate so that walking creatures would not get trapped, measuring water depth and current velocity at the mouth of the net before and after sampling (to estimate the volume of water passing through), removing the net, preserving the collected material in a flask with 95% ethanol, cleaning the net twice, putting it back in place, and finally measuring the water chemistry at the drift-sampling site. All this in five streams, with two nets per stream, every 3 hours over 2 days, in freezing water and an oxygen-deprived atmosphere at 4300–4600 m. These efforts revealed that in the fishless glacier stream, invertebrates drifted equally night and day, with a mean traffic of about 60 individuals per hour. But in the streams with fish, we observed about twice more drift traffic during the day than at night: the opposite result to what we expected. It was quite puzzling, but seemed to disprove the fish-avoidance hypothesis. We had two interpretations: either the introduction of trout was too recent and the organisms were still naïve, with no behavioral adaption, or trout density in the stream was too low and not considered a major threat. Our data did not allow us to confirm either of these two hypotheses, but Dean was of help to deepen our understanding, even if indirectly.

* * *

Students would come to our rescue. As Hutchinson wisely stated toward the end of his life: "In any effective university, you find mature scientists playing an enormous role as mentors of younger ones, which provides a major or even the most important aspect of their education."[130] During his years at La Católica, Dean was a strong proponent of mentoring. Redro, Pucho and Vero are living proof of

---

[130] Hutchinson (1993), p. xvii. Used with permission of John Wiley & Sons Books, from A treatise on limnology. Vol 4, The zoobenthos by GE Hutchinson; permission conveyed through Copyright Clearance Center, Inc.

Fig. 4.21 Round-the-clock fieldwork. At 2:00 a.m., undergraduate student Andrés Morabowen uses the beam of his head torch to take night-time conductivity measurements at a drift net in an Antisana glacier stream. Photo by Olivier Dangles

this—15 years after their undergraduate studies they are still dedicated to the study of streams. In the late 1990s, Dean supervised Andrea Encalada, who went on to pursue a successful research career in stream ecology. Andrea obtained her PhD at Cornell, where she worked on the effect of fish on stream invertebrates in collaboration with Alex Flecker. She returned to Ecuador and became a professor at the Universidad San Francisco de Quito, and since then has conducted countless studies in freshwaters throughout the country, including one that helped us interpret our drift results.

The study consisted of placing caged trout into a fishless stream in a remote part of the Guayllabamba basin and then sampling invertebrate drift downstream from the cages. It is fascinating how chemical cues constitute much of the "language of life"[131] under water. Just as piranhas can scent food from a few drops of blood, invertebrates can sniff out danger from chemical signals, probably emitted from fish-skin mucus. Andrea's results were incontrovertible: bug drift was about three times lower downstream of caged trout (Fig. 4.22). [132] While neither these invertebrates nor their ancestors are likely to have ever seen a trout, they still associated their odor with danger. So, this invalidated our hypothesis that they might be naïve to the presence of trout. Interestingly, Andrea also found that drift was proportionally higher at night, and that trout could influence this pattern. Normally, one-fourth of invertebrates drifted at night, but when fish were present, night traffic increased by 50%, a similar pattern as that observed by Alex in Venezuela. Why we found

---

[131] Hay (2009), p. 193.

[132] Vimos et al. (2015).

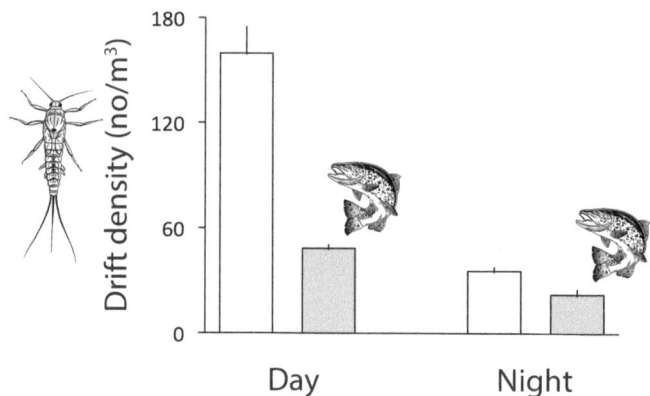

Fig. 4.22 The impact of trout on bug drift. Data from our experiment suggests that stream bugs avoid the presence of rainbow trout in an Andean stream in Ecuador. Insect drift was about three times lower with the presence of trout than without during the day, and twice lower during the night. Data from Vimos et al. (2015); Illustration: Olivier Dangles; Drawings by Paula Terán Ospina (mayfly) and Ivan Mahileuchyk—www.Dreamstime.com (trout)

conflicting results in our study might be related to the turbidity of milky waters, which handicaps visual predators, or it may be that higher meltwater flows during the day increase daytime drift. In any case, in the long term, the modification of stream color due to glacier retreat is likely to benefit hunters rather than prey.

So not only are humans threatening glacier-influenced ecosystems via global warming, but through introduced species. Of course, this is not only true for aquatic life, but for plants on land. Over a thousand introduced plant species have been reported as being naturalized in high mountain ranges worldwide, some of them with the help of global warming. A few introduced species such as dandelion and sorrel have already shown up near summits or glacier fringes. In the Andes, ironically, they are sometimes nursed by native *yaretas*, which help them colonize ever higher. And this rapid colonization of mountaintops is not limited to human-introduced species. As warming reduces the climatic barrier to reaching the highest elevations, it is becoming increasingly urgent to decipher how organisms' physiology has been shaped by temperature variations in what are now cold environments. For this we need to know about the relationship between heat and natural history in the high Andes.

4.6 Feeling the Heat

What is heat? When Joseph Fourier completed *The Analytical Theory of Heat* in 1822, he did not foresee all the consequences of his mathematics, but he knew that he had solved the heat equation.[133] This states that the flow of heat at a point goes from the hot side to the cold side and is proportional to the temperature gradient at that point. Fourier's equation defined heat as a *quantity of energy that flows* spontaneously from one body to another, an idea that would later be popularized by John Tyndall in his introductory book on the physics of heat.[134] And just as Fourier transforms are back in fashion in our high-tech world, the basic principles behind the heat equation have become fundamental to understanding how nature will respond in a warming world, from melting glaciers to suffocating organisms. Why? Because heat affects the amount of energy present in a body: what we commonly call its temperature. And in turn, temperature conditions every aspect of matter and life. As US thermal biologist Michael Angiletta puts it: "Nothing escapes its control."[135] In the living world, temperatures control the rate of chemical reactions, with profound impacts on the physiology, life strategies and, ultimately, the distribution of plants and animals worldwide.

* * *

A few decades before the heat equation, in 1789 Ramond de Carbonnières had observed that "the arrangement of plants on the slope of the mountains mainly obeys the temperature of their different zones."[136] Even more interestingly, he reported that "In those places where the vegetation should expire, these plants benefited from *particular climates* that interrupt the succession of the general climate and whose temperature is *less regulated by the relative situation than by accidental causes of heat.*"[137] In other words, microclimates: Ramond de Carbonnières observed that beyond mean air temperature, local climate can have a strong impact on plant distribution. A few years later, suspecting the importance of these local climates for understanding the links between the physical and living world, Humboldt made some measurements. On the slopes of Chimborazo, "The air was 2.3°R [2.87°C]. When placed three inches in the dry sand, the thermometer remained constantly at

---

[133] "In this work we have demonstrated all the principles of the theory of heat, and solved all the fundamental problems." From Fourier (1822), p. xvi.

[134] Tyndall (1863).

[135] Angilletta (2009), p. 1.

[136] Ramond de Carbonnières (1792), p. 283.

[137] Ibid., p. 286.

Fig. 4.23 **Seeing the heat**. A landscape view of the fast-retreating Carihuayrazo glacier (Ecuador) with a normal camera (**a**) and a thermal camera (**b**). What looks relatively homogenous to the human eye is revealed as heterogeneous by the camera. Ground temperature is key to understanding the dynamics of the colonization of life as the glacier retreats. Photos by Sophie Cauvy-Fraunié

4.7°R [5.87°C],"[138] while on the Orinoco, on "stones covered by an innumerable quantity of iguana and geckos, [. . .] the thermometer placed against the rock rose to 50.2°."[139]

Every good naturalist tries to see the world from his or her subject's point of view. For Humboldt, this took the form of taking physical measurements as close as possible to the conditions experienced by plant roots or animal feet. Today, modern ecologists can easily visualize these local conditions thanks to an instrument that Humboldt would surely have liked in his toolkit. Thermal cameras make the invisible world of temperature visible (Fig. 4.23). Thermography has a beauty and power beyond strict science, transporting viewers to a different world and helping them to sense the challenges faced by living organisms in a warming world. This invisible realm governs the biology of all the organisms that reside there and the natural processes they regulate. In recent decades, thermal images have increasingly helped to understand how local conditions can contribute to making predictions about the ecological consequences of a rise of 1–2 °C in air temperature. Just as fluid dynamics help to elucidate adaptations to water flow, the laws of thermodynamics are essential to understand how living organisms feel the heat. To see this energy is to be more aware of the fragility of living things in the face of climate change.

---

[138] Humboldt (2003), p. 220. °R = degree Réaumur. On the Réaumur scale, the freezing and boiling points of water are defined as 0 and 80 degrees respectively.

[139] Humboldt (1852–1853), vol. 2, p. 199.

Local variations in heat are particularly noticeable in high-altitude regions such as the Altiplano. French ornithologist Jean Dorst, in François Vuilleumier's words *"one of the most brilliant naturalists of the twentieth century,"*[140] an avid reader of Fabre and a tireless explorer of tropical mountains, sums it up well.

> As everywhere in the mountains, sunshine plays a very important role. The sun heats up the atmosphere considerably and especially the objects that radiate it. [. . .] It is frequent that in the streets of the towns of the Altiplano the ice does not melt on the side in the shade, while the passer-by is forced to undress on the side of the street exposed to the sun.[141]

No one can imagine the strength of this radiation until experiencing it. Living in La Paz, at 3650 m, I was stunned by the extreme effects of stepping into the shade or sun. As a warm-blooded mammal, thus with fairly good body temperature regulation, I still had to constantly change sides of the street, alternately chilling and sweating. This physical experience gave me an insight into cold-blooded animals—frogs, lizards or flies—that cannot regulate their own temperature and must constantly adapt their behavior to heat variations. Oliver Pearson carried out pioneering studies on this in the Altiplano. In 1973, during a family excursion in southern Peru, he measured the body temperature of free-ranging lizards and toads by inserting tiny automatic thermometers in their rectums.[142] He described how these animals alternately used warm stones or the shade of holes and nooks either to warm up or cool down and maintain an optimal temperature. As we have seen (Fig. 2.15), *yareta* cushions provide another way to escape the heat in these ecosystems. Pearson also pointed to the importance of wind and humidity on the thermoregulatory behavior of these animals.

Importantly, the challenge of coping with radiation and heat variation in the Altiplano increases with altitude. A few hundred kilometers north of Pearson's study site, at the foot of the Puka glacier at 5180 m, a close relative of the Titicaca water frog avoids temperature extremes by utilizing freshwater pools. While the air temperature is about 1.5 °C, with extremes between –9 and 22 °C, the frog's body temperature varies only between 4.6 and 8.0 °C. Amazingly, another frog species living in the same place but among the rocks has a body temperature varying between –3.5 and 44 °C, as it is strongly affected by the heat radiating from the substrate.[143] These frogs have evolved different strategies to survive the local climate. For one of them, the presence of water is vital in buffering temperature extremes. This begs the question, is it easier to face heat under water?

The answer is yes and no. It is true that aquatic habitats are more buffered from warming than the air. This is due to the high specific heat capacity of water (the amount of energy needed to change the temperature of 1 g of water by 1 °C), which is about 4.2 times higher than that of air. Water heats up less than air, and its temperature tends to be less variable over the day and seasons than on land. While

---

[140] Vuilleumier (2004), p. 1290.

[141] Dorst (1957), p. 6.

[142] Pearson and Bradford (1976).

[143] Reider et al. (2021).

a high-altitude fly has to cope with sudden and extreme temperature changes in the course of a day—typically frenetically active when the sun is shining, moving erratically in cloudy weather, and disappearing into crevices at night—its aquatic larvae have much more 'water-conditioned' lives. Also, the spatial variability in temperature is generally much lower in water than on land. The two frogs cited above are a good example of that phenomenon. However, while temperature buffering may seem an advantage for aquatic organisms in the face of warming, it also has its drawbacks. If the climate warms, while land creatures may be able to escape unfavorable conditions by taking advantage of the diversity of local climates at hand (Fig. 4.23b), those living in uniformly warming water may not have this option. In other words, they cannot escape the heat. As a consequence, it will be harder for them to manage their energetic demands and meet their respiratory needs. Although warmer temperatures modestly increase the supply of oxygen by increasing oxygen diffusion, they probably increase the demand for oxygen more. All this makes predicting the warming effects on aquatic biota quite complex, which has triggered a myriad of scientific studies over the last decade, from lab experiments to empirical studies along gradients of latitude and altitude. Of these, a few have focused on aquatic systems that depend on another source of heat than the Sun: the Earth.

* * *

Fourier's research led him to make predictions about how heat acted not only in the laboratory, but also in the Earth itself. The mathematician identified three different sources of heat on the surface of the Earth: the exposure to solar rays, the irradiation from other stars, and "primitive heat, which it contained when the planets were formed."[144] Fourier showed that this last heat source was negligible when compared to the Sun's heat, but his "admirable investigations"[145] on Earth's internal temperature gradients helped Humboldt lay the foundations of the study of groundwater temperatures. In his interest to understand this physical phenomenon, as complex as the limits of perpetual snows, Humboldt measured the temperature of close-to-boiling hot springs in Venezuela and Mexico and posited that hot springs owe their origin either to the nearby presence of magma or to the very deep circulation of groundwater.[146] Geothermal systems are quite common on Earth, even though their exploration from a biological or ecological standpoint is relatively recent. In 2010, Guy Woodward, a colleague from Imperial College London, published a paper that caught my attention. He used a geothermal stream system in Iceland to assess the ecosystem-level effects of warming.[147] With my fondness for natural experiments, I kept it in mind and hoped I might have the opportunity to perform a similar one. Guy's study had been conducted at sea level, where oxygen is

---

[144] Fourier (1827), p. 570.

[145] Humboldt (1846–1858), vol. 1, p. 166.

[146] "The temperature of springs which, for the last half century, has been so much an object of physical research, depends, indeed, like the limit of perpetual snow, on the concurrent influence of many and very complicated causes." Humboldt (1846–1858), vol. 1, p. 208. See also Davis (1999).

[147] Woodward et al. (2010).

not limited. Would we find similar patterns in a stream in the high Andes with regards to the trade-off between oxygen consumption and temperature?

I got my chance to test this in 2012, a few months after I had arrived in Bolivia. At my host lab in La Paz, the Universidad Mayor San Andrés, I met Antonio Daza (Tuco), an enthusiastic biology student. Tuco was never happier than in the field, preferably near water and looking for insects. In addition to his studies at the university, Tuco guided tourists to discover the natural wonders of Bolivia. At the time I met him, he was looking for an undergraduate project, preferentially on fresh water. I thought of Guy's research and asked Tuco if he was interested in studying high-altitude hot springs in the Bolivian Altiplano. He confirmed that he was and said he had brought tourists to hot springs in Tomarapi, at 4200 m on the slope of the Sajama volcano, about five hours by car from La Paz. As tourist sites are not necessarily the best place for scientific studies, Tuco suggested we could find another geothermal system near Sajama. However, he warned that we would need to talk to the native Aymara communities and get their agreement to work on their land. Like Dean, who had learned about Tibetan culture during his work there, I thought I should learn more about the Aymara. It turned out that the first anthropological study on these people, still relevant today, had been done by a Frenchman whose name was quite familiar to me.

* * *

The reason I'd heard of **Alcide d'Orbigny** was because of my kids. The French-Bolivian Lycée in La Paz bears his name. While he is widely known in Bolivia, until that point I knew nothing about d'Orbigny, which is difficult to understand because during his relatively short life d'Orbigny not only studied an amazing diversity of fields, including paleontology, geology, botany, zoology, archaeology and anthropology, but made major contributions in each of them. He developed the first geological time scale, named nine rock strata still in use today, and established the basis of micropaleontology, a discipline that would have crucial applications in geological engineering (from mineral exploitation to tunnel digging). He was also a precursor in biogeography, collecting more than 100,000 specimens, including fossils, and studying 24,000 species of plants, invertebrates, fish, frogs, birds and mammals. Moreover, he was a humanist, playing a crucial role in the development of anthropology and ethnology, and supported the recognition of human rights in Bolivia.[148] D'Orbigny was equally comfortable peering through his microscope or collecting samples in mosquito-infested rainforests, writing dry scientific publications or popular and poetic travelogues. His memoirs and treatises comprise at least 67 titles, "from which it would be easy to select half a dozen, any one of which might have constituted the lifework of any ordinary man."[149] His monumental work *Voyage in Austral America* is, in Darwin's words, "on a scale of magnificence

---

[148] Roux (2004).

[149] Heron-Allen (1917).

Fig. 4.24 A vestige of Alcide d'Orbigny in Paris. (**a**) Bust of the naturalist d'Orbigny on the Paris Natural History Museum's Gallery of Paleontology and Comparative Anatomy, as seen from a path in the Jardin des Plantes; (**b**) A close-up of the bust showing the wear of time. Photos by Olivier Dangles

which at once places him in the list of American travelers second only to Humboldt."[150]

Some years ago, convinced that there must be a place of honor for Alcide d'Orbigny in France, I looked for him in the neighborhood of the Natural History Museum in Paris. It took me quite a while before I finally discovered his bust, hidden behind the vegetation in the Jardin des Plantes, on one of the façades of the Gallery of Paleontology and Comparative Anatomy (Fig. 4.24). Though time and lichens had intervened over the years, I recognized his face by his characteristic chinstrap beard. I wondered why d'Orbigny was no longer present in French collective memory. Had Darwin, who wrote in English and made such a key discovery for science, taken all the limelight in the nineteenth century? Whatever the reason, d'Orbigny's memory, like Humboldt's, was much more alive in South America than in France. While near the end of his life d'Orbigny was offered the chair of Paleontology at the Natural History Museum, his true value would never be recognized. After the death of d'Orbigny in 1857, Louis Agassiz was contacted by the French Minister of Public Instruction and offered the chair left vacant. Agassiz refused, too attached to his independence and his projects in the United States, preferring "to build something new here than to go and fight among the coteries of Paris."[151] And so, France missed the chance to perpetuate the legacy of Humboldt and d'Orbigny among its scholars and citizens. On d'Orbigny's grave near Paris, the only sign of his scientific connection is hidden in the tiny shells embedded in the

---

[150] Darwin (1840), p. 110.

[151] Letter from Louis Agassiz to Charles Martins, 8 November 1858 (Harvard University, Houghton Library, bMs Am 1419 [140]), see Appel (1997).

calcareous headstone,[152] a tribute much less imposing than Agassiz's moraine in Mount Auburn Cemetery.

In fact, Alcide d'Orbigny owed his trip to South America to tiny shells. The son of Charles-Marie d'Orbigny, a naturalist himself, Alcide was just 11 when he started helping his father sort sand samples in search of microscopic fossils.[153] He soon became fascinated by the diversity and beauty of these ammonite-like shells and started drawing them. D'Orbigny observed that, unlike ammonites, these creatures' shells had holes. In 1825, d'Orbigny presented to the French Academy of Science the first classification of these microscopic animals that he would later place in a completely new group: foraminifera, Latin for 'hole-bearers.' He made such a strong impression with his "extensive knowledge, [. . .], activity, zeal and studious and wise conduct"[154] that 8 days later, he received a letter from the museum inviting him to lead a scientific expedition to the then poorly known South America.[155] While preparing his voyage, d'Orbigny undertook training in all branches of natural history with the best specialists in Paris and visited Humboldt, who asked him "a host of questions to solve" and made him "aware of the means of observation."[156] Humboldt helped him obtain barometers, the only instruments he brought with him.

d'Orbigny left France in July 1826. After visiting Tenerife on the voyage, he arrived in South America, where his exploits included going upstream the Rio Paraña, the continent's second-largest river, exploring the plains of Patagonia and making excursions in Chile. In May 1830, he reached the Peruvian coast. From there, he made his way up to La Paz, crossing about 100 km north of Sajama. His long walk in the Altiplano was demanding but fascinated him. Like Joseph de Jussieu, he observed the yareta, "of a beautiful green, [. . .] that only an axe can cut through." [157] Like Humboldt, he was interested in the color of rivers and "had

---

[152] Legré-Zaidline (2002), p. 229.

[153] Vénec-Peyré (2002).

[154] Legré-Zaidline (2002), p. 38.

[155] d'Orbigny's travels were complementary to Humboldt and Bonpland's. The region of Lima was the northern limit reached by d'Orbigny, while it was the southern tip of Humboldt's expedition. Also, d'Orbigny did not visit Brazil, which was already visited by the botanist Auguste Saint-Hilaire (1779–1853).

[156] d'Orbigny (1835–1847), t. 1, p. 5

[157] "The vegetation of this region [...] is quite particular, and appeared to me different from any that I had seen before. There are no trees, nor even shrubs; one sees there, with some rare grasses, only plants living in clusters and of a most singular aspect. None of them is high; all grow on the rocks to form a compact, rounded mass, often a few meters in diameter, of a beautiful green, but whose branches are so tightly packed in grass, that only an axe, so to speak, can cut them. Each mass represents a single plant, with a single root, and which, over several centuries, may not have acquired more than half a meter in height. These stumps are used as peat when they are dry." d'Orbigny (1835–1847), t. 2, p. 379.

very marked proof of the altitude-induced" *soroche*, in the form of "a very violent headache and great difficulty breathing." [158] Like Pearson and Dorst, he observed the habitats used by animals to adapt to the harsh climate: "a woodpecker that lives only in rocks, and two beautiful species of rodents that make their underground galleries in the wet plains."[159] Like Agassiz Jr, he took a boat on Lake Titicaca, where he "perfectly made out the bottom," though surprisingly missed the baggy-skinned frog.[160] But one of his most significant achievements on the Altiplano was the detailed archeological and ethnological studies of the Indigenous people who live there, the Aymara, with whom he spent several months. In his book *The*

---

[158]"Since my arrival at the top of the Cordillera, I had been suffering from the rarefaction of the air. I felt atrocious pains in my temples; I had nausea similar to that produced by seasickness; I breathed with difficulty. At the slightest movement, I felt strong palpitations and a general malaise, combined with discouragement that I could not overcome despite my efforts. I had very marked proof of what habit produces. While I was suffering in this way, I saw two natives, sent as couriers, nimbly climbing points, incomparably higher than those where I was, with ease to shorten their route. The shepherds there, as light-footed as the goats of the Pyrenees, were occupied in the valleys near the perpetual snow in guarding their flocks of llamas. They were, however, at an elevation equal to that of Mont Blanc. In the evening I experienced a major nosebleed, which relieved me a little; nevertheless, I spent a night all the more dreadful that I was without shelter, exposed to a sharp and biting cold, which transformed into ice all the water in the surroundings." Ibid. pp. 380–381; see also p. 387. Concerning the reference to the color of water, see Ibid. t. 1, pp. 182–184.

[159]Ibid., t.2, p. 384.

[160]It is really surprising, almost inexplicable, that d'Orbigny did not mention the Titicaca frog in his writings. Even though he was "forced by illness to discontinue [his] trip around Lake Titicaca" (d'Orbigny 1835–1847, t. 3, vol. 2, p. 243), d'Orbigny spent several days in the vicinity of the lake in early June 1833 (see his itinerary in the 'Carte générale de la République de Bolivia dressée par A. d'Orbigny, d'après des itinéraires relevés au cours des années 1830, 1831, 1832 et 1833.' (1852)). He had "the pleasure to collect on the shore several interesting objects of natural history" (Ibid. t. 3, vol. 1, p. 354) and the frog was supposed to be very abundant at this time. Moreover, the conditions were good for naturalist observations. "The waters are so clear that, at a depth of twenty meters or more, one can perfectly distinguish the bottom, covered in greenery by a species of aquatic plant found on all the coasts. I often saw fish flash underneath me with their silver reflections." (Ibid.) The frog was also important symbolically for local communities and could have been mentioned to d'Orbigny. While the professors with whom d'Orbigny corresponded were more interested in fish (Cuvier) and mammals and birds (Geoffroy Saint-Hilaire), d'Orbigny nonetheless collected 16 species of amphibians during his travels. One toad (*Bufo dorbignyi*) was named after him (see Lescure et al. 2002). A hypothesis to explain this mystery is that d'Orbigny did not go to Titicaca himself. It is known that he drew his 'Carte topographique du lac de Titicaca ou Chucuito et d'une parties du grand plateau des Andes (Bolivia et Pérou); dressée sur les lieux en 1833, 1839)' based on drawings and a map of the north of the lake made by English traveler John Christian Bowring (who accused him of plagiarism) (see 'Comptes rendus hebdomadaires des séances de l'Académie des sciences', tome 8, Paris: Bachelier, 1839, p. 364, 453, and 548). d'Orbigny described the Bolivian province of Caupolican without going there, based on the observations of his assistant François Rossignon (?–1846), who was never credited, see Loza (2005). So we can't be sure that d'Orbigny really observed the waters of Titicaca.

American Man, inspired by and dedicated to Humboldt,[161] d'Orbigny closes the chapter on the Aymara as follows:

> Would these deductions not establish for all, as they did for us, the near certainty that the Aymara are the first stock of the civilization of the Andean plateau? That they occupied the central point where agricultural and pastoral life first developed? Where social ideas germinated? Where the first monarchic and religious government was established, within this society, which reached perhaps long ago a degree of advanced civilization, and whose last splendor—the religion and industry transported from the banks of Lake Titicaca toward Cuzco—ended up creating the kingdom of the Incas, which, later, left its cradle completely forgotten?[162]

In his descriptions of his interactions with the Aymara, d'Orbigny presents them as gentle and peaceful but distrustful, unsubmissive and attached to their freedom.[163] I would soon have the opportunity to find out for myself.

* * *

Our first meeting with the representative of the Aymara community was like a scene from a Western. Strains of Charles Bronson's harmonica would not have been amiss. A dusty desert road, tumbleweeds rolling across the windy plain, two groups of men sat in their cars about ten meters from one another with a palpable tension in the air, at least for me. The meeting point was the entrance to the Sajama National Park, as if we needed to show our hands before we were allowed to go further. On the horizon, plumes of dust signaled the approach of another car: that of Rubén Lamas, the director of the park. He would help us to communicate and, hopefully, make a deal.

We launched the negotiation talking about *bofedales*, as I knew the Aymara's primary activity in Sajama was pastoral. Tuco and I presented the objective of our study, which was basically "collecting invertebrates in a geothermal stream to understand the future effects of climate change." Our hosts were skeptical. They explained that they already had indicators to help them predict climate variability and manage weather-related risks: they use star constellations to forecast the timing and quantity of precipitation that will fall in the rainy season; they listen to fox vocalizations to make seasonal forecasts for where and when to plant; they observe the moisture under rocks in the pre-dawn hours of the Fiesta de San Juan on 24 June to assess whether the following growing season will be rainy, and so on. I was to discover that these indicators are not only validated and sometimes explained by science, but are in fact superior to those currently provided by climate services in the country.[164] So the Aymara were not interested in our study about aquatic insects.

---

[161] "To whom would I think of dedicating [my work] if not to you, whose genius in a way inspired it; to you whom Europe has proclaimed the model of philosopher–explorers!" In d'Orbigny (1839), preface.

[162] d'Orbigny (1839), pp. 328–329.

[163] d'Orbigny (1835–1847), t. 2, p. 379, 391, 394, 554; d'Orbigny (1839), t.2, p. 172.

[164] See, for example, Orlove et al. (2002), Gilles et al. (2022).

What would they have to gain from understanding more about them? Our pitch, carefully prepared in advance, was a flop.

Changing tack, Rubén asked the community members to talk about hot springs in the area. The president spoke up. After welcoming us to his community, he said: "In front of the volcano, on the other side of the valley, there is a stream called Juntuma." Tuco whispered in my ear that *junt'u uma* means warm waters in Aymara: a good omen. "There is an area where you can find three types of waters: waters that are too hot for swimming, waters that are too cold for swimming, and waters with the perfect temperature for swimming." I thought to myself that we may have found the temperature equivalent for the "meeting of waters." (see Fig. 4.6) Instead of color, the site might be an example of a confluence of hot and cold streams. But the representative asserted that the differences in temperature occurred in only one stream, which crossed an area with steaming hot springs. Upstream from the hot springs the water was too cold for swimming, was too hot at the level of the springs, and further downstream was the perfect temperature until heat dissipated as the stream ran downhill. I appreciated this classification of waters through swimming characteristics and soon understood its purpose. The community had the idea of bringing visiting climbers to the site (many come to climb Sajama, which at almost 6550 m exceeds Chimborazo in elevation, see Fig. 3.2). Mountaineers would probably enjoy bathing in the hot spring after their cold ascent, as well as an extra night in the community. But the project had not yet materialized. The plan was to create *piscinas* (swimming pools) fed with water from Juntuma, but in order to plan where to build the pools, they needed to know more about water temperatures in different sites in the area and their variability over time. Tuco and I exchanged glances with the same idea in our heads. The community may not care about aquatic insects, but they might be interested in the water temperature data that we would collect. The Aymara are careful observers of nature and pay close attention to *pacha* (Earth, universe, time), but the fine-scale variations science can provide might be of use. We explained that we had temperature loggers that could measure the temperature of water every minute, for hundreds of days if necessary, and that we would be happy to share this data with them. This won their agreement, and Charles Bronson stopped playing.

* * *

After sharing a meal of *chuño* (freeze-dried potato) and *charqui de llama* (thin strips of dried llama meat), we all headed in the direction of Juntuma, following a dirt road winding through the high-elevation grasslands, or **puna**. I could feel the ice starting to break, but there was still not much talk. Suddenly, an ostrich-like bird crossed the road. *Suri*! exclaimed the Aymara. As I took a photograph to immortalize this encounter,[165] the story of the 'discovery' of this unusual bird came to mind. The first European description of the lesser rhea is ascribed to d'Orbigny, who named it

---

[165] See the picture at: https://www.inaturalist.org/observations/91442019

Rhea pennata in a footnote in volume 2 of his *Voyage in Austral America.*[166] However, the bird's colloquial name is Darwin's rhea. Why? Did d'Orbigny observe so many birds on his trip in South America, collecting more than 1500 specimens, of which 200 were new to science and five were named after him, that he left one for Darwin? [167] Certainly not, as Darwin also had his share of birds, collecting about 500 specimens, even if only two were named after him (of which not even a finch).[168] In fact, the rhea had been observed and described by the two naturalists at more or less the same time. In May 1829, Patagonian **gauchos** hunted a dozen rheas for d'Orbigny using traditional bolas, a weapon made of a cord with stone balls at each end, thrown to entangle the rhea's legs.[169] A few years later, in August 1833, Darwin first heard about this bird while he was conducting expeditions in northern Patagonia. The description of the rhea by d'Orbigny was probably only published in 1837,[170] and while Darwin knew that the French explorer had "made great exertions to procure this bird," thought he "never had the good fortune to succeed."[171] Darwin continued searching fruitlessly to sight one until one day in January 1834, when a rhea was shot, cooked and eaten by Darwin's company before he realized it!

> Fortunately, the head, neck, legs, wings, many of the larger feathers, and a large part of the skin, had been preserved; and from these a very nearly perfect specimen has been put together, and is now exhibited in the museum of the Zoological Society. Mr Gould, in describing this new species, has done me the honor of calling it after my name... [172]

The name *Rhea darwinii* chosen by the famous ornithologist John Gould was later supplanted by d'Orbigny's *Rhea pennata*. But the name Darwin's rhea remained as a souvenir of the discovery of this bird by European naturalists. I told my companions that the first scientist to have described the *suri* was a Frenchman who had traveled near Sajama. I also mentioned that during his trip, he had observed parakeets, which he named after the Aymara (*Psilopsiagon aymara*). Impassive, they barely nodded their heads. But the cool atmosphere seemed to be thawing.

* * *

When we arrived at Juntuma hot springs, it brought to mind Earth in Precambrian times: the ground steaming with geothermal heat, gas bubbles rising through water holes and emitting popping sounds, orange carpets of bacterial mats, a sulfurous smell filling the air (Fig. 4.25). In the past, environments like these hosted conditions that may have triggered the initial chemical reactions linking together simple molecules in a first step toward complexity. At its biochemical core, the recipe for

---

[166] "In this species the acrotarsus [fore part of the tarsus] is covered with small feathers; this is why I have given it the name Rhea pennata." d'Orbigny (1835–1847), t. 2, p. 67.

[167] Berlioz (1933).

[168] Frith (2016).

[169] d'Orbigny (1835–1847), t. 2, p. 194.

[170] Dickinson (2019).

[171] Darwin (1839) p. 110.

[172] Ibid. p. 108.

Fig. 4.25 Hot springs in the Altiplano. Like a primitive soup, a hot spring (75 °C) surfaces in the Sajama National Park, Bolivia (4300 m). Some of these springs are connected to the Juntuma stream, making it an ideal place to study the ecological effects of warming water. Photo by Olivier Dangles

life relies on only a few ingredients: chemical elements, water and an energy source to power reactions. All of these ingredients exist in hot springs.

But Tuco and I were less interested in using hot springs to travel back in time than to predict the future—namely, the effects of global warming. We toured the site with the Aymara and determined that Juntuma's configuration was perfect for our study. A dozen pools of water, at about 70–80 °C, were scattered over a salty plateau. Only submerged sheet-like growths of orange bacteria and deep green algae grew there. While half of the hot pools were unconnected to streams—"hot spring islands"—the other half flowed directly into the Juntuma. And because each spring had different outflow volumes and temperatures, the downstream waters were heated to different levels. Our thermal camera allowed us to quickly visualize the site's thermal heterogeneity (Fig. 4.26). This brief reconnaissance mission confirmed the potential of the Juntuma as an open-air lab of the ecological effects of heat on aquatic biota: the site provided a mosaic of average temperature conditions, from about 9 °C about 100 m upstream from the first hot springs to 36 °C at the confluence with the hottest spring. Between the two extremes, we could find sections of the stream with basically all intermediate temperatures. Collecting data here would not only be useful for our study, but also for the Aymara's project.

Juntama was such an exceptional site that we decided to broaden the scope of our study. We would not only investigate aquatic insects in relation to water flow, temperature, oxygen and minerals, but also in relation to algae and fish. In addition

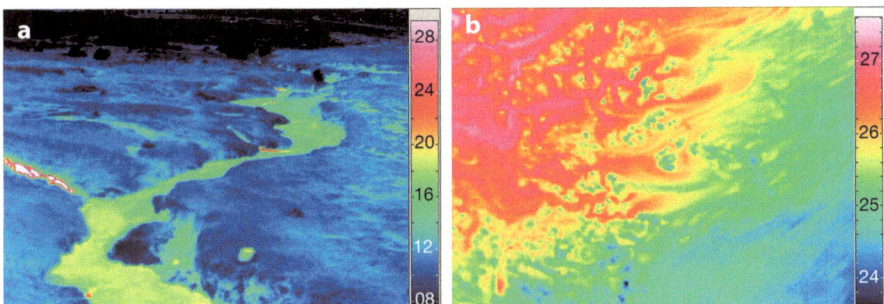

Fig. 4.26 When waters meet (3/3). (**a**) Thermal image of the confluence of waters with two temperature regimes in the Juntuma stream, Sajama National Park, Bolivia. On the left, a jet of warm water originating from a hot spring joins the stream, modifying the temperature (the deeper yellow downstream from the confluence). Water from other hot springs joins the stream at several other localized sites. (The temperatures are indications, as water quickly cools at the surface due to evaporation, making measurements biased.) (**b**) Close up of the confluence of hot spring and stream waters. The temperature scale is in °C. Photos by Olivier Dangles

to Dean and Sophie, three more people joined our team: Jorge Molina and Estefania Quenta-Herrera, students at Universidad Mayor San Andres, and Xavier Lazzaro, a hydrobiologist. They would help take measurements over 20 sites along the Juntuma over a period of one year. We monitored the stream physics with dozens of flow and temperature loggers and recorded over 30 different families of invertebrates, certain species of which had not been described before. A fluorometric probe, an instrument that looks like the Ghostbusters' 'proton pack,' allowed us to measure the different kinds of chlorophyll-producing algae at the base of the stream's food web. Our nets also offered up some nice surprises, since unlike at Antisana, trout had not infiltrated the remote Sajama. We found a small (5–7 cm long) native fish closely related to Humboldt's volcano catfish and belonging to the genus *Trichomycterus*.[173] This name "conceived by Humboldt,"[174] literally means 'hairy nostrils,' in reference to the fish's barbels. We were unable to identify the species, as many *Trichomycterus* fish have been described recently and many more may exist.

And just as importantly, our work allowed us to truly experience the puna ecosystem. Facing temperature variations of more than 50 °C in 24 h, observing plant blooms in response to rare precipitation, feeling the presence of a puma or Andean mountain cat roaming nearby made us enter into true communion with the place. During the nights spent by the Juntuma stream, I experienced some of the most profound connections to nature in my life, a feeling comparable to what d'Orbigny expressed at the sight of the Cordillera Real from the Altiplano: "*never has such a grandiose and majestic view been offered to me.*"[175] For me, this majestic

---

[173] https://www.inaturalist.org/observations/91759183

[174] Humboldt (1811–1833), vol. 2, p. 348.

[175] d'Orbigny (1835–1847), t. 2, p. 38.

Fig. 4.27 Connecting to the cosmos. The Milky Way observed from Sajama National Park, Bolivia, with a queñua tree in the foreground. The Aymara call the galaxy a 'river' which connects the stars across the sky. **Photo by Olivier Dangles**

view was the Milky Way (Fig. 4.27). The clear winter skies of high-altitude deserts like those around Sajama are some of the best places on Earth to observe stars. In total darkness, without the glare from the lights of human civilization, a canopy of endless glittering stars made me acutely aware I am a tiny creature in the cosmos, a small part of an immense universe. Knowing that the light of the stars had taken many hundreds of years to reach me, I knew I was looking into the fathomless past. But like d'Orbigny, I had work to do: "*In the midst of my contemplation, I had quite forgotten myself, and when, abandoning this magical picture, I remembered my existence, I lowered my eyes and looked around me.*"[176]

* * *

So, what findings came out of the Juntuma? A key result of our study was to reveal the maximum water temperature at which each invertebrate, algae and fish

---

[176] Ibid., p. 389.

species could survive in the stream. Overall, we found that most invertebrates were not affected by warming water until the temperature reached about 24–25 °C.[177] Above this threshold, species disappeared one after the other, until no invertebrates were left at 36 °C. As with glacier retreat, warming water would have its losers (many) and its winners (few). Mayflies, stoneflies, crustaceans, blackflies and many other groups simply disappeared when the water warmed above 20 °C. Other insects didn't mind the heat, in particular two groups of flies: non-biting midges and shore flies (Ephydridae). But while the former abounded at all temperatures, from 9 to 34 °C, shore flies were true heat lovers; in fact, they were the only insects that *preferred* hot water. They felt at home only when the stream became 22 °C and well beyond, up to 32 °C. To the naked eye, shore fly larvae look like those of other flies: minute moving specks of grey or white (Fig. 4.3b). However, their minuteness conceals a complex structure and a wide diversity of forms.[178] Thanks to their forked breathing-tube tails (Fig. 4.3d), shore fly species have evolved solutions to master extreme conditions. One of the most extraordinary cases is the petroleum fly, a species living in the natural crude oil pools of California. It breathes by means of its posterior spiracles thrusting above the surface of the oil and feeds on the remains of small animals that fall into the petroleum. If shore flies can adapt to crude oil, hot water shouldn't be a challenge, right? In fact, it is: thermal adaptation is such a serious problem for insects that they have systems for this deep in their body cells.

At low temperatures, insects must avoid ice crystals forming in the cells and body tissue freezing. Pierre Moret's favorite—ground beetles—are good at this. One Alaskan species can survive temperatures as low as –87 °C, and some fruit flies can survive temperatures below 100 °C! Their main strategy is accumulating anti-freeze molecules such as glucose or sucrose in their circulation systems. Dealing with high temperatures is a similar biochemical story. Heat unfolds the three-dimensional structure of various types of proteins, including vital enzymes on which most life forms rely. To imagine this denaturation, think about poaching an egg. Before cooking, it is fluid and transparent; as it cooks it becomes hard and opaque.[179] Denaturation can occur when insects are subjected to heat, but they can also die before their proteins become affected. Heat can directly affect vital functions such as respiration, **osmosis**, or the nervous system.

How can some shore flies develop in water too hot for almost all other insects? We don't really know, but it may be due to the production of heat-shock mechanisms, which assist in the refolding and stabilization of damaged proteins. Some insect species have acquired these through natural selection, probably because living in hot environments provides some advantages. First, plenty of food is available and

---

[177] Quenta-Herrera et al. (2021).

[178] Ephydrid flies in hot springs were probably observed by Humboldt during his journey from Quito to Lima. In Cajamarca, northern Peru, he observed that the Tragadero hot spring (46°R = 57.5°C) "was filled with small insects that do not yet appear to be developed, and in this state resemble Nereids with a suction tube for air." Humboldt (2003), p. 270.

[179] See Wharton (2002), p. 134.

there are few competitors. In the Juntuma, we found that cyanobacteria, which live in colonies, were dominant, providing a food supply for fly larvae. And predators such as the hairy nostril fish, known to be fond of fly larvae, were not found in waters above 25 °C. In fact, this fish's body size decreased by about 1 cm every 5 °C, which may mean that it cannot meet its higher metabolic demands with the available prey in the stream. The only predators were just outside the stream. In the vegetation along the banks, we observed thousands of orb-web and money spiders (I measured up to 144 webs per meter of vegetation!) waiting for fly larvae to emerge.

A second objective of the study was to assess whether altitude, and thus limited oxygen, affects how aquatic life responds to warming. Our reference was Guy Woodward's Icelandic study. He had found a completely different response of aquatic life to temperature than what we had observed in the Altiplano: in Iceland, Guy found that invertebrate diversity increased from 5 to 15 °C and then sharply declined, non-biting midges almost disappeared above 25 °C, and fish (trout) mass increased as water warmed. This illustrates the diversity of ecological patterns found in similar ecosystems worldwide. The only result our studies shared concerned the shore fly. Whether on the coasts of Iceland or in the highlands of Bolivia, these flies enjoy hot springs. There is even one Icelandic species, the aptly named *Scatella thermarum,* that can live in water up to 48 °C! All this suggests that many local factors affect life in warm water, and that the evolutionary past of each place is key to understanding species' adaptation to heat.

Unfortunately, the history of the colonization of life on the Bolivian Altiplano is still fragmentary, so we did not glean much about how the lack of oxygen there may jeopardize the resistance of aquatic insects to warming. However, we did learn that many aquatic species in the Bolivian Altiplano are confronted with extreme temperatures, which may allow them to adapt to future climate change. The situation might be different closer to the equator. There, many species are not exposed to extreme heat, so they may be locked into relatively narrow temperature ranges at specific altitudes. If they are already living close to their upper thermal limits, this may doom them in a warming world.

What about the water's 'swimmability?' I should admit that we had only sporadic exchanges with our Aymara hosts throughout the study. On one occasion they assisted us in securing the temperature loggers; on another occasion we helped them pull out a llama that had sunk deep into a muddy bofedal. Once, while they were observing us sampling a hot spring, they explained that these springs connected the earthly world (*akapacha*) where people, animals and plants reside and the inner world (*manqhapacha*), the place of death and birth, conveying their cyclical sense of life. These were rare opportunities to get to know each other better and exchange information and viewpoints, yet the cultural barrier was not easy to overcome. One day, an Aymara woman approached Tuco to ask him what he was up to with his bucket and nets (Fig. 4.28). Everyone in the community had been informed about our survey, but the insects were probably not part of the story. The woman understandably suspected that Tuco was fishing in the stream. When he told her that he was fishing for insects, this did not lessen her skepticism. So Tuco gave her additional information about aquatic insects: their diversity, habits and their

Fig. 4.28 Cross-cultural encounter. Hydrobiology undergraduate student Antonio Daza talking with an Aymara woman curious about our study on the banks of the Juntuma, Sajama National Park, Bolivia. Photo by Dean Jacobsen

importance for fish as a food resource and as an indicator of water quality. Nor did this appear to convince her of anything other than that Tuco was *loco*.

Despite our communication problems, we were able to provide our hosts with relevant information for their hot spring project. Near the end of the study, Tuco met with a dozen community members to present our results, discuss key data and distribute a written report. We gave them a detailed map of the area with the water temperature fluctuation over a year in about 20 sites, of which over half were swimmable. We also shared two other results of potential interest for the community project. First, the farther the water moves from a hot spring, the less variable the temperature within the moving body of water. Building a *piscina* too close to a hot spring was not recommended, as a current of scalding water may occur unpredictably. Second, the central part of the moving water mass retained its elevated temperature longer. If the community decided to deviate some water from the stream, they should place the pipe accordingly in the channel. And I had my own recommendation for the community. *"Think about developing stargazing tourism! A hot spring in a cold desert looking up at the Milky Way would be an unforgettable experience for people living in artificially lit cities empty of stars."* Our Aymara hosts slightly nodded their heads.

So, in conclusion, what is in store for high-altitude stream flies given climate change predictions? As ecologists always say when they are invited to make projections, this is a complex problem: even from the perspective of a fly. The future of water creatures is even less clear than that of land plants and animals. On land, the escalator to extinction is in motion, and is known to be pushing species to higher elevations. But in water, temperature, turbidity, flow, oxygen and introduced species all simultaneously interact and make future predictions more chaotic. Warmer waters push flies higher, turbidity pushes them lower, fish push them higher, oxygen pushes them lower, meltwater flow can push them higher or lower depending on the cycle ... For flies, finding the optimal threshold on the mountain is likely to depend on their physiological and behavioral adaptations. However, for many aquatic bugs yet to be discovered, we are far from having this information. Our quest to understand the future of Andean stream flies links back to the legacies of earlier scientists, from Fourier, Humboldt and Ramond de Carbonnières to Tyndall and Agassiz to Needham and Hutchinson, and forward to what tomorrow's students will discover. Technological advances will help the latter better grasp the physical and biological worlds and predict how they may evolve.

But beyond science and the search for knowledge, it is the curiosity and sense of wonder defended by Anna Botsford Comstock that pushes future naturalists to wade in rivers as children and later as adults, allowing new discoveries. While the naturalist Comte de Buffon ironized that *"A fly should not occupy more space in a naturalist's head than it does in nature,"* [180] there is perhaps no better parable than a seemingly futile hunt for flies to illustrate the human thirst for understanding. In *The Art of Travel*, Alain de Botton philosophizes about this (in reference to the quote of Humboldt that introduces this fourth chapter):

> How does a person come to be interested in the exact height at which he or she sees a fly? [...] Curiosity might be pictured as being made up of chains of small questions extending outwards, sometimes over huge distances, from a central hub composed of

(continued)

---

[180] Buffon (1856–1857) Tome 6, p. 164. This sentence is more meaningful than it sounds. It was written by the naturalist Georges-Louis Leclerc de Buffon (1707–88), director of the Jardin du Roi (which became the Jardin des Plantes), for the attention of (without naming him) the entomologist Réaumur. Buffon and Réaumur had contradictory approaches to natural history, which, from a retrospective point of view, illustrate the values that Western culture attributes to nature. Buffon thinks big, perceives the world from a macroscopic point of view, creates a hierarchy placing humans at the top of creation, followed by quadruped animals, then birds, etc. He doesn't include flies and other insects, whose study he leaves to the budding science of entomology. Europeans dominate all: mastering and improving nature and taking a moralizing approach to animals. In contrast, Réaumur (see Sect. 4.4) is not only interested in the 'great' facts of nature, but is curious about all of them, without distinction. His approach to nature is more objective, attentive and open but, unfortunately, has remained a minority in Western culture.

a few blunt, large questions. In childhood we ask, 'Why is there good and evil?' 'How does nature work?' 'Why am I me?' If circumstances and temperament allow, we then build on these questions during adulthood, our curiosity encompassing more and more of the world until at some point we may reach that elusive stage where we are bored by nothing. The blunt large questions become connected to smaller, apparently esoteric ones. We end up wondering about flies on the sides of mountains.[181]

I would add that fly hunting encapsulates a reciprocal exchange between the role of nature in piquing human curiosity, and the increasingly urgent role of humans in preserving the world's remaining wild places. A passion for even the most 'insignificant' things in nature is vital not only for science, but for all of us on Earth. Sharing this passion about our intricate and warming planet with the public is an essential task for scientists.

---

[181] The Art of Travel by Alain de Botton (2004), Penguin Random House, p. 116.

Chapter 5
Telling Stories

> *The descriptive genre, simple and scientific, will constantly be*
> *mixed with the oratorical genre. Such is Nature herself. The*
> *glittering stars delight and inspire; nevertheless, everything*
> *that moves on the celestial vault is subject to*
> *mathematical laws.*
>
> *Alexander von Humboldt (1860, p. 65)*

The new main protagonists of this chapter are shown in Fig. 5.1.

5.1 The Heart of the Andes

My two boys were amazed by the enormous jaw suspended from the ceiling of the hall at the American Museum of Natural History: about 2.4 by 3.3 m, its size was equivalent to the height and width of several human adults standing next to one another.[1] The original owner of the jaw was a megalodon shark, which Louis Agassiz first described in 1843 based on fossilized tooth remains. I could see on my children's faces that the sight of this jaw was worth a thousand words—engaging their imagination in an immediate way and bringing to life these giant sharks whose teeth could make short work of prey in ancient seas. While gazing at the jaw provided clear clues about the mechanics of the shark's bite and diet, it also raised questions: Why did these sharks grow so big? What kind of animals did they feed on? Why did they become extinct? It was hard to prise them away from the jaw, but we still had a lot to see during our family weekend in New York in July.

Leaving the museum, we crossed the street toward Central Park and came face to face with a massive bronze bust perched on a 3.6-m granite pedestal; it was Humboldt, a memorial erected some 150 years ago, on the occasion of the centennial of his birth, 14 September 1869. In the year the AMNH was created by a one-time student of Louis Agassiz, many thousands of people assembled for a whole day of

---

The original version of this chapter was revised: a sentence has been updated on page 194. The correction to this chapter is available at https://doi.org/10.1007/978-3-031-39528-4_7

---

[1] In fact, this famous reconstruction is considered too large, as it was made using only the largest teeth. See Prothero (2015), p.105.

O. Dangles, *Climate Change on Mountains*,
https://doi.org/10.1007/978-3-031-39528-4_5

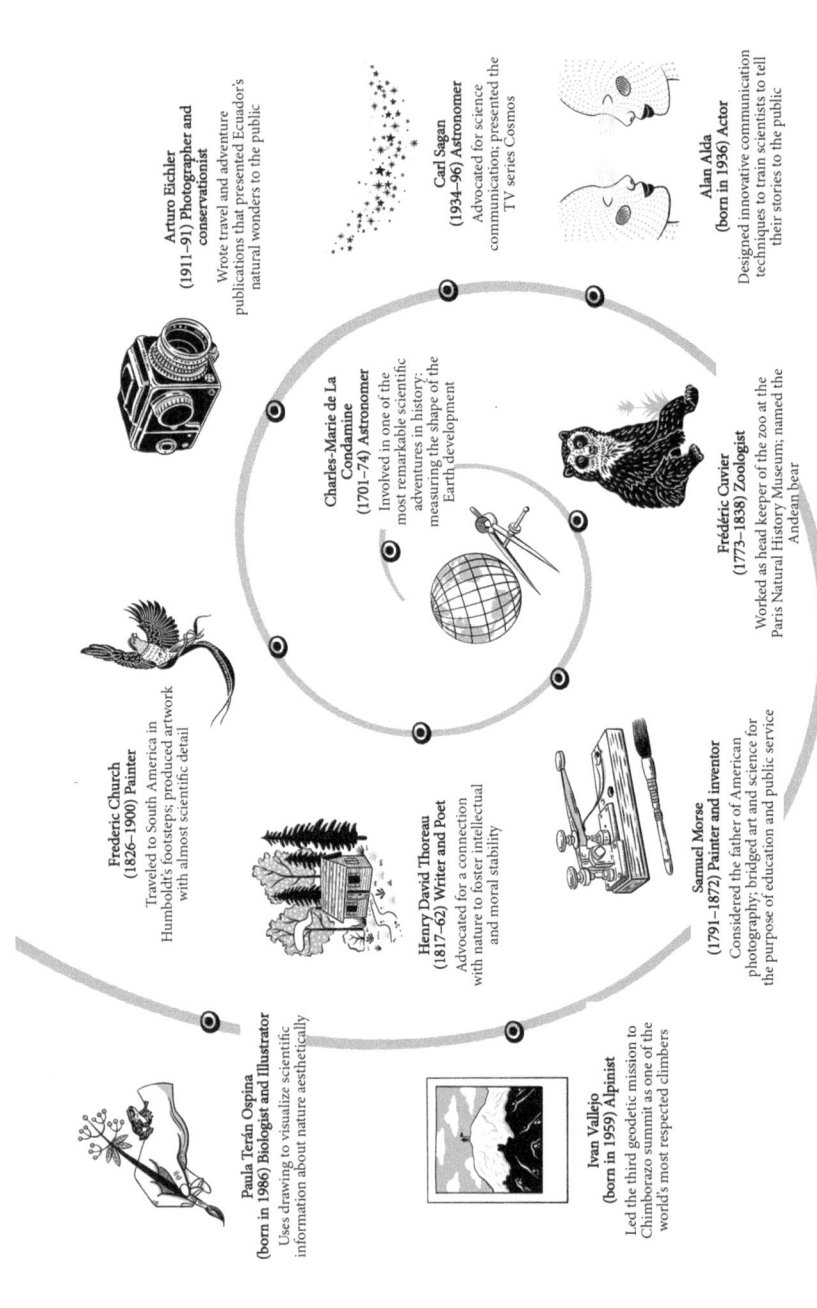

Fig. 5.1 Central figures in this chapter and Chap. 6. This spiral timeline shows ten historical or contemporary figures who contributed important ideas mentioned in this section. Design and drawings by Paula Terán Ospina

celebrations to commemorate Humboldt, with a parade, a banquet and a torchlight procession. I wasn't sure how much interest current New Yorkers had in this bust of Humboldt, but my sons found it far less exciting than the megalodon jaw. I, on the other hand, was happy to observe this monument I had vainly sought in Paris's Jardin des Plantes. We carried on through the woods of the Naturalist's Gate to see the next wonder I had promised my kids, which was on the other side of Central Park, in the Metropolitan Museum of Art.

Binoculars in hand, our stroll through the park was the occasion of a few bird observations, including a brown creeper climbing up a big elm and a gray catbird, loudly 'meowing' near the intersection of Terrace Drive and 5th Avenue. I had purposely made a detour on our way to the MET to pay tribute to another illustrious nineteenth-century savant, the "American Leonardo,"[2] **Samuel Morse**. The statue depicts the famous inventor with one hand holding a strip of code and the other on an electromagnetic telegraph. But before making his name in the world of science, Morse was a master in the world of the arts,[3] and this was the context in which Humboldt first met him, in Paris in 1831.

Morse—one of the United States' leading portraitists—was in France as an artist, reproducing dozens of European masterworks in miniature on a single canvas, the *Gallery of the Louvre* (Fig. 5.2). Like Humboldt,[4] Morse was deeply interested in building networks of knowledge interconnected across large distances. With his miniaturized Louvre, he wanted to transport the best of European art across the Atlantic to inspire the American public. Completed in 1833, the *Gallery of the Louvre* did not attract the audience he expected, but it was influential in inspiring his discovery of other means to capture images and to transmit information. In 1838, at the age of 47, Morse put aside his brushes and palette to dedicate his time to the development of the telegraph, and was warmly acclaimed by Humboldt during his presentation before the French Academy. One year later, Morse became the first American to witness Louis Daguerre's invention of photography and, like Humboldt,[5] was filled with admiration for the process. Morse succeeded in taking

---

[2]Mabee (1943).

[3]When Morse patented the telegraph, Humboldt acclaimed him in a congratulatory note as "illustrious in two worlds:" science and art. See Harvey (2020), p. 202.

[4]In Humboldt's books, images play a salient role in enhancing his texts, offering Europeans their first views of South American landscapes. For Humboldt, landscape was the key interface between nature and culture and between science and art; he felt that landscape painting was a key way to express the unity of knowledge and transfer it to others. Humboldt was influenced by the exotic landscapes painted by William Hodges (1744–97), the official artist on Cook's second voyage. Written descriptions, landscape paintings and plant and animal illustrations interacted to convey a vivid sense of the multiplicity of nature and the naturalist's worldview to a broader public.

[5]Alongside the physicists and mathematicians François Arago and Jean-Baptiste Biot (1774–1862), Humboldt was one of the experts asked to give his opinion on the process invented by Louis Daguerre (1787–1851)—the daguerreotype, the antecedent of photography—before the Chamber of Deputies decided if the French state should acquire it to offer it to the world for free. In a letter to the German naturalist Carl-Gustav Carus (1789–1869), dated 25 February 1839, Humboldt describes his enthusiasm for this prodigious invention. See Recht (1989).

Fig. 5.2 A compendium of masterpieces. *Gallery of the Louvre* by US artist and scientist Samuel Morse (oil on canvas, 187 × 274 cm, Terra Foundation for American Art, 1992.51) Like Humboldt's book, Cosmos, Morse's painting aimed to serve as a compilation of knowledge, through a miniaturized Louvre that could be appreciated by the American public. Photo: https://commons.wikimedia.org/wiki/File:Gallery_of_the_Louvre_1831-33_Samuel_MorseFXD.jpg, Samuel Finley Breese Morse, Public domain, via Wikimedia Commons (as reported by Eleanor Jones Harvey, senior curator at the Smithsonian American Art Museum, in her masterful book *Alexander von Humboldt and the United States* (p. 190), there are other connections between Humboldt and this painting. First, Humboldt frequented the Louvre with Morse: "Sometimes the great explorer would seat himself beside Morse as he painted at the Louvre, and discourse with the utmost charm from his vast store of observation and thought." Also, Humboldt was a source of inspiration for the woman seated in the back corner of the salon painted by Morse: Susan Fenimore Cooper (1813–94). Her parents were close friends of Morse, who gave Susan painting lessons. Born in Cooperstown, 130 km east of Ithaca, she was an artist, naturalist and one of the first environmentalists in the US. Described as "a very clever woman" by Charles Darwin (in Barrett and Freeman 1992, p. ii), Susan Fenimore Cooper refers specifically to Humboldt in her *Rural Hours* (p. 123 and p. 482), anonymously published "by a lady" in 1850. This book predates similar messages in Thoreau's *Walden*. Humboldt is also cited in her *Introduction to Rhyme and Reason of Country Life*. See Kurth (2016), p. 38; Cooper (2002))

portraits with the daguerreotype and started teaching this technique to a whole generation of future American photographers.[6] In a sense, photography allowed him to pursue his interest in art through a different means.

---

[6]Since the 1850s, Morse has been recognized in the United States as the 'father of American photography' for having been the first American witness to Daguerre's invention, see Brunet (2004).

Both Humboldt and Morse were convinced that science and the arts were complementary pathways for the elevation of the human soul.[7] I imagined the monuments of both savants in Central Park like stepping stones connecting the AMNH's world of science to the MET's world of arts. After all, is there so much difference between these two worlds? In his article 'The naturalist as an art critic,' Evelyn Hutchinson analyzed the dichotomy between natural history and art museums, and asked why some objects are put into one and some into the other. *"If at first the answers seem obvious, there will, I think, prove to be enough difficulties to lead us into interesting if obscure regions of the human mind."* For Hutchinson, the objects in both types of collections serve *"to arouse admiration and pleasure in their beauty, wonder at their strangeness or history, envy or awe at their costliness or rarity."*[8] In other words, the basic motivations and objectives behind science and the arts may fundamentally be the same: they both seek to tell stories about the world.

* * *

On our arrival at gallery 760 in the MET's American Wing, the guard was somewhat wary when he saw my son Matias using binoculars to study the painting we had come to see (Fig. 5.3a). But Matias was only following the recommendation on the original announcement for the exhibition of the painting: *"Visitors are requested to bring opera glasses."*[9] Back in 1859, American landscape painter **Frederic Church** had the bizarre idea of recommending that visitors bring a magnification device to fully enjoy his magnum opus *The Heart of the Andes*, the most famous representation of the tropical Andes ever painted. Why binoculars?

The most obvious reason is that they help take in the fine degree of detail in this oversized canvas that would be invisible from a distance: a blooming orchid and a bromeliad,[10] a resplendent quetzal perched on a dead branch, two fluttering orange butterflies, a small waterfall, the artist's name carved on a trunk (Fig. 5.3b). Looking at a painting with binoculars also accentuates the pleasure of observation. As if viewing a real landscape in Ecuador, Matias moved vertically from the frozen summit to the teeming cloud forest, taking in the changes in climate and habitats as if he was traveling from the poles to the equator. Moreover, observing a painting with binoculars is a symbolic metaphor for connecting artistic and scientific modes of perception and methods. Indeed, both scientists and artists generally observe their

---

[7]In *Cosmos*, Humboldt describes the attitudes toward natural phenomena among painters, poets, and students of nature. He argues that in science, philosophy, poetry, and the fine arts, "the primary aim of every study ought to be an inward one, that of enlarging and fertilizing the intellect". Humboldt (1846–1858), vol. 1, p. 39. See also Briffaud (2006). In *Views of Nature*, sixty-nine plates play a key role in accompanying the text, making the book appear, as Humboldt acknowledges, like 'a picturesque Atlas'. See Boelhower (2013).

[8]Hutchinson (1963), p. 100.

[9]Avery (1993), Fig. 24, p. 35.

[10]Likely *Wallisia pretiosa*, a species endemic to Ecuador that is now red-listed as 'vulnerable:' https://www.inaturalist.org/observations/76177473

Fig. 5.3 Observing *The Heart of the Andes*. (**a**) My son Matías observes the details of F. Church's painting at the Metropolitan Museum of Art in New York. The painter recommended that viewers bring 'opera glasses' when his painting was released to the public in 1859; (**b**) close-up of lower left corner: the many details of the painting include a resplendent quetzal, a pair of golden-winged butterflies, and several epiphytic plants, among which a possible bromeliad of the genus *Wallisia (white arrow)*. Photos by Olivier Dangles

subjects carefully, looking for deeper and hidden patterns underneath observable phenomena. Then they communicate these patterns symbolically, whether in scientific hypotheses, graphs or pictorially. For an artist or a scientist, the meaning of nature lies in the underlying pattern. By looking at *The Heart of the Andes* through his binoculars, Matias was sharpening both his acuity of observation and his artistic sense. And, unconsciously, he was acquiring a deeper knowledge about the interconnectedness of the natural world. As **Henry David Thoreau** encouraged, the body and senses "must conspire with the mind".[11]

Like Morse's *Gallery of the Louvre*, Church's *The Heart of the Andes* is a massive, overwhelming and impassioned compendium that portrays the unity of knowledge, and it too has a link to Humboldt. Frederic Church grew up on Humboldt's writings and *The Heart of the Andes* was his illustration of the geological, climatological and botanical knowledge that Humboldt had unified in the *Tableau Physique*.[12] Church's painting is not only an evocation of Humboldt's

---

[11] In Walls (1995), p. 153

[12] Using Humboldt's writing as a guide, Frederic Church traveled to Ecuador in 1853 and 1857. In *Cosmos*, Humboldt had challenged artists to portray the unity of interconnected landscapes, and Church had taken up the gauntlet. Following the success of *The Heart of the Andes* in America, Church planned to take it to Berlin, wanting Humboldt, his intellectual mentor, to see his masterpiece. On 9 May 1859, Church wrote of this desire to the poet and translator Bayard Taylor (1825–78), who had met and spent time with Humboldt. "[The] principal motive in taking the

scientific ideas, but also of his immersive approach to understanding nature: a desire
to link scientific accuracy and **aesthetics**, from details of birds and plants to the
nuance of cloud forms, from the color of rocks to the play of water in motion.[13] In
Church's paintings, art and science are interlinked.[14] For this reason, *The Heart of
the Andes*, as well as several of his other paintings such as *Niagara* (1857), *Cotopaxi*
(1862) and *Icebergs* (1863) can be considered some of the most creative communi-
cations of scientific phenomena for the general public.

In 1859, *The Heart of the Andes* was a blockbuster, more popular than any
painting previously unveiled in the United States, despite its foreign subject matter.
In New York, more than 500 people per day lined up, sometimes for hours, to see
this singular artistic achievement.[15] The following year in Boston, the painting
would draw over 30,000 citizens in 6 weeks.[16] Why was it so interesting to the
public? As people today line up to go to the cinema or theatre, the crowd that stood
on Broadway wanted to witness the spectacle and the story of the tropical Andes
with their own eyes. Church succeeded in captivating people who did not know
much about the Andes, satisfying their appetite for wonder by showing them some of
the most surprising features of the region.[17] His paintings used the canvas to tell a
scientific story about the natural world, allowing the eye to trigger both sides of the
brain: the analytical, which stimulates understanding, and the creative, which taps
the imagination. Church also added a third key ingredient to tell his story: emotion.
To do this, he took certain liberties and created an idealized picture, as Humboldt had
done decades before with the *Tableau physique*.

The Heart of the Andes does not depict a real place, but assembles various scenes.
From a topographical viewpoint, it is incorrect. The mountain is not *"Cayambe or*

---

picture to Berlin is to have the satisfaction of placing before Humboldt a transcript of the scenery
which delighted his eyes sixty years ago—and which he has pronounced to be the finest in the
world." In Gould (2000), p. 28. But Humboldt died on 6 May 1859, a week after *The Heart of the
Andes* first went on public view in New York. Church felt "as if [he] had lost a friend." See Avery
(1993), p. 39.

[13] Church's rigorous scientific approach to painting and attention to detail involved long hours of
observations and studies. In 1856, for example, Church went to Niagara Falls where he drew almost
100 detailed sketches of water in motion: plunging waterfalls, foaming rapids, swirling eddies, and
the vapory mist formed by the falls. See Harvey (2020), p. 338.

[14] See Gould (2000), p. 28.

[15] In a mere 3 weeks, more than 12,000 people paid 25 cents each to view Church's painting when it
was first unveiled in New York City. For the closing day, more than 6000 people bought tickets,
with a 1.6-km line of visitors forming between lower Broadway and the Tenth Street Studio.
Manthorne (1989), pp. 31–32.

[16] Avery (1993), p. 41.

[17] Church's paintings were enormously popular in the mid-nineteenth century. His greatest paintings depict the physical reality of nature: the raging waters of Niagara in 1857, the luxuriant
vegetation in the tropical Andes in 1859, the fiery fury of Cotopaxi in 1862, and monumental
icebergs in 1863. Behind these paintings, scientific debates were taking place about their subjects.
Cotopaxi was painted at a time of debate between plutonists and neptunists; *The Icebergs* during the
scientific controversy about erratic boulders between Louis Agassiz and Charles Lyell. See
Mitchell (1989).

Chimborazo, or any other peak of the equatorial group. It is each and all of them, and more than any."[18] Also, the most prominent bird in the painting, the resplendent quetzal, is Central American not Andean, and was probably observed by Church in Panama. Yet Church's painting remains true to the underlying science (the assemblage of tropical ecosystems along an altitudinal gradient), and his astonishing birds and mysterious icy summits also convey a strong emotional impact. He surely knew that scientific details alone don't persuade the public: stories and emotions do. This is why *The Heart of the Andes* could demand *"far more than a vague confidence that we can safely admire without committing ourselves.*"[19] The natural landscapes painted by Church are not mere representations of the world. They are ideas transmitted to the viewer. Beyond his paintings, Church contributed to raising awareness of nature through his role in establishing Niagara Falls as an international park, founding the MET and serving on the Central Park commission. It is by telling stories that scientists and artists can engage the public and spur it to action.

* * *

Church, Réaumur, de Saussure, Humboldt, d'Orbigny, Agassiz, Morse, Tyndall, Comstock and Fabre were all extraordinary science storytellers. They all engaged with the broader society in which they worked and lived and were important figures in placing science on the cultural and educational map.[20] All recognized the need not only to perform rigorous research, but also, in Humboldt's words, to *"fulfill their commitment to the public*"[21] and communicate their knowledge. Tyndall, an admirer of Church's paintings and an unparalleled popular writer on physics, claimed in his preface to *Heat* (1863): *"My aim has been to rise to the level of these questions from a basis so elementary that a person possessing any imaginative faculty and power of concentration might accompany me.*"[22] His books sometimes ran to ten editions.

However, even at the time of Humboldt and Tyndall, many scientists (such as, for example, the zoologist George Cuvier) were already so professionalized that they

---

[18] Winthrop (1869), p. 17.

[19] Ibid.

[20] In particular, Réaumur designed books to encourage people to look at nature for themselves, to make their own experiments and collections, and to report back to the author on anything noteworthy. In this way, his books created a conversation instead of a monologue, an early form of citizen science. See Terrall (2014), p. 80. See also Drouin and Bensaude-Vincent (1996).

[21] In the introduction of his '*Relation historique*,' Humboldt asserts: "The difficulties I have experienced since my return, in the composition of a considerable number of treatises, for the purpose of making known certain classes of phenomena, insensibly overcame my repugnance to write the narrative of my journey. [. . .] I also perceived that such a preference is given to this sort of composition, that scientific men, after having presented in an isolated form the account of their research on the productions, the manners, and the political state of the countries through which they have passed, imagine that they have not fulfilled their engagements with the public until they have written their itinerary." Humboldt (1852–1853), vol.1, p. xix. For a thoughtful analysis of travel narratives over the 1760–1840 period, see Stafford (1984). See also Lubrich (2022) on Humboldt's skills in science communication.

[22] Tyndall (1863), p. ix. For a discussion of Tyndall's ability to keep audience members or book readers engaged, see Dry (2019) pp. 44–45.

considered public interest in science as a possible danger to the advancement of knowledge. When scholars aim to reach a popular audience, scientific descriptions may seem too vague; guarding the legitimacy of academia often involves setting aside the importance of public interest in science.[23] Today, a stigma still exists in academia regarding scholars who actively engage in outreach efforts with the public: spending too much time communicating one's research is seen as necessarily at the expense of research quality and rigor, a syndrome described as the Sagan effect.[24] In the early 1980s, planetary scientist **Carl Sagan** brought Humboldt's masterwork *Cosmos* to life in a television series, which recounted fascinating scientific stories from the origin of life to humans' place in the universe.[25] Seen by at least 500 million people across 60 countries, the show made Sagan a superstar. But this fame damaged Sagan's standing in the scientific world. The US National Academy of Sciences rejected him as a member, even though later analyses of Sagan's publications suggest that his academic career was comparable to Academy members who were less involved in public outreach. Yet the world has changed since Sagan's times: ever more urgent environmental challenges and the accelerating diffusion of fake news (which spreads farther and faster than real news)[26] are inciting an increasing number of scientists to climb down from their ivory towers and share their scientific findings with the public. Inspired by past naturalists, there is a pressing need for them to again *"fulfill their commitment to the public."*

For the ordinary scholar, scientific knowledge is hard to communicate. Researchers are still devoted to rational arguments and scientific facts, convinced that what was persuasive for them will be the best way to change others' minds. But as US marine biologist-turned-filmmaker Randy Olson points out, science on its own is not interesting to the public: it is complex and distant and needs to include a more human element.[27] Fascinating life stories and experiences, spectacular paintings and images, artistic performances, facts related to familiar places, or metaphors of things that matter to us are examples of these more human elements. Although this lies outside 'hard' science, reviving past naturalists' approach of communicating scientific knowledge, at least partly based on the senses, perceptions and emotions, has the potential to make a profound impact on modern science communication. In two of my experiences in Ecuador, I saw at first hand the power of stories to communicate science to ordinary people. It is key to find a frame of reference for the message and to find common ground with others from sometimes diverging viewpoints. Communication relies on establishing trust and coming up with ways to help people understand and feel concerned about what can be maddeningly complex issues, such as climate change.

---

[23] See Drouin and Bensaude-Vincent (1996).

[24] Ecklund et al. (2012); Martinez-Conde (2016).

[25] Walls (2016) argues that Sagan's TV show bore "striking parallels to Humboldt's masterwork" (p. 118).

[26] Vosoughi et al. (2018).

[27] Olson (2009), p. 71.

5.2   A Mountain Higher than Everest

"Tell a story." This is solid advice for scientists who want to communicate about their work in a way that is human, compelling and relatable to everyday life. From TED Talks to university media offices to three-minute thesis challenges, all over the world scientists are honing their skills in narrative crafting to engage with society. However, modern scientific stories can seem quite dull compared to the adventures of the first scientist–explorers. Of these, the story of the first geodetic mission to the equator in is certainly one of the most remarkable scientific sagas ever told in history.

The mission had all the ingredients of a blockbuster science adventure of the time.[28] First, there was a compelling hook. The French scientists were charged with a huge task: to discover the true shape and exact measurement of the globe or, as said at the time, the "figure of the Earth." It was known in the eighteenth century that the Earth was not a perfect sphere. Isaac Newton's theory of gravitation postulated that the Earth's rotation caused it to bulge at the equator and flatten at the poles, while Descartes' vortex theory of planet motion supported the idea of a lemon-shaped Earth. Yet no one had ever been able to determine empirically whether the Earth was long like a lemon or squat like a pomegranate. To settle this matter, the Royal French Academy of Sciences commissioned two expeditions, one to Lapland, close to the North Pole, and the other to Quito, at the equator. Measuring the length of a meridian arc (the curve between two points on the Earth's surface along the same longitude) for one degree of latitude at each of these two sites and comparing the results would resolve the issue.

The story also had a strong cast of characters. Three wig-wearing Royal Academy mathematicians led the mission to Quito: the Casanova-like Louis Godin, the adventurous and charming **Charles-Marie de La Condamine**, and the versatile genius Pierre Bouguer. At the time, every educated person had heard of them. They departed from the old continent with a band of companions, including the reclusive physician and naturalist Joseph de Jussieu, and two Spanish scientists, Antonio de Ulloa and Jorge Juan y Santacilia, who were supposed to both learn from and keep an eye on the French. When they reached Ecuador, they interacted with a cosmopolitan world of politicians, Jesuits, Spanish and creole elite, Indigenous peoples, brigands, slaves and charming ladies. The geodesists' closest Ecuadorian associate was Don Pedro Vicente Maldonado, who would become a member of the French Royal Academy of Sciences in 1747.

Finally, there was exciting action with myriad twists and turns. After arriving at the Pacific port of Manta on March 1736, the never-resting Academicians climbed up and down mountains to determine the length of a degree (see Fig. 5.4). But field

---

[28]There are several, more or less fictionalized, accounts of the expedition. See, for example, Trystam (1979), Whitaker (2004); BBC (2021) 'Voyages of Discovery,' episode 4. For a complete account of the scientific complexity, physical difficulty and portraits of all the principal protagonists, see Ferreiro (2011).

Fig. 5.4 Measuring the shape of the Earth. During the first geodetic mission in Ecuador (1735–45), Academicians used triangulating equipment to measure a height angle with a **quadrant**. In front of a tent (as the team often had to camp in the wild to take measurements), two members of the mission use a quadrant to make a sighting, while two others (in the background) ascend the slope of an active volcano to install a signal on another point of the triangle. The Andes offered a great context for triangulation, with two parallel mountain ranges to cross-section the entire area and a lower inter-Andean valley. Drawing from La Condamine (1751, p. 3)

circumstances proved far more daunting than anticipated, and instead of achieving their measurements in a year or so, it ended up taking them almost 10 years. They faced cloudy weather that impeded observations for long periods, altitude sickness and blizzards, life-threatening earthquakes and erupting volcanoes. They ran out of funds, as Godin spent the budget on his mistresses and the Academy cut off their credit. They got entangled in Spanish colonial politics and the ripple effects of European wars. One companion died of yellow fever and another was murdered at a bullfight. Over the period, the purportedly close-knit team became a bickering band of rivals, putting in danger the very purpose of the mission. Yet the savants persevered through all these odds, emerging with a remarkably precise measurement that proved Newton right, allowed more accurate world maps, and established the foundation of modern geodesy. They also made valuable chance discoveries along the way: rubber—one of the most important raw materials in industrial societies, and quinine, which would save millions from malaria.

Today, the geodetic mission's fame has faded in collective memory, in France as elsewhere. But this remarkable scientific story had a long-lasting and not-just-scientific legacy in Ecuador. In fact, the very name of the country was given in honor of the "equatorial science" of the mission,[29] and museums, statues, a geodetic

---

[29] In the 1820s, Venezuelan military and political leader Simón Bolívar (1783–1830) followed Humboldt's call to honor the equatorial science of the previous century, resulting in the renaming of the Quito region as the district of Ecuador in the newly constituted Republic of (Gran) Colombia.

pyramid, street names and Quito's French school still commemorate it today.[30] This first-ever international scientific collaboration was also one of the first examples of contact between the 'Global North' and 'Global South' based on scientific, cultural and humanistic objectives, as opposed to power, territorial control or exploitation.[31] These goals were revived in 1899, when the French Academy of Sciences sent a second geodetic mission to Ecuador to confirm the results of the first mission.[32] Across the centuries, the mission had remained a powerful icon, epitomizing the role of science to advance society and to engage the public in scientific questions. Much later still, the geodetic mission would become an opportunity for me to use story-telling to communicate climate change science to a broad audience.

* * *

The opportunity came from François Gauthier, the French ambassador to Ecuador, in 2015. At that time, I had been named head of the French Research Institute for sustainable Development's cooperation team in Ecuador, overseeing a dozen researchers, including geophysicists, medical entomologists, geographers, ecologists and an archaeologist. This position involved weekly meetings with the French Embassy staff in my role of scientific adviser and spokesperson for French–Ecuadorian academic collaboration. The ambassador was more interested in economic than scientific diplomacy, and I had difficulty explaining our research in ways he would find interesting. Yet Gauthier had a history background, which is probably

---

[30] For a critique of the commemoration of the French geodetic expedition in Ecuador, see Capello (2018).

[31] For more than two centuries after the discovery of South America by Europeans, the Andes remained the exclusive territory of Spanish conquistadors. Unwilling to share the wealth of these new regions, Spain excluded all foreigners in order to conceal information about their natural resources. In 1735, the geodetic mission, perceived as a group of harmless savants, was allowed to enter the Andes. As the physician Casimiro Gómez Ortega (1741–1818) argued to José Gálvez, minister of the Indies, "twelve naturalists [...] spread over our possessions will produce as a result of their pilgrimages a profit incomparably greater than could an army of 100,000 strong fighting to add a few provinces to the Spanish empire" (see Cañizares-Esguerra 2006, p. 122). By 1815, a French embassy had opened in Brazil in a spirit of budding diplomatic relations with the Kingdom of Portugal. For the first time, the embassy included cultural and scientific teams, whose activities would support diplomacy. Artists and scientists also became ambassadors of their countries. Today, scientific relations are an integral part of diplomatic activity.

[32] The second geodetic mission was assisted by the Ecuadorian army and equipped with the latest instruments for making triangulation measurements, including the theodolite, which can simultaneously measure angles on both vertical and horizontal planes (unlike the quadrant, used by the first mission, which could only measure horizontal angles). This objective was greatly expanded by the archaeological and ethnographic works of French ethnologist Paul Rivet (1876–1959), who maintained his ties to Ecuador and South America throughout his life. Rivet became interested in the Indigenous peoples of the region and gathered fascinating data that made him a recognized anthropologist and ethnographer on his return to France. His stay in Ecuador would eventually motivate him to create the Musée de l'Homme (Museum of Mankind) in Paris in 1938. Rivet was a great admirer of Alcide d'Orbigny, who, in his opinion, stood up to comparison with Humboldt. He was "less lyrical, but more precise [...]; less of a generalist, but more of an analyst; less of an encyclopedist, but more scrupulous." From Rivet (1933), p. 26.

what led him to read La Condamine's *Journal du voyage fait à l'Équateur* (1751), a lively narrative of the first geodetic mission. The story gripped him, and 1 day, during a cocktail at the Alliance Française in Quito, he declared: "*In 2016, France will celebrate with Ecuador the 280ᵗʰ anniversary of the first geodetic mission.*" While 280 seemed a bit of an odd choice for an anniversary, I thought it could be an entertaining event. The ambassador planned to organize a concert of Baroque music, a dinner with French and Ecuadorian food from the time, and a symposium on the history and science of the mission. And then he stated: "*And we will launch the third geodetic mission in Ecuador!*" I almost dropped my glass of wine. Was he serious? It was the first I'd heard about it. After his declaration, I approached the ambassador and told him that a third mission required a scientific objective, and if we wanted to honor the memory of the illustrious Academicians, it had to be a really big deal. He answered, "*I trust you to find a third geodetic mission that will tell a good story and make headlines in the national and international press!*" I began to think about how to accomplish this mission.

I was lucky, as I had a direct heir of La Condamine at hand. Jean-Mathieu Nocquet, a brilliant geophysicist, is a world specialist in geodesy applied to the study of earthquakes, and at that time, he was based in Quito. His specific skill is the use of high-precision GPS data to detect the faintest movements of the Earth's crust. As he once explained to me: "*Every year, like your nails grow a few tens of millimeters, the Earth's crust moves. But unlike fingernails, it also deforms, and as a consequence, stress increases. Earthquakes are the sudden and sometimes catastrophic release of the accumulated stress. Geodesy allows us to see how the Earth is preparing the next earthquake.*"[33] Ardently curious, with a genuine love of science, meticulous but pragmatic, Jean-Mathieu is a present-day example of the historic Academicians, minus the wig. The day after the ambassador's surprise announcement, I visited Jean-Mathieu in his office at the Instituto Geofísico del Ecuador to present my problem. We had to find a mission that commemorated the first, that was interesting from a scientific viewpoint, 'sexy' enough to appeal to public interest, relatively quick to organize and perform (we had several months, not 10 years), and cheap.[34] Jean-Mathieu thought for a moment, and finally suggested: "*Why not measure the exact altitude of Chimborazo?*" After some discussion, I began to understand how great this idea was.

* * *

Humans have long been interested in measuring the height of mountains. In 200 BCE, the Greek polymath Eratosthenes calculated the altitude of Mount Olympus (2930 m). Geodesy was born. But the discipline did not progress much further

---

[33] Technological advances in latest-generation GPS have led to a true revolution in the field of seismology. For example, they have allowed the discovery of 'silent' processes (not emitting seismic waves) that release stress on faults and are crucial to understanding earthquake dynamics.

[34] Jean-Mathieu's first suggestion was for the deployment of a complete geodetic network: "€500,000 would put us on a par with the best networks in the world," he argued. The ambassador responded: "Please come up with an idea that will only require a few thousand euros!"

until the development of navigation gave measuring height a practical interest. Peaks rising from the ocean and visible by navigators several days before arriving provided valuable information about the position of the ship. Because of its strategic location in the Atlantic, Mount Teide on Tenerife Island (see Fig. 3.2) was the first mountain whose height was closely scrutinized. From his first voyage in 1492, navigator Christopher Columbus wrote, *"it is held to be the highest [mountain] that can be found,"*[35] and the peak long retained its world altitude record in the eyes of Europeans, with height estimates up to 10,000 m, and even 100 km high![36] It was not until the first geodetic mission that a successor to Teide was found: Chimborazo. The French savants did not use the imposing mountain in their triangulations,[37] but to test another prediction of Newton's theory of gravitation: gravitational pull. Bouguer thought that this massive mountain should be enough to measurably deflect a pendulum from true vertical. In December 1738, he set up camp with La Condamine on the flanks of Chimborazo to carry out the experiment. They took the opportunity to measure its height: 3220 toises or 6276 m. Bouguer thought it was *"perhaps the highest [mountain] in the world."*[38]

When Humboldt reached South America, the mountain height record set by the French Academicians was still in effect. Humboldt had been fascinated by La Condamine's journal, which literally served him as a travel guide.[39] Humboldt's accounts contain dozens of explicit references to the geodetic mission, and he constantly compared his own scientific observations with those of La Condamine.[40] It is even likely that Humboldt's celebrated *Tableau physique* was inspired by La Condamine's collage-like juxtaposition of texts, images and maps in his journal. But there was at least one thing Humboldt was skeptical about in La Condamine's

---

[35] In Jouty (1989).

[36] In 1724, advances in trigonometry allowed the Franciscan explorer Louis Feuillée (1660–1732) to measure the peak with more precision. He calculated it to be 2213 toises (4313 m), which still overestimated its height by 600 m. While Teide is actually lower than many peaks in the Alps, no other mountain in Europe had time to supplant its reputation as the highest peak in the world before South American summits were measured by the French geodesists. See Cajori (1929).

[37] The French geodesists initially thought that reaching the highest points of mountains would ensure the best measurements. However, they soon realized that climatic conditions were particularly difficult on mountain peaks, which were often shrouded in mist and subject to heavy snowfall and violent storms. This made it impossible to see from one mountain triangulation station to the other. They no longer ascended so high in the mountains, settling for an altitude where the two neighboring signals were visible.

[38] Bouguer (1748), p. 55.

[39] La Condamine (1745) is the first source for the geodetic mission. The book is based on a rich collection of maps, manuscripts and letters that La Condamine brought back to France and also stories that he had gathered from local informants.

[40] "The testimonies collected by M. de la Condamine are very remarkable; he has published them in the greatest detail. And I'd like to add that if this traveler passed for the man of the most ardent curiosity in France and England, he is considered in Quito, in the country he described, as the most sincere and truthful man." From Humboldt (1814–1825), Tome II, p. 486 (this sentence is absent from the English translation).

findings: his measurement of Chimborazo's height. He wrote: "*La Condamine had found Chimborazo to be almost 3,220 toises high. My own trigonometric measurements, which I repeated twice, gave me 3,267 toises [6,544 m], and I may have some confidence in my own observations.*"[41] Later, dozens of explorers would give different estimations of its altitude: 6247 m, 6352 m, 6284 m, or even 6586 m. In today's geographic atlases, the most common figure, dating back to Reiss and Stübel (1873), is 6310 m, although a British survey team in 1993 measured 6268 m.[42] In short, no one could ever quite agree on Chimborazo's exact height. It was time to clear up this matter once and for all.

* * *

The idea of this quest to measure Chimborazo excited the French ambassador. The scientific objective of the third geodetic mission was approved, and an official press release announcing the measurement was sent out. Soon after, my phone started ringing incessantly. The news aroused interest from all sides, from journalists and mountaineers to ministries and schools to international diplomats and local politicians. Napoleón Cadena was the mayor of Riobamba, the tenth largest city in Ecuador, located at the foot of Chimborazo. He called to invite the third geodetic mission to Riobamba, as a stopover on the team's way to Mt Chimborazo. Napoleón wanted to organize a welcome celebration at the town hall to place the mission under the best possible auspices. He reminded me that Riobamba was the true host city of past geodetic missions. It was visited on a regular basis by La Condamine (who found love there) and was the city Maldonaldo was from. Riobamba even hosted a geodetic pyramid during the second geodetic mission. For Napoleón, welcoming the third mission was a matter of prestige, as was the mountain's altitude—the prestige of a city and the prestige of a mayor. The mission had turned political.

In total, 30 people from Ecuador and France, including diplomats, scientists, technical assistants, guides, journalists, and the armed forces, were involved in the organization of the third geodetic mission. The core team for the altitude measurement was made up of three French geophysicists—Jean-Mathieu, Frédérique Rolandone and Matthieu Perrault—seven members of the Ecuadorian armed forces, one photographer and four Ecuadorian guides. The latter included two national heroes: **Ivan Vallejo**, the first Ecuadorian and third in the Americas to crown the 14 summits over 8000 m without the use of supplemental oxygen; and Carla Pérez, the first Ecuadorian and woman in the Americas to climb both K2 and Everest without supplemental oxygen. (To give an idea of their achievement, while more than 4000 people had summited Everest at that time, fewer than 200 had done so without oxygen, and only seven had been women). With such an experienced team, the mission was off to a good start.

On 2 February 2016, I accompanied the mission to Riobamba, where Napoleón was awaiting us with much fanfare. We made our way toward the town hall in a

---

[41] Humboldt (1803), p. 330.

[42] Reiss and Stübel (1873), p.34; for more on the other altitude estimations of Mt Chimborazo, see Meyer (1907), pp. 69–70 and Horrell (2019), p. 110.

crowd of local celebrities, political representatives, journalists, photographers and dozens of small children, in a friendly and colorful atmosphere heightened with tambourines and folk dances. In an emotional welcoming speech, Napoleón stressed the highly symbolic importance of Chimborazo for the Ecuadorian people. *Taita* ('Father') *Chimborazo*, as the native people call the mountain, was a place of pilgrimage for the pre-Incas, and local communities still make offerings to it.[43] This 'historical mountain' was also honored by '*El Libertador*' Simón Bolívar (1783–1830) who, during his campaign to create the nation of Gran Colombia and drive the Spanish armies out of Peru, ascended Mt Chimborazo "in the footsteps of La Condamine and Humboldt."[44] In short, Ecuadorians take great pride in their highest mountain, which is even featured on their national flag. Wishing the geodesists good luck, Napoleón then invited Jean-Mathieu to give more details about the mission, shifting the ceremony from the sensory to the scientific (Fig. 5.5).

In presenting the aim of the mission, Jean-Mathieu underlined the importance of precision. The first geodetic mission achieved astonishing precision considering the means available at the time. Years of field measurements, months of calculations, thousands of pages of notes and figures allowed the Academicians to find a meridian distance value of 110,576 m. This turned out to be only 58.7 m more than the 'exact' distance, which was not established until 1924: a precision of 0.05%![45] With equal scientific rigor, our third mission aimed at measuring Chimborazo's altitude with the highest precision available. Metrology instruments today are much more precise than the GPS chip of a smartphone. To monitor planet Earth, modern geodesists have developed satellite-based technologies with accuracy to the centimeter. This has revolutionized our ability to study changes on the globe, from the centrimetic thinning of glaciers, to the tiniest movements of faults in the Earth's crust, to the slightest changes in sea levels. This is where geodesy meets climate change science. Measuring Chimborazo's altitude to the centimeter would set a solid baseline for monitoring any future decrease in glacier depth. This new geodetic mission was not only a way to celebrate the past, but a way to prepare for the future.

---

[43] In 2017, during a field trip to the western flank of Chimborazo with archeologist Alden Yépez and historian (and Humboldt specialist) Segundo Moreno, I had the opportunity to observe at about 4800–5000 m Inca stone buildings and hundreds of spherical holes in the rocks, seemingly used for sacred purposes. Under the rocks, we also found traces from modern Indigenous pilgrims: food and prayers written on paper, candle stubs, old photographs, fruit, flowers, hair, jewels, dolls and even two human skulls. For more details see: Yépez (2022).

[44] "Historical mountain," as written in the Legislative Decree s/n (Registro Oficial 1272, 5-XII-1900) that regulates the Escudo de Armas y Pabellón Nacional of the Republic of Ecuador. In his famous poem 'Mi delirio sobre el Chimborazo' (My Vision on Chimborazo) from 1822, Simón Bolivar narrates his dreamlike arrival in the area of the volcano, the "dominator of the Andes:" "I searched for the footsteps of La Condamine and Humboldt; boldly I followed them, nothing stopped me; I reached the glacial region, the ether suffocated my breath. No human plant had trodden the diamond crown placed by the hands of Eternity on the exalted temples of the dominator of the Andes."

[45] See Francou (2013), p. 31.

Fig. 5.5   Science meets politics. Mayor Napoleón Cadena (center, with the microphone) welcomes the third geodetic mission at the Riobamba city hall in Ecuador, 2 February 2016. Photo by Juan Sebastián Rodríguez @sebas_foto

We left the still festive Riobamba around noon to arrive in the early afternoon at our refuge, in a grassy valley beneath the southwest face of Chimborazo. For 2 days, the team would make preparations here, doing acclimation walks at 4000–5000 m, getting the equipment ready, and perfecting the altitude measurement procedure. Jean-Mathieu thought several team members should know how to measure, in case he was unable to make it to the top. He had done a lot of preparation for the past 2 months, even climbing Kilimanjaro in Tanzania, but his experience of ice climbing was limited, and he wanted everything to be foolproof. He explained the procedure step by step to the other geodesists: *"First, you need to deploy the geodetic antenna and mount it on this mast that you've previously anchored into the ice. Make sure it's at the highest point of the mountain! Then you have to tighten this tiny screw, which is a bit tricky to manipulate, to give the device extra stability. It's important to attach it carefully because if the weather gets windy, this would ruin the measurement. Then, connect the antenna to the GPS receiver, plug it to this battery, and connect the battery to the solar panels. Batteries die quickly in low temperatures."* (Fig. 5.6).

Matthieu asked how long the measurement would take to conduct. *"Well, to obtain a very precise measurement, the antenna has to stay in place for quite a long time so that many observations can be collected and the error reduced."* How long is

Fig. 5.6 The third geodetic mission. French geodesists Matthieu Perrault, Frédérique Rolandone and Jean-Mathieu Nocquet (from left to right) practicing the installation of the GPS antenna the day before the ascension of Mt Chimborazo. In the background, guides Ivan Vallejo and Julio Mesias sharpen the spikes of their ice crampons with a metal file. Photo by Juan Sebastián Rodríguez, @sebas_foto

"quite a long time?" *"Two hours should be enough. Pray for good weather!"*[46] Seeing darkening clouds on the horizon, I thought to myself that Jean-Mathieu's protocol should have perhaps started with an offering to *Taita* Chimborazo.

* * *

February 4, 11:00 p.m. The company gathered at the start of the ridge route; it would take an estimated 8–10 h to cover the 5.6 km to the top of Chimborazo. The measurement equipment was divided between the climbers, with Ivan and Carla carrying the heaviest loads. I gave the expedition flag to plant at the top to Ivan, wished everyone good luck and watched the string of people move away under a luminous starry sky. Not adept at ice climbing, I would not go with them, so all I know about what happened next comes from Matthieu.

The company arrived in staggered groups at the base of the glacier, at some 5500 m in altitude, where everyone put on their spikes. From there, Ivan and Carla

---

[46] A 2-hour measurement is needed for accuracy to the centimeter. This allows: (1) accumulating a number of observations sent by GPS satellites (more data allows an error-averaging effect); (2) obtaining data from different satellite positions in the sky (this permits triangulation with different configurations and improves precision); and (3) estimating and correcting for the delay experienced by the GPS signal while crossing the atmosphere.

led an advance group with Matthieu, Juan Sebastian Rodríguez, the photographer, and a few soldiers (with some of the equipment), while the other two guides stayed with Jean-Mathieu and Frédérique to ascend more slowly. But after only a few hundred meters up the glacier, Jean-Mathieu slipped. Because of an unusually dry period, the glacier was in a dangerously unstable condition, and the guides, preferring not to take risks, thought it would not be safe to climb without a high level of experience. Thus ended Jean-Mathieu's and Frédérique's adventure, leaving the success of the scientific mission on Matthieu's shoulders. He would have to represent the whole team, tackling the long ridge and the obstacles that awaited him at the top.

February 5, 6:00 a.m. Ivan and Carla's advance team reached the end of the ridge first. Matthieu and Juan Sebastian had managed to stay the course without requiring roping, but the soldiers had dropped behind. In front of the remaining team lay an enormous field of ice stretching across the summit plateau. But instead of being smooth and compact as Matthieu was expecting, the glacier surface was "cracked like dry skin." The glacier field was full of holes and icy blades jutting up three meters high, like giant spires: **penitentes** (Fig. 5.7a). The name comes from their resemblance to a crowd of kneeling religious brothers doing penance in their pointy hoods. First described by Darwin in Patagonia, penitentes are snow or ice formations found throughout the Andes, resulting from a combination of the fusion and sublimation of snow and ice. At the top of Chimborazo, they are really impressive. The reason is that frequent and violent eruptions of the neighboring volcanos, Tungurahua and Cotopaxi, can cover the snow surface with thick ash layers, decreasing snow albedo (its reflectivity), which causes snowmelt and the formation of tall blades. In Matthieu's mind, the penitentes resembled the frozen waves of a stormy sea, blocking the way to the summit. They looked daunting to cross. But Ivan and Carla were already leading the way, "jumping like gazelles" from one penitente to the other.

In less than an hour, the party reached the summit, named after the English alpinist Edward Whymper, who successfully achieved the first ascent of Chimborazo, and the Matterhorn in the Alps (where he was engaged in a race with John Tyndall to reach the top).[47] After taking a few minutes to observe the stunning view, they started to take the equipment out of the bags. This was their first pang of worry: the battery was missing. It was in the backpack of one of the soldiers! They blamed themselves for making the mistake of separating the pieces of the GPS instrument. They would just have to wait—it was a beautiful day and there was no immediate rush. But after an hour, they had a second pang: what if something had happened to the soldiers? The minutes started to drag ... Finally, they spotted the silhouettes of the three soldiers on the other side of the sea of penitentes and, relieved, Matthieu and

---

[47] In December 1879, Edward Whymper went to Ecuador with two guides to climb Chimborazo. Less than 1 month later, he reached the summit. Beyond climbing, Whymper studied the human body's reaction in rarefied air and widely collected plants and insects. See Whymper (1892). See Schaumann (2020), pp. 174–208 for a thoughtful analysis of the influence of de Saussure, Agassiz and Humboldt on Whymper's narratives and his rivalry with Tyndall.

Fig. 5.7 At the top of Chimborazo. (**a**) The 'sea' of penitentes with Whymper's summit in the back; (**b**) Alpinist Ivan Vallejo searches for the highest penitente, (**c**) The team planting the expedition flag on 5 February 2016 (from left to right: Matthieu Perrault, Carla Pérez, Juan Sebastian Rodríguez and Ivan Vallejo). On the left, you can see the deployed antenna used to make the measurement of Chimborazo's precise altitude (6263.47 m). Photos **a**, **b** by Matthieu Perrault, **c** Juan Sebastián Rodríguez, @sebas_foto

Ivan started to make preparations for the measurement. Most importantly, they had to look around for the highest spot to plant the antenna. That caused another concern. The penitentes were so irregularly shaped that it was not easy to decide which was the highest. Ivan began to examine the penitentes grouped at the summit one by one (Fig. 5.7b). After 20 min of exploration, measurements and discussions, the team finally agreed on the tallest blade. They planted the mast, attached the antenna and... then came a new cause for anxiety. Jean-Mathieu's "tiny screw, which is a bit tricky to manipulate" escaped from Matthieu's fingers and fell down a hole, two meters below. After a fruitless search, they had to face the facts: the screw was lost. They would just have to hope that the wind would not pick up during the measurement.

Finally, the soldiers arrived, completely exhausted, but in one piece—and with the battery. Once the device was plugged into it, the measurements could finally start. The soldiers' contribution to the mission was vital, but proved ephemeral. Soon after, one got altitude sickness and his companions had to take him back down the mountain. During the next two hours, while waiting for the data to be collected, the rest of the company contemplated the magnificent view to the north of Cotopaxi, Cayambe and Antisana. The weather was perfectly windless, and the incidents with the screw, the battery and the dangerous climb were already forgotten. The team planted the flag (Fig. 5.7c) that would make headlines in the newspapers. Mission accomplished!

After packing up all the equipment and starting on their way back, the next problem arose: Matthieu got altitude sickness and thought he would never be able to make the return trip. Zigzagging through the penitentes was arduous, but once he got through them, he got his strength back for the descent. This was quite incredible as Chimborazo was Matthieu's first summit ever. It seems that *Taita* Chimborazo looked favorably on him. The team returned to the refuge at 5:00 p.m., concluding the expedition.

* * *

The mission of the geodesists was far from over, however. Jean-Mathieu's GPS does not display the results on a digital screen, but stores large amounts of data that have to be analyzed. It was time to do some calculations, and this was where Jean-Mathieu's skills would be vital. The hardest part of the analysis was handling the measurement errors, a challenge also faced by his illustrious predecessors. In his publication on measuring the meridian, La Condamine dedicated 18 pages to describe "*the various causes that may affect the accuracy of the observations*" and the corrections needed to fix errors arising from telescope parallelism and quadrant divisions, or the estimation of the distance to the stars.[48] In our case, Jean-Mathieu had to correct for modern-day phenomena affecting satellite orbit and clock errors, GPS signal delay and its curve when entering the atmosphere. But after weeks of calculations, assisted by his Ecuadorian student Paul Jarrín, he finally obtained the exact height of Chimborazo with respect to sea level: 6263.47 m. This demonstrated that La Condamine's estimate had been the closest, no matter what Humboldt and others had said after him.

Importantly, this precise measurement lays a solid foundation for studying variations in Chimborazo's altitude in the future, especially in the context of global warming. But beyond the cold data, the moving testimonials of the mission members struck a deeper chord. As Ivan expressed to a crowd of journalists on his return from the expedition: "*Reaching the summit with this historic mission filled us with joy, but also with sadness. We could see the effects of global warming on the glaciers of*

---

[48]La Condamine (1751), pp. 141–158.

Chimborazo."[49] In the time that had passed since 1978, the first time Ivan climbed Chimborazo, dramatic changes had adversely impacted the glacier: the ice was darker and wetter, the snout of the glacier was higher in elevation, the pockets and cracks were filled with meltwater. His emotion was palpable: he spoke about Chimborazo as if it were a dying friend. At the timescale of a human life, glaciers once seemed eternal. Not anymore.

This emotional response to the visual reality of glacier retreat would prove just as significant as science in conveying our mission. This led me to wonder how the aesthetic perception of nature might change in a world affected by global warming. Will mountains without ice be considered beautiful in the future? Will people adapt to these new landscapes and value them? Given our responsibility in global warming, is it morally acceptable to appreciate nature that has been disfigured by climate change?[50] While Chimborazo may still be beautiful and a symbol for Ecuadorians if its summit becomes bare rock, aspects of its appearance and role in the landscape would be lost forever. First of all, its massed snow penitentes, those formidable but sublime guardians of its summit. The ice at the summit of Chimborazo is estimated to be 54-m thick, and most of the melting occurs at the glacier's edge, not at the top, but an extreme event such as a volcano eruption could change the game and potentially accelerate the effects of global warming. For Ivan and many other Ecuadorians, losing Chimborazo's icecap would strike a blow to their national identity.[51] They would have to modify their flag to erase the white peak. Climate change will also mean cultural change.

* * *

But does it matter beyond Ecuador? At the time of Humboldt's trip to South America, Chimborazo was considered the highest mountain in the world. Today we know that Chimborazo is not even in the top 100 highest mountains worldwide. Yet what does that even mean today in a context of rising oceans when altitude is measured in relation to average sea level? This will inevitably make all Earth's mountains lower. Why then not take an invariable point as a reference, such as the center of the Earth?[52] In fact, that would change everything. To explain why, we need to go back to the Academicians' "figure of the Earth."

Recall that the first geodetic mission proved that the Earth bulges slightly around the equator. 'Slight' is a manner of speaking: the Earth's radius at the equator is some 21 km greater than at the poles. This means that mountains closer to the equator are a greater distance from the center of the planet than peaks that are farther from the

---

[49] El Comercio, February 13, 2016, p. 7.

[50] These questions are addressed by Brady (2014) and Auer (2019).

[51] For a discussion of the importance of landscape to Ecuadorian national identity, see Borchart de Moreno and Moreno Yánez (2012).

[52] Another way of defining a mountain's height is by measuring it from its base to its top. In this case, mountains in the sea are the tallest. Interestingly, measured by this benchmark, Teide is the third tallest mountain in the world, at 7500 m from its base on the seafloor to its peak. Only Haleakala (9100 m) and Mauna Kea (more than 10,200 m) in Hawaii are taller.

Fig. 5.8 Chimborazo dethrones Everest. The story of the third geodetic mission attracted much interest from national and international media as on this over of the Ecuadorian newspaper *El Telégrafo* on 10/04/2016. An article entitled "The Mountain That Tops Everest (Because the Earth Is Fat)" was also published in *The New York Times* on 16 May 2016. The media coverage was huge including, to name a few, the BBC (England), El País (Spain), and Le Monde (France)

equator. This has profound implications for Chimborazo. At 6263.47 m above sea level at 1° latitude, using complex geodetic models, Jean-Mathieu estimated that Chimborazo was 6384.415 km away from the Earth's center. [53] Measured from this reference point, this makes it higher than Everest, which at more than 27° latitude is 6382.605 km from the Earth's center. Our geodetic mission thus revealed that the summit of Chimborazo is the furthest point from the center of the Earth, and therefore, in a way, the 'highest,' dethroning Everest. When this announcement was made public, it made news—I received calls from the press around the world (Fig. 5.8). The French ambassador was ecstatic that his initial idea had resulted in such a powerful message and successful French–Ecuadorian collaboration. For Riobamba, it was also great news: tourism in the region increased by 30% in July 2016 as people came to see and climb Chimborazo. Our story had made science appealing for a broad audience.

* * *

There is an epilogue to this story. The contest for the highest mountain in the world has been repeated throughout history. In Humboldt's time, no sooner had he returned from his trip to the Americas than rumors from British surveys of mountains in India threatened the altitude record of Chimborazo. In 1808, he firmly defended its top position: *"English envoys were carried on litters through northern India to Tibet. They came from Calcutta where barometers are very common, and yet we still know nothing about the elevation of this country. I believe that in Europe one has very exaggerated ideas of it, and that Tibet is much lower than the plateau of Quito."*[54] This of course only delayed the inevitable. Observations around the globe accumulated over the following decades, and Himalayan and South American peaks repeatedly changed heights in geographic atlases (see the Aconcagua and Sajama overpassing Chimborazo in Fig. 3.2).[55] Chimborazo ended up losing its record,

---

[53] We can trace the origin of these geodetic models back to the first geodetic mission, which resulted in a precise calculation of the length of a degree of latitude, as well as Bouguer's revelation of gravitational anomalies at Chimborazo. We know today that the surface of the planet is uneven. The Earth is neither a pomegranate nor a lemon, but rather a potato with surface undulations of ±100 m. After complex calculations, Jean-Mathieu obtained a measurement for Chimborazo with precision to ±3 cm, a huge difference to that of Huascarán, which has a precision of ±5 m.

[54] Humboldt (1808), p. 105.

[55] The first European expeditions to the Himalayas started when Bengal was colonized by the British in the 1770s. In the 1790s, English orientalist and mathematician Henry Thomas Colebrooke (1765–1837) observed Mt Jomolhari in Tibet, whose altitude he estimated at 7930 m (the actual altitude is 7326 m). The Ganges Valley was explored in 1807–1808 by the new Surveyor-General of Bengal, Robert Colebrooke (1762?–1808), Henry's cousin, who observed from the city of Gorakhpur two peaks whose elevation he estimated to be 8045 m above the plain. He was convinced that the peaks were "as high if not higher than the Cordilleras of South America." He died in 1808, and his assistant Lieutenant William Spencer Webb (1784–1865) continued the search for the source of the Ganges. On the Nepalese border, the latter measured Mt Dhaulagiri, obtaining an altitude of 8190 m (close to the present 8167 m). Until around 1815, these discoveries were not published or known in Europe. For a history of early travelers and adventurers in the Himalayas, see Mason (1987). Concerning South America, in 1826 and 1827, the Irish naturalist Joseph Barclay

although never to Humboldt. *"All my life I have prided myself on the fact that, of all mortals, I had reached the highest point on Earth: I mean on the slopes of Chimborazo!"*[56] Humboldt would certainly have been interested in the results of our geodetic mission. They returned Chimborazo to the highest point on Earth. This means that at the summit of Chimborazo, Humboldt was in fact not only the highest man in the world, but also the closest to the cosmos.

Today the contest is still on, with global warming now setting the rules. The status of Chimborazo as the farthest point from the center of the Earth is rivaled by a close runner-up: Mt Huascarán in the Cordillera Blanca in Peru at 9° latitude and 6768 m in altitude places it 6384.372 km from the Earth's center. This is a mere 43-m difference between the two mountains. Whether Chimborazo will keep its record or not depends on the rate of glacier retreat on both peaks. If this is equivalent, Chimborazo should still be the highest when all its ice is gone, as its glacier seems to be thinner than that on Huarascán. But due to global warming, many mountains may lose their altitude record in the future. Some have already lost it. In 2019, Mt Kebnekaise (2095.6 m), which lies 150 km inside the Arctic Circle in Sweden's far north, lost its title of the country's highest peak by 1.2 m. In the past 10 years, the melting rate of its glacier has been 1 m per year. Climate change will increasingly change the "figure of the Earth." We will have to be prepared to continually update our atlases.

What does this adventure teach us about storytelling in climate change science? Communicating scientific knowledge only through data and facts is not enough. More is needed to reach people and try to persuade, rally, inspire or change attitudes to environmental questions. A number of elements triggered strong interest in our mission: locally, the proximity of the subject and its

(continued)

Pentland (1797–1873) conducted an 18-month survey of the Bolivian Andes and measured six peaks higher than Chimborazo, including the highest, Sajama (6542 m). Eight years later, in 1835, Captain Robert Fitzroy sailed his ship the *HMS Beagle* with Charles Darwin on board up the west coast of South America. Fitzroy's surveyors saw a few mountains rising above the coast of Chile that proved to be much higher. One of these was Aconcagua (6959 m), the highest mountain in South America (see Humboldt 1854, note 86, pp. 542–546). For an illustration of the changes in mountain height in nineteenth-century world atlases, see Bailly et al. (2014), pp. 60–61 (for the Himalayas) and pp. 192–193 (for the Andes).

[56]Letter to Heinrich Berghaus on 25 November 1828, in de Terra (1955), p. 128. In fact, Humboldt did not establish an absolute altitude record for any living mortal at this time, as pre-Columbian remains have been discovered on many peaks of the Andes. In 1999, the frozen remains of three children were unearthed near the summit of Volcán Llullaillaco, Argentina, at an elevation of 6739 m (Reinhard and Ceruti 2010). In 1854, Humboldt obtained backing from the King of Prussia to support an expedition to Tibet and central Asia by the Bavarian Schlagintweit brothers. According to their barometric measurements, they reached an altitude of 6785 m on Mt Kamet, which would have made them the first Europeans to climb higher than those who had ascended Llullaillaco.

importance in Ecuador and the team's openness to the surrounding society; internationally, the surprising result that Everest is not the highest mountain if the reference point is changed and the involvement of well-known alpinists and scientists, as witnesses of global warming, in the expedition. Reflecting on why this scientifically modest work had more impact on society and more exposure in the media than most of his other work, Jean-Mathieu ventured: "It provided an answer to a simple question that everyone wants to know: What is the highest mountain in the world? And for Ecuadorians, the added interest: What if the highest mountain is in our country? What if climate change takes away our record?" For Jean-Mathieu, the lesson is clear: "Communicating on a question of public interest is much more effective than first having to explain the question we are asking as scientists, and then its results, and finally its consequences for society." When scientific knowledge generates media coverage and creates conversations in the community, it raises the possibility that it can begin to change views. Framing the Chimborazo scientific mission as a modern adventure gave it the capacity to reshape the way people perceive and imagine a rapidly changing world. After all, the meaning of height goes beyond science—whose measurements are constantly called into question and revised as more sophisticated tools become available—to touch on the symbolism of summits in human culture.

5.3   Lessons from a Bear

At first, James Painter was not convinced that my story would be a good fit for the BBC. As the head of the Journalism Fellowship Program at Oxford University, he is an expert in science communication, especially about climate change. In January 2011, James had contacted me while reporting on glacier retreat in the Andes, and I had taken him to see the Antisana streams. He found the confluence of the café-au-lait and black-tea rivers a striking sight (see Fig. 4.6), but he was unsure about the communication power of my study subject, fly larvae, to talk about climate change. I suggested a metaphor: *"You might say some of these invertebrates are the polar bears of high Andean streams."* In James's article about the effects of glacier retreat, he used my proposal, albeit in a tongue-in-cheek approach: *"Of all the things to worry about, it is not immediately obvious why anyone would choose some tiny bugs living in a stream some 4,500 m up in the Ecuadorean Andes. It is hard to see how something you can only see properly under a microscope—and with complicated names such as Parochlus (Chironomidae, Diptera)—could rival the polar bear as a symbol of climate change.*"[57] James was right. If polar bears have become the ultimate icon of the fragility of nature in the era of global warming, it is because they have appealing attributes that bugs do not.

Humans feel empathy and affinity with animals that resemble them in certain ways. Bears are mammals with a rounded head and small round ears; they can stand upright on two legs and use their forelimbs to grasp objects. Female bears have a highly developed maternal instinct. And bears and humans are keen on similar foods, including salmon, berries, corn and honey.[58] Clearly, fly larvae are much harder to identify with. They cannot compete with bears as environmental icons, no matter how threatened they are or important to the ecosystem. To appeal to a large audience about the fragility of high-altitude Andean wildlife, I would need a different focus. Fortunately, there are bears in the high tropical Andes. But my first encounter with a living bear was actually in Paris.

*　*　*

When I was about ten, I was taken to the zoo at the Jardin des Plantes near Paris's National Natural History Museum. There, a Syrian brown bear was confined in a bear pit. As I approached the high rim of the pit, I couldn't see over it, and it was only when my parents lifted me that I discovered the bear at the bottom. Surrounded by solid walls with vaulted niches on the sides, the bear walked a barren cobblestone floor with a small basin in the middle. From time to time, he stood on his hind legs against the wall to sniff up at zoogoers, hoping that someone would throw down a peanut or treats. This depressing pit was from another age; built in 1805, it had not changed much over the last two centuries (Fig. 5.9). While human life had completely revolutionized since that time, for captive bears it remained unchanged.[59]

The bear pit was built a decade after the establishment of the Ménagerie du Jardin des Plantes, the first public zoo in the world. After the French Revolution in 1789, bears and other wild animals were not uncommon in the streets and squares of the capital. There were many fairground menageries, where animals were exhibited or made to perform in wrestling demonstrations. In 1793, the Commune of Paris

---

[57] Painter J (2011) 'Ecuador's highland bugs key to biodiversity. Antisana Glacier, Ecuador'. From BBC News at www.bbc.co.uk/news, Published 30 April 2011.

[58] While today's humans tend to be sympathetic to bears because they look like them, this wasn't always the case. See Pastoureau (2007) for a history of relations between humans and bears, and in particular the description of Buffon's hatred of this animal (pp. 306–308).

[59] Since the beginning of zoos, bears have been contained in pits, as they have long had a reputation for ferocity. To ward off danger from bears, Catholics used to call on Saint Martin, protector of the poor and 'bear hunter.' In the nineteenth century, bears began to gain the sympathy of the public, who came in great numbers to see them in zoos. In the Paris menagerie, one polar bear became particularly popular with children and lived for 33 years. When the bear died in May 1901, he was mourned. An affectionate obituary was written for him in *Le Petit Journal (no. 549, Supplément du dimanche, 26/05/1901, p. 166)*: "The children who frequent the Jardin des Plantes have just suffered a great sorrow." The bear was taxidermized and joined the galleries of the Natural History Museum. The use of pits to keep bears at the menagerie was stopped in 2014.

Fig. 5.9 Urban encounters with bears through the ages. The bear pit built at the zoo of the Jardin des Plantes in Paris was built in 1805 and remained in use until 2014. Photograph by Pierre Petit, 'Fosse aux ours', around 1885, Muséum National d'Histoire Naturelle, Paris

outlawed performing wild beasts, as this was deemed a public danger. The police confiscated all such animals and took them to the Jardin des Plantes. As no facilities were available to keep them, the menagerie was created, turning the problem into an opportunity: a collection of living animals would serve public education and the study of natural history, or as the naturalist Etienne de Lacépède elegantly put it: *"spread lasting and easy instruction under the appearance of a temporary and light satisfaction."*[60] Two polar bears were among the very first live animals to arrive in the menagerie.

From 1803 onwards, the menagerie was directed by **Frédéric Cuvier**, the younger brother of noted naturalist and zoologist Georges Cuvier. Most specimens were received from private donations and traveling explorers—Humboldt, in fact, sent a capuchin and a night monkey from Venezuela.[61] Until his death, F. Cuvier not only organized the daily life of the menagerie and the exhibition of the animals, but also developed a more comprehensive assessment of animal behavior than anyone had yet attempted. This is illustrated in his reaction when in 1825 he received a bear specimen of a previously unknown species brought back by a royal ship from Chile: *"Unfortunately the fatigue of the voyage had weakened it to such an extent that it was not possible for us to preserve it, whatever care had been taken for it, and it was*

---

[60] Lacépède (1804), p 21. For more information about the first days of the Ménagerie, see Burkhardt (2001).

[61] Humboldt (1811–1833), vol. 1, pp. 313–314.

Fig. 5.10 The misnamed 'bear of Chile.' In this illustrated plate from his *Natural History of Mammals*, Frédéric Cuvier labels this species as "the bear of the cordillera of Chile"—which is surprising as Andean bears are not known to live there. It was later clarified that this specimen originated from the mountains near Trujillo in Peru. From Geoffroy Saint-Hilaire and Cuvier (1824–1842)

still very young. If it had lived, we would have been able to study it better, and to determine with more exactitude the nature of its distinctive characteristics." However, F. Cuvier was confident that this South American bear would not remain unknown for long: "*It will surely happen for this species what has happened for so many others: once indicated, it will be sought by travelers who will know exactly its origin, and soon it will be common enough to complete its history.*"[62] His predictions would prove to be wrong.

With this first encounter, F. Cuvier became the first to describe an Andean bear (Fig. 5.10), naming it *Ursus ornatus* ('decorated bear') in reference to the light-colored rings around its eyes (the species is also called the spectacled bear). But for the rest of the century, only a handful of Andean bears were shown in European zoos and only for short periods. It was not until the early 1910s that zoologist William Hornaday, director of the New York Zoological Society, finally obtained from Quito "a fine specimen of the species so long desired," after more than a decade of intensive searches in South America.[63]

Before this, many explorers in the nineteenth century had sought wild Andean bears in vain. Alcide d'Orbigny actively searched for it in the wild in Bolivia but did

---

[62] Geoffroy Saint-Hilaire and Cuvier (1824–42) *L'ours des Cordillères du Chili*, tome 5.

[63] Hornaday (1911), p. 748.

not see "the slightest trace of it." Surprisingly, he saw a captive bear in Cochabamba "moving the bellows of a blacksmith."[64] The French chemist Jean-Baptiste Boussingault, during his decade-long journey in South America (1822–32), had an even stranger encounter: he saw a bear cub in Bogotá on the bed of the Ecuadorian revolutionary heroine Manuela Saenz, intimate of Simon Bolívar; she had adopted it as a pet.[65] Humboldt may have seen one on his expedition near Riobamba.[66] But two naturalists who wrote about Andean bears at length, the Swiss Johann von Tschudi and the Italian Antonio Raimondi, never actually saw a wild bear.[67] So in the 200 years that separated the travels of Charles-Marie de La Condamine and Jean Dorst (neither of whom reported seeing a bear), naturalists had learned almost nothing about this elusive animal. In fact, much of the information disseminated about Andean bears was not based on first-hand observation, but came from second-hand testimonies. Not far off from the French expression casting doubt on those who repeat hearsay: *l'homme qui a vu l'homme qui a vu l'ours* ('the man who saw the man who saw the bear').

* * *

So who was the first naturalist to write about the Andean bear based on first-hand observations? It may have been the Riobamba-native Juan de Velasco, who, in 1789, described in the region of Quito the **Ucumari** bear, which "is very daring and lives in cold climates, where there are many of them and [they are] seen frequently." But he also asserted that there was another type of bear, the *Iznachi*, which is "only seen in very hot climates, and is all black without fur of any other color."[68] In his writings, von Tschudi claimed that the cold-habitat bear was carnivorous, feeding on deer, vicuñas, guanacos and cows, while the hot-habitat bear was a separate species (which he named *Ursus frugilegus*) that was vegetarian and fed mainly on palm fruits and maize crops. Adding to the confusion, in 1911 Hornaday described a subspecies of the Andean bear (*Ursus ornatus thomasi*) from southern Colombia that

---

[64] D'orbigny *(1835–47), tome 2*, p. 463, note 3. The use of animals, especially dogs, in blacksmithing was common in Europe until the twentieth century. See an illustration in Diderot and d'Alembert's Encyclopedia (1751–70), Plate 1 of the "Cloutier Grossier" (nail maker).

[65] Boussingault offered a lively description of Manuela Saenz's bear: "Manuelita adored animals; she had an unbearable teddy bear that had the privilege of roaming around the whole house. The ugly beast liked to play with visitors. If you pet him, he would scratch your hands terribly, or cling to your legs: it was difficult to get rid of him. One morning I visited Manuelita. As she was not yet up, I had to enter the bedroom, where I saw a frightening scene: the bear cub was lying on his mistress, its horrible claws posed on her breasts. Seeing me enter, Manuelita said calmly: 'Don Juan, go to the kitchen and bring a bowl of milk that you will place at the foot of the bed: that devil of a bear does not want to leave me.' The milk arrived. The animal slowly left its victim and got down to drink." In Boussingault (1892–1903), Tome 3, pp. 214–215.

[66] "We were very happy with our stay in Riobamba. We had worked a lot, measured trigono [metrically] Tungurahua and Chimborazo, made excursions to both volcanoes, and in Riobamba drew maps of the Condor and the Preñadilla, saw wolves, bears." Humboldt (2003), p. 223.

[67] Raimondi (1862), pp. 148–149; von Tschudi (1847), p. 420.

[68] Velasco (1841–1844), Tome 1, p. 85.

was completely black, with no clear marking around its eyes. Then in the 1970s, the Andean bear was ultimately declassified from the genus *Ursus* and put into another genus, *Tremarctos*, which had been described in 1855, but only for prehistoric bears whose fossils had been found (the current Latin name of the Andean bear is *Tremarctos ornatus*).[69] How confusing! So much for Cuvier's prediction.

Of course, while the Andean bear may have been new to science, Andean people certainly knew of its existence. In pre-Columbian times, the bear was closely linked to the spiritual world, and celebrations and legends involving bears persist today. During the Qoyllorriti festival in Cuzco, the largest pilgrimage of Indigenous people in the Americas, dancers dressed like bears (*ukuku*) protect pilgrims from dead souls (*condenados*) in a ritual climb to the Ausangate glacier (6385 m) to celebrate the magical properties of melting water. Through water, bears are seen as symbolic mediators between earth and sky. The *ukukus* used to break off blocks of glacial ice, which they carried down the mountain to their villages, where they would use the meltwater as medicine. Due to glacial melt, this practice has been forbidden, and today *ukukus* are only allowed to bring meltwater in bottles.[70] From rock carvings in the state of Merida in Venezuela to the Oruro Carnival in Bolivia, bears have long been important in Andean culture. However, contact with Spanish colonizers would transform the image of the Andean bear in colonial times through the lens of European myths. Its resemblance to a creature between man and beast reinforced a belief that male bears kidnap humans, especially women, to sexually pair with.[71] Thus, Old World myths, legends and customs have mixed with native cosmology to give rise to an intercultural folklore of Andean bears that still has a strong presence in local cultures. The *oso andino* ('Andean bear') exists in the mythic space of the culture of the Andes (*Lo Andino*), not just as an ecological entity.[72]

* * *

Today, the elusive Andean bear is still little known by scientists. Some basic facts have been established: of the eight existing bear species, the Andean bear is the only that is native to South America. It is endemic to the tropical Andes, ranging from

---

[69] Although the genus *Tremarctos* was established by Gervais (1855, pp. 20–21), it was not used before the 1970s, as the clear distinction between the Andean bear and the other *Ursus* species had not been confirmed. The vernacular name of the Tremarctinae is short-faced bears due to the relative shortness of the skull in comparison to other bears. They are also differentiated by a hole in the humerus (*trema* is Greek for 'hole'), whose function is still subject to debate.

[70] For more information about bears in Andean cosmology, see Paisley and Saunders (2010); Gade (2016), pp. 217–238; Stensrud (2020); Torres (2021).

[71] Humboldt recounts that a tale of a savage (*el salvaje*) pursued him everywhere "during five years from the northern to the southern hemisphere" and he invited "the travelers who will visit after [them] on missions of the Orinoco [. . .] to examine if it is some unknown species of bear [. . .], which could give rise to such strange tales." Humboldt (1852–1853), Tome 2, pp. 270–272.

[72] See the masterful series of essays on the relationship between landscape, plants, animals and people in the Andes by Gade (1999). Gade discusses the meaning of "*lo andino*" and provides a list of 20 material symbols, including the Andean bear, together with the queñua (*Polylepis*), the mountain tapir and Inca ruins.

Venezuela to northern Argentina through a long and narrow mountainous strip (more than 4000-km long but only 200–400-km wide). It is typically found in a wide gradient of altitudes across the Andes, from the cloud forests in the foothills to the snow line in the páramos. It is a rather small bear (130–190 cm in height; 35–70 kg for females and 130–175 kg for males), yet it has gigantic relatives. In fact, the Andean bear is the only living relative of the short-nosed bears (hence its classification in the genus *Tremarctos*), a group that lived exclusively on the American continent. Some were similar in size to today's brown bears, and, if Humboldt had penetrated deeper into the *Cueva de los Guacharos* (see Sect. 3.2), he would have found the fossilized remains of one of them.[73] Others were truly giants: the largest bears known to have existed on the planet. The South American giant short-faced bear reached an estimated size of 3.5 m in standing height and exceeded 1500 kg, which makes it a strong candidate for the largest carnivorous land mammal that ever lived.[74]

As for its lifestyle, the Andean bear's habits are known mostly from inferential evidence, such as hair, scat, claw marks, tree nests and feeding sites; most sightings are from camera recordings rather than encounters with the animal itself. This is not unusual for most bear species and is how scientists tend to obtain information on them worldwide. For example, during my year in Ithaca, I joined Catherine Sun, a doctoral student in Cornell's Department of Natural Resources, during one of her surveys of black bears in the Connecticut Hill Wildlife Management Area. She had set up 260 trail cameras and analyzed hundreds of hair samples caught in strategically placed barbed wire to estimate how the species' density is changing in the region. Black bears are on the rise in New York state, and one of the reasons is that they cope quite well with habitat modification: there are not fewer bears in areas with agriculture and human development compared to forested areas. One of the objectives of Catherine's study was to obtain information that would help to manage potential bear–human conflicts, which she achieved through calling on citizen science, collecting hundreds of observations via an app. From the early days of the university when John Henry Comstock would wrestle with a tamed bear,[75] to the Big Red Bear mascot of Cornell sports teams,[76] Catherine was writing a new page in Cornell's long history with bears.

---

[73] See Rincón and Soibelzon (2007), Stucchi et al. (2002).

[74] Soibelzon and Schubert (2011).

[75] The first bear to come to the Cornell campus was bought for 15 dollars in 1872 from a passing menagerie by zoology professor Burt Green Wilder (1841–1925), an outstanding pupil of Louis Agassiz. It was lodged in the basement of McGraw and cared for by John Henry Comstock, whom Wilder described as an "inspired anthropomorphic squirrel." This is maybe why Comstock grew fond of the bear and performed wrestling spectacles with it for the amusement of students. The bear eventually became a specimen in the university's museum. See Bishop (1962), p. 131.

[76] The Big Red Bear (or 'Touchdown') is the unofficial mascot of Cornell University. The first mascot was a live black bear introduced in 1915 by the Cornell University Athletic Association. Until the end of the 1930s, three more bears were used as mascots until costumed students replaced them. A bear statue was erected outside Teagle Hall in 2015. See Foote (2008).

Unfortunately, much less is known about the bears in the Andes than in New York State. However, since the pioneering studies by Bernard Peyton, a biologist and adventurer (and artist[77]) sent to Peru in 1977 by the New York Zoological Society—but this time to learn about bears, not to capture them—an increasing number of field studies have uncovered some of their secrets. The Andean bear is mainly active during the day and moves along established trails in search of food and potential mates, which it finds through its acute sense of smell. Though classed as carnivores, these bears are omnivorous and opportunistic feeders, and in reality are largely vegetarian. They mainly feed on plants, flowers and fruits, and occasionally on carrion, eggs or small animals (insects, worms, rabbits and rodents), or tapirs and cattle. Due to the large amount of fruit it consumes and its high mobility through the landscape, the bear plays a very important role as a seed disperser.

Adults are generally solitary and form pairs during the mating season, which can occur any time throughout the year. Females give birth to one or two cubs, which measure less than 20 cm and weigh only a few hundred grams. Females provide maternal care for over a year. Today Andean bear distribution is mainly constrained to pockets of habitat that have escaped human development. Yet even these territories are shrinking rapidly due to the expansion of agriculture and the construction of power lines, pipelines, roads, mines and quarries. There may be only 20,000 individuals left in the wild, with an estimated population of 5000 in Ecuador. The bear is expected to move faster toward extinction than any other carnivore in the region.[78] To prevent this, efforts are underway to create natural corridors to connect small populations.

Highly endangered, endemic to the tropical Andes, the only living relatives of giant prehistoric bears, mysterious, shy and poorly known yet omnipresent in popular folklore and culture[79]—not to mention adorable—the Andean bear is the perfect face for the protection of Andean wildlife. As polar bears are the tragic icons of climate change, Andean bears could become the ambassador for myriads of mountain creatures endangered by changes due to global warming and land use: aquatic flies, fish, ground beetles, diademed sandpiper-plovers, *Azorella* cushions, and so on. The plight of the bears could raise public awareness for all these threatened communities.

* * *

So what was my plan? I decided to use wildlife photography to capture images of Andean bears in their habitat that would make a visual argument for mountain wildlife protection, inspire interest, stimulate dialogue, and ultimately make people care. A striking image can help science deliver its message by engaging other ways

---

[77] Bernie Payton is a skilled origami artist who uses paper sculptures to convey a sense of the fragility of creatures, including bears. https://www.youtube.com/watch?v=eotfuZvmIVk

[78] See Iturralde-Polit et al. (2017).

[79] The image of the Andean bear is incorporated on the back of the 50 bolivar banknote in Venezuela, adorns the logo of the Colombian National Parks agency, and is the mascot of the Pontificia Universidad Católica del Ecuador in Quito.

of knowing and can have more resonance than academic texts or scientific reports. Bears, which are often anthropomorphized, produce strong emotional responses and empathy; I was convinced they could help raise awareness of the fragility of the tropical Andes. Now I just had to find a bear.

In the same way that it was once believed that there were two species of Andean bears, *iznachi* and *ucumari* (this has now been disproven, the face markings of Andean bears are highly variable), there are two different settings to choose from: the cloud forest or high in the páramo. Despite their dense vegetation, cloud forests offer ample opportunities for photographing *iznachi*, either through camera traps on well-defined tracks or during the blooming of certain tree species around which the bears gather in specific places and periods. In March 2009, the news circulated that bears could be easily seen in the lowland cloud forest of Maquipucuna reserve, a two-hour drive from Quito. The ripening of the olive-shaped fruit of the **aguacatillo** had attracted some 20 bears in search of one of their favorite foods. This occasion allowed my first observations and photographs of Andean bears: feeding, playing, building nests, resting, even posing for a family portrait. But if I wanted the bear as an icon of the páramo, I would need a picture of its high-altitude cousin.

Photographing an *ucumari* is another matter. In the páramo, bears are so elusive, shy and dispersed over large territories that most published pictures are of captive, reintroduced, radio-collared animals. Others are either blurry or taken from a long distance. How would I get up close to an *ucumari*? On serendipitous walks, luck had led me to Andean foxes, condors, owls, and even to the almost impossible to see mountain tapir. But over the 10 years I had lived in the Andes, I had never seen any bears in the páramo. Not even the slightest clues of their presence, despite intensive scouting. It was no different for me than for all those naturalists who had vainly searched for the bear through the decades. This just made me more obsessed with seeing an *ucumari*.

* * *

As by now you won't be surprised to hear, a clue to finding an *ucumari* was to be found in Humboldt's narratives. While Humboldt may or may not have seen wild bears during his travels, on his way from southern Colombia to Ecuador at the end of 1801, he commented about the bear's food sources in the highlands. In his description of the beautiful waterfalls of the Rio Vinagre, Humboldt writes: "*In the foreground of the sketch is a group of* Pourretia pyramidata, *[...] known in the Cordilleras by the name of* achupallas. *The stem of this plant is filled with a farinaceous pith, which serves as food to the great black bear of the Andes, and in times of scarcity, even to men*" (Fig. 5.11).[80] Humboldt was so fascinated by the singularity of the Andean bear diet that he used it to support his romanticized view of the tropics: "*Unconcerned, the ewe watches a group of bears wade by. Nature has assigned to each its food. [...] The young leaves of the crown [of* achupallas*] attract the bear more than sheep and cattle, and only in the case of raging hunger have*

---

[80]Humboldt (1810), p. 221.

Fig. 5.11 Humboldt and an *achupalla*. Detail from the *Waterfall of the Rio Vinagre* near the Puracé volcano in Colombia showing Bonpland (kneeling and collecting plants) and Humboldt (with hat) gazing at a waterfall near an *achupalla*, a bromeliad that grows in the páramo and is a favorite food source of Andean bears. Drawing by JA Koch from a sketch by Humboldt. In Humboldt (1810), plate 30

these to fear. Thus, in the tropics everything takes on a milder, more peaceful form and customs." [81] I realized that to find bears, I first had to find their food, the relative of the pineapple, the ***achupalla***.[82]

In the páramo of Ecuador, *achupallas* alone fill the *ucumari*'s 'simple bare necessities,' as the song goes. More than half of their diet is made up of these plants, which are particularly abundant in certain humid moorlands. Where there are

---

[81] Humboldt (2003), p. 162.

[82] The new genus name for *Pourretia* is *Puya*.

numerous patches of *achupallas*, it is quite common to find evidence of feeding by the bear. In the summer of 2016, as I visited some potential study sites for a new research project on water conservation in the Cayambe-Coca National Park, I discovered a zone with many *achupallas*. And this would indeed prove to be where I finally had the opportunity to see my first *ucumari*. The encounter was purely by chance, as the bear practically stumbled upon me as it crossed my path. It was not very shy, and I was able to take several photos at a relatively close distance. But most importantly, this encounter served as training so I could pick out the bears' markings, so well camouflaged in the páramo grasses. As Thoreau advised: "The question is not what you look at—but how you look and whether you see."[83] With practice, my eyes could spot what they could not before, and now I knew how to find what I was looking for.

After this initial encounter, I spent almost all my free time in search of bears in the Cayambe-Coca National Park. I only had a year before I left for my sabbatical at Cornell, and I was concerned it would take a while to meet my second bear. But I was wrong. In 3 months, I observed three different individuals, all feeding on *achupallas* (Fig. 5.12a, b). There was a shy black-faced youngster, a mother with a very young cub, and a large male that surprised me in the high grass, and approached much closer than I initially expected. When I took his portrait, I could barely fit his head into the camera frame (Fig. 5.12c). The bears ignited my curiosity in a way I had never experienced, and bear-watching became a bit of an addiction. I wanted to visit them every day and see what was going on in their *achupalla* haven. I thought about bears, talked about bears, dreamt about bears. My family's patience tired. In the documentary *Grizzly Man* (2005), the bear biologist Larry van Daele talks about the "siren song" of spending time in the bear's world. Bear-watching revealed my inner Thoreau: I enjoyed being far from civilization and almost preferred the company of bears to people. I longed for a picture that would show my connection with the bears. This day would come soon.

On the last day of 2016, I was again out in the national park. Andean bears do not hibernate—given mild temperatures and plenty of food, they have no reason to. As is often the case in the páramo, the weather was rainy and foggy. There was an earthy smell of moss in the air and the piping whistle of the tawny antpitta. Above my head, dozens of megawatts pulsed through a power line, reminding me of the bigger story of the human impact on the planet. I scanned the landscape, alert for a black hump in the *achupallas*. A thin mist gave the páramo a ghost-like beauty, but it soon began to thicken, and I found myself shrouded, with no visibility. I shifted to another sense, hearing, and listened to the sound of a stream nearby, then walked up it to get back on the main trail.

The fog and rain seemed to be conspiring to make me leave, but I enjoyed being out bear-spotting whatever the weather. My tenacity paid off. Suddenly, a small silhouette crossing the path materialized out of the mist and stared at me: a bear cub, about 8 months old. Soon, a second silhouette joined it. They sensed my presence

---

[83] Thoreau, Journal 5 August 1851, in Walls (1995), p. 127.

Fig. 5.12 Close encounters with the Andean bear. (**a**) A female tearing off the central leaves of an *achupalla* to take it to her cub (**b**, white arrow). To the untrained human eye, individual bears may look the same. However, each Andean bear's facial markings are highly specific, similar to human fingerprints: see **c** and **d**. All photos were taken by Olivier Dangles in Cayambe-Coca National Park, Ecuador

and sniffed the air. The bears lifted their heads, looked in my direction and moved away. I suspected their mother was nearby and stayed quiet. She finally emerged from a bush, and all three went off in the direction of the cliff. Despite their lumbering appearance, Andean bears are actually pretty spry. They walk fast and are excellent climbers, able to scale trees and rocky walls in steep terrain above 4500 m in altitude. I decided to follow them. The fog dissipated as I climbed. I slid on the slippery slope and dropped my lens. After 20 min going uphill without seeing the bear family again, I took a break to scan the landscape. There they were, their heads barely sticking out of a *Loricaria* bush, 50 m above me. The low angle did not allow a good shot; I needed to get closer. This would require moving carefully. I concentrated on every step on the steep slope, my heavy camera on my back.

Fig. 5.13 Connecting with a bear. A perfect photo opportunity of an Andean bear in the páramo of Cayambe-Coca National Park, Ecuador, and a reciprocal exchange between human and non-human. Photo by Olivier Dangles

Arriving at the edge of a gully, now 20 m from the bears and at their level, I found a stable spot for my tripod and waited. One after the other, three curious heads popped up from the bush. The two young started to play pawing games under the watchful eye of their mother. I began to take pictures, and one of the cubs, intrigued by the sound of the camera shutter, turned to me and lifted up its right paw, as if in a wave (Fig. 5.13). This was the picture I was looking for. I stayed with the family some 15 min until the mother yawned and looked away. It was time to leave them alone.

Surprise, awe, tenderness, humility, concern: these are the reactions that people shared on social media when they saw the picture, in addition to the hundreds of likes. Few messages can be designed to produce the same emotional response in everyone, but the bear cub was a unanimous success. It was communication that needed no language, a direct appeal to make people care about the vulnerability of nature. With this human-like gesture, the bear provided an opportunity for reflexive awareness, blurring the modern divide between nature and culture, the wild and human worlds. The bear had acknowledged and responded to my presence: I did not only see a bear; I was seen by the bear. Thoreau described this feeling in an encounter with a muskrat: "While I am looking at him, I am thinking what he is thinking of me. He is a different sort of man, that is all."[84]

---

[84] Cited in Walls (1995), p. 143.

* * *

But while this image was an ideal hook to capture attention about the urgent need for páramo conservation, I did not want to only romanticize and anthropomorphize the bears. By spring 2017, I had observed a total of 11 bears and collected enough pictures to share my experience of their world with a broad audience. In collaboration with the Quito Water Company, the Ecuador Fund for Water Protection, and Ecuadorian artist Belén Mena, we designed a book, *OSO* ('bear' in Spanish), aimed at the "iconization" of these animals in order to protect the páramo.[85] The images captured during my walks provided a window into a bear's life in the páramo (Fig. 5.14). Some zoomed out to show the bears as part of their landscape. Others revealed the dangers faced by these animals in their day-to-day lives, mostly due to human disturbances in the heart of their habitat. The images sought to highlight the role of the Andean bear as a keystone species, whose protection equally protects other species with similar space and habitat needs, and its reliance on the preservation of crucial ecosystem services, such as water provision. This immersion in the lives of the bears might reenchant humans with nature, might trigger their concern.

So did these images have any impact on páramo conservation? They at least contributed to the effort by several stakeholders in the 2010s to make the Andean bear an icon of environmental protection in the Quito region. For example, the Ecuador Water Fund used the sale proceeds from *OSO* to support research grants on páramo ecology and deployed a campaign using photographs from the book. The campaign's objective is to maintain a sustainable water cycle in Quito by informing and assisting all stakeholders in water conservation and wildlife protection. Over the last decade, this campaign has led to a gain in popularity of the Andean bear among Quito's citizens—it is now viewed as a steward of water quality and conservation.

This was exemplified in the rebranding of Ecuador's most popular water brand, Güitig. The brand is so pervasive in Ecuadorian culture that instead of sparkling water, most Ecuadorians say "Give me a Güitig." Its water comes from the glaciers and páramos of the Cotopaxi volcano after passing through underground streams where it absorbs minerals. Since 1940, a polar bear had featured on the bottles as the icon of the brand (Fig. 5.15a). In 2021, when the brand decided to redesign its bottle, its customers suggested replacing the polar bear by an Andean bear. A contest was launched, asking local artists to use *"their creativity to imagine the Andean bear in its habitat to raise awareness of its importance and conservation."*[86] Many of the submissions presented the Andean bear in its environment, with Cotopaxi and its snowcapped peak in the background and native páramo plants in the foreground (for example, the *achupalla* and *Azorella* cushion, as in the design in Fig. 5.15b). The Andean bear showed it could rival the polar bear as an icon of the wonder and importance of nature.

---

[85] I borrowed the term 'iconization' from the study on polar bears by Born (2019).

[86] *El Universo*, 'Güitig plantea tres opciones para nuevas etiquetas que incluyan al oso de anteojos', 27 April 2021.

Fig. 5.14 Powerful conservation symbols. Photographs from the book *OSO* (Dangles and Mena 2017), which aims to promote the protection of the páramo. (**a**) Andean bear in its habitat; (**b**) Mountain tapir (*Tapirus pinchaque*), a species with similar habitat requirements to Andean bears; (**c**) Andean bear near a water source, a crucial resource for humans and animals alike; (**d**) Andean bear crossing sign along a road in the páramo. All photos were taken by Olivier Dangles in Cayambe-Coca National Park, Ecuador

Fig. 5.15 The Andean bear becomes an icon. In 2021, Ecuador's Güitig brand of mineral water decided to change its bottle labels to promote páramo protection. (**a**) The original design with a polar bear; (**b**) one of the four new designs with an Andean bear, this one showing the bear in its habitat (designed by Ecuadorian artist **Paula Terán Ospina**)

While the Andean bear was increasingly entering the consciousness of Ecuadorians, it had also fundamentally altered my way of seeing things. My quest for photographing these bears had taught me unexpected new lessons in communication. Bears have extraordinary facial expressions and communicate with each other through body language. In my time photographing these bears up close, I became alert to these nonverbal cues, which allowed me to interpret their moods and emotions. I found that each bear, recognizable by its specific face markings, had its own personality: shy, curious, irritable or indifferent. Recognizing nonverbal cues is also critical in human communication. In **Alan Alda**'s book *If I Understood You, Would I Have This Look on My Face?*, he emphasizes this aspect of communication that relies on senses other than just hearing. *"The person who's communicating something is responsible for how well the other person follows him. [. . .] Synchrony can't occur without acute observation [. . .] that enables people to see and respond to the slightest shift in another person's behavior."*[87] Reading signals from others is an ability as vital to photographing bears as to communicating science. A vital conclusion I drew from my experience with these bears would be to focus more on how others receive the message when sharing my scientific work.

How you frame your message is as important in climate change science communication as in any other area. Storytelling—including metaphor, imagination, multisensory impressions—can be a bridge that spans the distance between science and human interest. This requires remaining true to the underlying science, while tailoring the message to the attitudes, values and perceptions of different audiences. As mountain environments alter under human pressure and a warming world, what can be done to safeguard their ecosystems, plants and animals? How can scientists make their work make a difference? Changing public opinion and spurring people to action will not be driven by science alone, but by stories that appeal to the emotions. In my work, the Andean bear and Chimborazo have served as powerful ways to communicate to the public about páramo conservation, glacier retreat, warming and other global changes. Natural science is not based solely on the rational and logical, but on the senses, perceptions and emotions. Like naturalists in the Age of Enlightenment, scientists need to go beyond science-splaining and find ways to express the wonder and threat in the world.

---

[87] If I Understood You, Would I Have This Look on My Face? by Alan Alda (2018), Penguin Random House, p. 30 and p. 37

Chapter 6
Conclusion: Humboldt Tenured

*I shall spend a few days in Sanssouci [a palace in Potsdam],
where I shall reach—alas!—my seventy-fifth birthday. My
only reason for regret is that, in 1789, I thought the world
would have solved several more problems. I have seen many
things, but, measured by my hopes, still little.*

Alexander von Humboldt (1860, Berlin, 13 September 1844,
p. 122)

The new main historical protagonist of this chapter is shown in Fig. 5.1.

It was a cold April morning in the high sierra of Ecuador. The icy wind stinging our hands and faces was a reminder that we had climbed higher than 4500 m. Each step was an effort and the thin air that entered our lungs left us breathless. It felt like our hearts might burst out of our chests. Periodically resting and advancing up the steep slope, we finally reached the spot we were looking for. This time, we were not searching for a particular plant, animal, cave, stream or glacier or interested in reaching the top of a mountain. We were trying to find a lost landscape. Very close to 4800 m, in the saddle between the twin peaks of the Illinizas, we took out the book we had brought and compared one of its photos with the view lying before us. There was no doubt: this was the place.

I was with Juan Diego Pérez (Juandi), an Ecuadorian photographer, documentary filmmaker and writer who had first discovered his country in the book we were holding, *Nieve y selva en Ecuador* ('Ecuador: Snow Peaks and Jungles'), the book that had brought us to the Illinizas. It was published in 1952 by **Arturo Eichler**, a German-born writer, photographer, mountaineer, university professor and conservationist, who arrived in Ecuador in the mid-1930s. The chronicles of his adventures appeared periodically in a column in the Ecuadorian newspaper *El Comercio*. With its photographs and wonderful cover by the Ecuadorian painter Oswaldo Guayasamín, *Nieve y selva en Ecuador* contributed to a generation of Ecuadorians discovering a facet of their country that was little known at the time: its wilderness. The natural landscapes of the high Ecuadorian Andes, the dense jungles and people

O. Dangles, *Climate Change on Mountains*,
https://doi.org/10.1007/978-3-031-39528-4_6

who lived there, and the enchanting Galapagos Islands became accessible through the images and stories of this intrepid traveler.

In early 2016, Juandi had come to talk to me about his new project. His idea was to use the photographs in Eichler's book as points of reference for the "relatively pristine" (postcolonial) state of Ecuadorian landscapes in the past and compare them with the present day.[1] Most of us forget to what extent the landscape is different from the one that existed decades ago, as the changes from year to year are gradual. Jared Diamond refers to this as landscape amnesia: the incremental loss of what our surroundings look like from one generation to the next.[2] Juandi thought that comparing 'before' and 'after' pictures might act as a wake-up call to the rapid anthropogenic changes to the Ecuadorian sierra. I decided to join Juandi in his quest for these lost landscapes. While it was relatively easy to find the exact view from the Illinizas saddle (Fig. 6.1a, b), the task was harder for other photographs in the book, such as the Conocoto road, in the southern suburbs of Quito (Fig. 6.1c, d). The tops of the Cotopaxi and Sincholagua volcanoes, far in the background, were our only landmarks. As for the bucolic view of the countryside around the little church of Mulalo, a small village at the foot of Cotopaxi, it had simply disappeared (Fig. 6.1e, f). Concrete walls had been erected where the photograph had been taken decades before us, and we had to climb a half-built terrace littered with rebar to get the church and Mt Cotopaxi in the frame. The Andes have not only been warming since Humboldt's times, the whole region has been perceptibly disfigured.

This transformation signifies a brutal shift in the sensory experiences of the environment today. Past sounds, smells, sensations and sights have been lost: the bleating of sheep, the sweet, earthy smell of agave, snow crunching underfoot, cobblestone paths. These have been replaced by uglier, more urban experiences disconnected from nature: the honking of cars, exhaust fumes, gray concrete, cold steel. For scientists or artists, it now seems impossible to study, write or paint nature without showing the human domination exercised over it. Since Humboldt, the exponential increase in human activities has ravaged planet Earth, and in the absence of past information on historical conditions—cultural, ecological, aesthetic—this results in a creeping normality. 'Before' and 'after' pictures of the Andes offer stark insights into the nature of these changes. Hopefully, they also provide a catalyst for the construction of a sustainable future; a future in which natural places are understood as part of the vast interconnected system of the Earth.

* * *

Solving problems. The quote introducing this conclusion shows that Humboldt explored not only to advance knowledge, but to find the larger meaning of life; he felt humble in the face of so many unanswered questions. Swiss philosopher Benjamin Constant wrote that Humboldt *"did not seek [in science], as many scientists do, a means of rendering himself indifferent to the interests of humanity and of relieving himself of the responsibility of having opinions and showing*

---

[1] Pérez Arias (2022).
[2] Diamond (2005), pp. 425–426.

Fig. 6.1 Resisting landscape amnesia. Before (**a**, **c**, **e**) and after landscapes (**b**, **d**, **f**) in the Ecuadorian Andes from Eichler's book *Nieve y selva en Ecuador* (1952) and 2016. Comparing past scenes with those of today may help combat collective landscape amnesia and enlist the aesthetic appreciation of nature to promote conservation. (**a**, **b**) The Illinizas saddle; (**c**, **d**) The road to Conocoto near Quito; (**e**, **f**) Mulalo church in Cotopaxi province. Photos by Arturo Eichler (**a**, **c**, **e**), Olivier Dangles (**b**), and JuanDiego Pérez Arias (**d**, **f**)

courage.[3] At the beginning of the twenty-first century, 200 years after Humboldt's travels in South America, the emergence of sustainability science has triggered a

---

[3]Benjamin Constant (1805) Journal intime, 12 February 1805. In Melegari (1895), p. 99.

reexamination of the 'Grand Challenges' with a transdisciplinary approach, and a new social contract for academia.[4] For decades, a scholar was expected to be detached and apolitical, but many feel it is time for a more socially responsible way of doing research, bridging objective study with moral values that aim to shape policy and social action. To respond to the challenges of our times, climate change in the first place, it seems crucial to rethink the way knowledge is produced, the way research is conducted, who researchers conduct it with and for, and how results can inform concrete solutions. In part, this means being more like Humboldt in our thinking.[5]

What if Humboldt was around today? What would his plan be to save the planet, to fight against landscape amnesia? It is hard to imagine what he would think about the current realities of the modern world. Perhaps G. Evelyn Hutchinson could give us a clue, as he was alive to see the challenges of the Anthropocene. In his last piece for American Scientist, entitled 'What is Science For?', Hutchinson wrote that it was *"useless to complain passively that destructive tendencies are human nature,"* claiming that *"the most pressing need for mankind is to learn to produce a technology of love."*[6] What he meant, I think, is that the environmental crisis is also a spiritual crisis, and that empathy, sensitivity and humility are key to sustainable behavior. Beyond finding solutions to our problems, we need to reconnect to nature, to others, and to ourselves.[7] It is time to reconsider our relationship with the non-human and acknowledge the social and political role of sensitivity in protecting the Earth.

In practical (or ideal?) terms, a 'university for a sustainable world' in the spirit of Humboldt would bring together different disciplines in communities (of thinking and practice) in which science, the humanities and the arts actively engage with wider society to design and share their studies. These communities should be centered on sustainability goals anchored in specific places and local needs, considering all the dimensions of humanity's modes of relating to the Earth.[8] An idea of 'community' based on a broader conception than in Western culture, like the Andean *Pachamama*, which includes non-human beings as well as the non-living—mountains, forests, oceans.

---

[4] See Lubchenco (1998); Urai and Kelly (2023).

[5] For a discussion on the modernity of the Humboldtian concept of science—at once transdisciplinary, intercultural and globalized—see Ette (2012), p. 215 and Ette (2014), pp. 122–123

[6] Hutchinson (1983).

[7] In his seminal book *A Sand County Almanac* (1949), early American environmentalist Aldo Leopold (1887–1948), was one of the first to stress the importance of considering both the intellectual and emotional processes in the preservation of the Earth. "A thing is right when it tends to preserve the integrity, stability, and beauty of the biotic community. It is wrong when it tends otherwise," (Leopold 1949, pp. 224–225).

[8] The term 'communities' was used in this sense by Leopold (1949, p. viii): "When we see land as a community to which we belong, we may begin to use it with love and respect."

To create such communities, we first need to relearn to see the non-humans that surround us, to know their names[9] and habits, to appreciate their beauty and mysteries. Respecting nature and feeling wonder at its creative power can help to cultivate a more sustainable way of life. To imagine a more desirable future and innovate the change needed to get there, we also need communities that nurture the imagination: a quality that should be as prized as intellect. We need observers, thinkers and practitioners that, like Humboldt,[10] use the authority and credibility of science in tandem with aesthetic appreciation and emotional appeal. Science advances through logic, but also through surprises: conceptual leaps, unplanned detours and unexpected encounters.

These new communities need to recognize the right of different forms of knowledge to coexist. They can be fused to tell a richer story: data and poetry, art and experiments, cutting-edge technology and traditional knowledge, myth and math are all legitimate windows on the world. Diversifying how we represent nature can reflect its complexity and help design solutions to wicked problems. Science can describe the characteristics of plants, animals and their habitats, but the historical, emotional and aesthetic qualities that make these unique are more likely to make people feel attached to them, to care for them, and to transmit these values to future generations. A holistic approach is our only hope of responding to the anticipated transformation, unforeseen risks and increasing uncertainty of the future we face. For artists, knowledge transforms sensitivity; for scholars, sensitivity allows better understanding.[11] By contextualizing human experience as part of nature, a 'university for a sustainable world' would integrate the reflexivity of the humanities, the pragmatism of science and the sensitivity of art: Humboldt would find his place there.[12]

* * *

At the close of this book, I realize that the process of writing it has worked its own transformation on me. Where once I stood contemplating the majestic Cotopaxi from the slope of Antisana and marveling at this perfect volcano, today I view not just a landscape, but all the interrelated strands—scientific, cultural and sensory—that weave the natural tapestry of the place. Boundaries are blurred; between me, the ground, the air, the water (Fig. 6.2). All is continuous with the vast mountain and its innumerable dwellers, its creatures, plants and people. I am connected to the whole. Today, I perceive the invisible features of the landscape, what is unseen, the "*more than human*:"[13] the ice dragons, the thermal patterns and boundary layers, the

---

[9]In the words of Thoreau: "With the knowledge of the name comes a distinctive knowledge of the thing. That shore is now describable, and poetic even." In Walls (1995), p. 147. See also Bate (2000) about the importance of naming places, objects, beings, etc. to "reawaken [their] pre-scientific magic" (pp. 168–175).

[10]See Dettelbach (2005), p. 58.

[11]See Zhong Mengual (2021).

[12]For a summary on how artists engage in sustainability, see Blanc and Benish (2016).

[13]Abram (2012).

Fig. 6.2 Immersed in the mountain. Silhouetted against the sky, I stand in the shadows of the páramo facing Mt Cotopaxi, Ecuador, at one with the mountain, in all its physical, biological, aesthetic and spiritual dimensions. Photo by Dean Jacobsen

meltwater stream cycles, the fish spewed from volcanoes, the diamond-headed pins of oxygen, the spirits of *ukuku*, the narratives of La Condamine and the ghost of Humboldt. I feel an abstraction of space and time: Everest superimposed on Chimborazo, the contrasting colors of Ithaca's McLean bogs and the Altiplano's bofedales; the past in the giant *Azorellas*, the future climate predicted by foxes' yelping. Today, my left and right brain are in greater synchrony, allowing a deeper, more holistic understanding of the place.

My intention with this book was to take Humboldt's approach to present a collection of stories that enhance our understanding and perception of climate change on mountains. The way the stories link humans and non-humans, epochs and places, imaginaries and facts is often experiential rather than descriptive. Imagination may help us to face a future without cushion plants, Andean bears or glaciers. While I am a scientist, I believe in the importance of weaving knowledge into a compelling narrative that transcends pure reason and instills wonder, which has the power to change behavior.[14] Past naturalists have much to offer, and their subterranean networks remain to be discovered by other authors. As for Humboldt, the best tribute is to link his work to contemporary ideas, put his approach into practice and make his exceptionally prescient perspective resonate with our world today. It is one of the most exciting times to be a naturalist in Humboldt's sense—to forge a different way of producing knowledge that helps us reconnect to nature and solve complex problems. This requires placing natural history within a broader context: intellectual, moral, cultural, artistic and emotional. Our future depends on it.

---

[14] See Brodie et al. (2016), p. 11.

Correction to: Telling Stories

Correction to:
Chapter 5 in: O. Dangles, *Climate Change on Mountains*,
https://doi.org/10.1007/978-3-031-39528-4_5

This book was inadvertently published with incorrect sentence on page 194. The last sentence has now been corrected as "Finally, there was exciting action with myriad twists and turns. After arriving at the Pacific port of Manta on March 1736, the never-resting Academicians climbed up and down mountains to determine the length of a degree".

The updated online version of this chapter can be found at
https://doi.org/10.1007/978-3-031-39528-4_5

Epilogue

It came and touched the ear like a cold finger. It was a long, muffled breath, a throaty purr, a deep noise, a long, monotonous song from an open mouth. It filled all the tree-clad hills with its presence. It was in the sky and on the earth like rain; it came from all sides at once and it surged up slowly like a heavy wave, rumbling in the narrow corridors of the dales.

Jean Giono[1]

Spring 2023. Just as I was finishing this book, I received an unexpected email from the composer and music artist Ricky Chaggar. He said that he had been inspired by a scientific paper I co-authored on Andean wetlands and had composed a piece of music based on it. His piece had just been selected to be performed at the Cambridge Science Festival in Britain. When I listened to it, I found that his artistic project was the result of an attentive reading of our study and deep thinking about how to translate the data and findings into music.

The composition succeeds in unifying the physical, ecological, cultural and mystical worlds of the Andes: the sound of a celesta suggests the icy environment, extreme climate events are conveyed with a rumbling bass drum and clashing cymbals, flutes evoke gentle bird calls, the energy of lifeforms is expressed through the pacing and interlinked melodies, the rhythm conjures Bolivian folk music. The piece is in four movements—landscape, ecosystem, impact and damage—progressing from the initial tranquility of the wetlands to the upheaval of climate change (Fig. 1). In the beginning, you hear a recurring melody, like the heartbeat of the *Pachamama*, but by the last movement, some of the musical notes are missing, like dying patches of wetlands after glacier retreat. Ricky had written the music of mountains in a warming world, a different form of narrative for the theme of this book. This was a new way to represent climate change, an inspiring and hopeful invitation to expand the definition of science, the humanities and the arts.

---

[1] Giono (1934), pp. 15–16.

Fig. 1 The music of climate change. Extract from the third movement of 'Andean wetlands: a signal of climate change' composed by Ricky Chaggar. Here, the flowing rhythm and melody of the wetlands is suddenly interrupted by an extreme weather event. In Ricky's words, "The music becomes busy and more complex, to communicate the chaos that extreme weather brings and the panic for survival. The wildlife tries to return after the storm and we hear the birds once more. However, this time there is a feeling of sadness from the negative experiences suffered from climate change. This is communicated in the change of musical harmony." https://www.scimusic.org/

Glossary

Achupalla Local name of a South American plant of the Bromeliaceae family. It has spiny leaves and its stem is a common food resource for the Andean bear.

Aesthetics A set of principles concerned with the nature and appreciation of beauty.

Aguacatillo Local name of a native tree species (*Nectandra acutifolia*) found between 100 and 2300 m a.s.l. in the tropical Andes. Andean bears commonly climb them to feed on the fruit.

Altiplano High-altitude plateau at about 3700 m a.s.l., stretching almost 1000 km from southeastern Peru to the southeastern tip of Bolivia and fringing Chile to the northwest.

Aymara Indigenous people in the Andes and Altiplano of present-day Peru, Bolivia, Chile and Argentina. Their cosmovision is an interrelated whole, guided by the moral logic of *Suma Qamaña* or Living Well Together, in which natural, economic and social worlds are in harmony.

Bofedal High-altitude wetland in the arid grasslands of the central Andes. They are dominated by peat-forming plants of the Juncaceae family and provide crucial services to people such as water regulation, forage provision and carbon storage.

Boundary layer A layer of fluid formed in the immediate vicinity of a surface and in which fluid velocity is modified. In most cases, the boundary layer is very thin compared to the rest of the fluid, but it can have a significant effect on tiny organisms living at the surface of a river substrate.

Chronosequence A series of sites that differ in age but otherwise occur in similar environmental conditions. Chronosequences are used to study the changes in natural communities (plant, animals, microbes) during successive stages: for example, in recently deglaciated areas.

Chuquiragua A species of flowering plant (*Chuquiraga jussieui*, Asteraceae) growing above 2500 m in the tropical Andes, used by local people for its antioxidant and antiinflammatory properties. Hummingbirds (e.g. *Oreotrochilus Chimborazo*) sip nectar from its bright yellow-orange flowers.

© The Author(s), under exclusive license to Springer Nature Switzerland AG 2023
O. Dangles, *Climate Change on Mountains*,
https://doi.org/10.1007/978-3-031-39528-4

Cuchilla In Spanish, the term *cuchilla*, or 'knife edge,' is used to describe a narrow ridge with steep cliffs on either side.

Diatom A single-celled alga with cell walls composed of transparent silica, mainly found in aquatic environments. They are widely used to monitor past and present changes in environmental conditions.

Diptera Any member of an order of insects containing two-winged or so-called true flies. It includes many common insects such as mosquitoes, midges and the house fly.

Divergence In evolutionary biology, divergence refers to the discrepancy in biological features (genes, organs, body structure, physiological processes, behavior) between different species or populations of the same species. The longer populations or species are separated, the greater the degree of divergence.

Drumlin Elongated, egg-shape hill formed by glacial ice acting on underlying unconsolidated ground.

Endemism In biology, endemism refers to the distribution of an organism limited to a particular geographic area (e.g. island, mountain, country).

Esker A long ridge of gravel and other sediment, typically having a winding course, deposited by meltwater from a retreating glacier or ice sheet.

Facilitation An interaction in which the presence of one species modifies the environment in a way that benefits a second, neighboring species.

Gaucho Nomadic horseman and cowhand of the Argentine and Uruguayan Pampas (grasslands). They are folk figures similar to cowboys in western North America.

Glen A long, deep valley formed by a glacier or river, with steep or vertical sides.

Guácharo Local name of the oilbird (*Steatornis caripensis*), a bird species found in Guyana, Trinidad, Venezuela and along the tropical Andes. In old Castilian, *guácharo* means 'one who wails or laments.'

Harmonics In physics, a wave with a frequency that is a fraction of another reference wave.

Hydrograph A graph showing the variation in water flow over time at a specific point in a river.

Hypoxia Low or depleted oxygen in a water body.

Island syndrome This concept explains how organisms on islands differ from their continental counterparts in an array of morphological, behavioral and ecological features (e.g. reduced wings, gigantism, loss of defensive behavior).

Kaira Local Aymara name for the Titicaca frog (*Telmatobius culeus*).

Kame Highly stratified irregularly shaped mound of debris deposited after glacier melting.

Kettle Deep depression left behind by a buried block of ice.

Keystone species An organism that has a disproportionately large impact on its environment relative to its abundance.

La Católica Pontifical Catholic University of Ecuador (PUCE) founded in 1946 in Quito.

Lifer Birder vocabulary for a first sighting of a bird species.

Limnology The study of biological, chemical and physical features of inland aquatic systems.

Lineage In evolutionary biology, a temporal series of populations, organisms, cells or genes connected by a continuous line of descent from ancestor to descendant.

Microclimate Climatic conditions (temperature, precipitation) in a small or restricted area, especially when this differs from the climate of the surrounding area.

Moraine Accumulation of unconsolidated debris that forms on the surface, sides and front of a glacier.

Nevado In some Latin American countries, a mountain with a permanent cover of snow and ice.

Niche In ecology, the range of resources and conditions allowing a species to maintain a viable population.

Osmosis Physical process in which molecules of a solvent, such as water, migrate across a barrier from the side containing a lower concentration of a particular solute (such as sodium) to one containing a higher concentration of it.

Pachamama Quechua word meaning 'Earth mother' and relating to both a goddess and the cosmovision of Indigenous Andeans. The *Pachamama* designates both the Earth as a living planet of interrelated beings and the deity that governs it. The equal rights of *Pachamama* (or nature) are enshrined in the constitution of Ecuador and Bolivia.

Páramo In the tropical Andes, high mountain ecosystem above the timberline and below the permanent snowline.

Penitentes Hardened snow/ice formations that take the form of closely spaced elongated, thin blades pointing towards the general direction of the sun.

Phenology In ecology, study of the timing of recurrent natural events (e.g. plant flowering, bird migration) in relation to seasonal climatic changes.

Preñadilla Local name of catfish of the genus *Astroblepus* that are able to climb short distances. These are restricted to streams in the tropical Andean mountains. Many species have restricted distributions at elevations above 1000 m, so endemism is common.

Puna A biome of montane grasslands and shrublands at about 3500–4800 m that stretches southward from central Peru through western Bolivia into northern Chile and Argentina.

Quadrant An instrument used for taking angular measurements of altitude in astronomy and navigation, typically consisting of a graduated quarter circle and a sighting mechanism.

Quadrat A square or rectangular plot of soil/sediment used to mark off a physical area to isolate a sample.

Quechua Indigenous people of South American highlands who speak the Quechua language, which originated among the Indigenous people of Peru. Although most Quechua speakers are native to Peru, there are some significant populations in Bolivia, Ecuador, Chile, Colombia and Argentina.

Queñua Trees of the genus *Polylepis* (family Rosacea) endemic to the mid- and high-elevation regions of the tropical Andes. The name *Polylepis* ('many scales' in Greek) refers to the characteristic shredding of the multi-layer bark that protects these trees against low temperatures.

Radiation (1) In evolutionary biology, a process in which organisms diversify rapidly from an ancestral species into a multitude of new forms.

Radiation (2) In physics, energy coming from a source in the form of invisible particles or waves (e.g. heat, light, and X-rays).

Radiocarbon dating Method that provides age estimates for carbon-based materials that originate from living organisms by measuring carbon-14. When plants or animals die, they stop absorbing carbon, but the carbon that they have accumulated continues to decay. The amount of carbon-14 present in the sample indicates the age.

Respounchous Local name in the dialect of Aveyron, France, for black bryony (*Dioscorea communis*), a climbing herbaceous plant that looks like a tropical vine and is common in hedges. The plant is poisonous except for the young shoots, which resemble wild asparagus and are commonly eaten in southern France.

Rotifer A microscopic, many-celled freshwater aquatic invertebrate that has a ring of cilia resembling a wheel.

Sierra Name of the highlands region of Ecuador that extends from north to south through the Andes.

Soroche Quechua word for altitude sickness, which is caused by rapid exposure to the low concentration of oxygen at high altitude. This can affect those sensitive to it from altitudes 2000 m and up. Symptoms include headache, vomiting, tiredness, confusion, trouble sleeping and dizziness.

Speciation The formation of new and distinct species in the course of evolution.

Species turnover The magnitude of change in species composition along predefined gradients (e.g. island size, distance to a glacier margin).

Stochasticity In environmental science, stochasticity refers to unpredictable fluctuations in space and time in environmental conditions. These random events can affect the dynamics of natural populations (e.g. birth, mortality) and communities.

Taita An informal term for 'father' in Latin American countries. In Andean mythology, several mountains/volcanoes are considered as divine fathers and are called *Taita* (e.g. Taita Chimborazo, Taita Imbabura). Other mountains are considered female (e.g. Mama Cotacachi).

Transect In ecology, a standardized path along which an observer counts and records occurrences of the objects of study (plants, birds).

Ucumari Local name given to the Andean bear in several high-altitude regions of the tropical Andes.

Ukuku Young men dressed as bears during ritual processions to glaciers in the Andes. Ukukus are considered mediators between the pilgrims and the mountain spirits. They climb up to the glaciers to retrieve the sacred ice that will bring health and fertility to their communities.

Yareta Local name of the cushion plant *Azorella compacta* found above 3200 m in Peru, Bolivia, northern Chile and western Argentina.

References

Abram D (2012) The spell of the sensuous: perception and language in a more-than-human world. Vintage, New York

Agassiz L (1840) Etudes sur les glaciers. Aux frais de l'auteur, Neuchâtel

Agassiz L (1842) La théorie des glaces et ses progrès les plus récents. Tiré de la bibliothèque universelle de Genève, Genève

Agassiz L (1847) Nouvelles études et expériences sur les glaciers actuels: leur structure, leur progression et leurs actions physiques sur le sol. Masson, Paris

Agassiz L (1869) Address delivered on the centennial anniversary of the birth of Alexander von Humboldt, with an account of the evening reception. Boston Society of Natural History, Boston

Agassiz AE (1875) Hydrographic sketch of lake Titicaca. Proc Am Acad Arts Sci 11:283–292

Agassiz EC (ed) (1885) Louis Agassiz: his life and correspondence. Houghton Mifflin, Boston

Agassiz GR (ed) (1913) Letters and recollections of Alexander Agassiz with a sketch of his life and work. Houghton Mifflin, Boston

Agassiz AE, Garman SW (1875) Exploration of Lake Titicaca. I. Fishes and reptiles. Bull Mus Comp Zool 3:273–278

Agassiz L, Vogt C (1839) Histoire naturelle des poissons d'eau douce de l'Europe central. De Petitpierre, Neuchâtel

Alda A (2018) If i understood you, would I have this look on my face? My adventures in the art and science of relating and communicating. Random House, New York

Anderson K (2006) Does history count? Endeavour 30:150–155

Angilletta MJ Jr (2009) Thermal adaptation: a theoretical and empirical synthesis. Oxford University Press, Oxford

Anthelme F, Buendia B, Mazoyer C et al (2012) Unexpected mechanisms sustain the stress gradient hypothesis in a tropical alpine environment. J Veg Sci 23:62–72

Anthelme F, Cavieres LA, Dangles O (2014a) Facilitation among plants in alpine environments in the face of climate change. Front Plant Sci 5:387

Anthelme F, Jacobsen D, Macek P et al (2014b) Biodiversity patterns and continental insularity in the high tropical Andes. Arct Antarct Alp Res 46:811–828

Anthelme F, Meneses RI, Valero NNH et al (2017) Fine nurse variations explain discrepancies in the stress-interaction relationship in alpine regions. Oikos 126:1173–1183

Anthony P (2018) Mining as the working world of Alexander von Humboldt's plant geography and vertical cartography. Isis 109:28–55

Appel TA (1997) L'Anatomie philosophique, l'évolution et les muséums: les relations entre le Muséum et Harvard. In: Blanckaert C, Cohen C, Corsi P et al (eds) Le Muséum au premier siècle de son histoire. Muséum national d'Histoire naturelle, Paris, pp 649–671

Armstrong RA (2004) Lichens, lichenometry and global warming. Microbiologist 5:e35

Arrhenius S (1896) XXXI. On the influence of carbonic acid in the air upon the temperature of the ground. Lond Edinb Dublin Philos Mag J Sci 41:237–276

Arrhenius O (1921) Species and area. J Ecol 9:95–99

Auer MR (2019) Environmental aesthetics in the age of climate change. Sustainability 11:5001

Avery KJ (1993) Church's great picture, the heart of the Andes. The Metropolitan Museum of Art, New York

Bailly JC, Besse JM, Palsky G (2014) Le monde sur une feuille: les tableaux comparatifs de montagnes et de fleuves dans les atlas du XIXe siècle. Fage Editions, Lyon

Baldwin IT (2004) A most productive passion for natural history. Science 303:958–959

Barrett PH, Freeman RB (1992) The works of Charles Darwin: v. 1: Introduction; diary of the voyage of HMS Beagle. Routledge, London

Bate J (2000) The song of the earth. Picador, London

Bell S (2010) A life in shadow: Aimé Bonpland in Southern South America, 1817–1858. Stanford University Press, Stanford

Berghaus HKW (1852) Physikalischer Atlas, oder Sammlung von Karten, auf denen die Hauptsächlichsten Erscheinungen der anorganischen und organischen Natur nach ihrer geographischen Verbreitung und Vertheilung bildlich dargestellt sind. Justus Perthes, Gotha

Berlioz J (1933) D'Orbigny, ornithologiste. In: Publications du Muséum d'Histoire Naturelle, 3, Commémoration du voyage d'Alcide d'Orbigny en Amérique du Sud 1826–1833, Masson et Cie, Paris, pp 67–74

Beutel RG, Friedrich F, Leschen RA (2009) Charles Darwin, beetles and phylogenetics. Naturwissenschaften 96:1293–1312

Bishop M (1962) A history of Cornell. Cornell University Press, Ithaca

Blackwell SH, Johnson K (eds) (2016) Fine lines: Vladimir Nabokov's scientific art. Yale University Press, New Haven

Blanc N, Benish BL (2016) Form, art and the environment: engaging in sustainability. Taylor & Francis, London

Blankenstein D (2014) À qui appartient le cosmos? In: Savoy B, Blankenstein D (eds) Les frères Humboldt, l'Europe de l'esprit. PSL Research University, Paris, pp 165–175

Boelhower W (2013) Views of the cordilleras and monuments of the indigenous peoples of the Americas: a critical edition; and Alexander von Humboldt and the Americas. AAG Rev Books 1:189–193

Bonpland A (1773–1858) Manuscrits d'Aimé Goujaud, dit Bonpland, Cote: Ms 53–54, Paper. 122 et 121 feuillets. 320 × 220 mm. Reliure en peau verte. Classement: Coupé en trois morceaux, le journal botanique du voyage de Humboldt et Bonpland est à reconstituer comme suit: N° 1–690 : Ms 1332, N° 691–1215 : Ms 1333, N° 1216–1591 : Ms 1334, N° 1592–2257 : Ms 2534, N° 2258–3698 : Ms 53, N° 3699–4528 : Ms 54

Borchart de Moreno C, Moreno Yánez SE (2012) From País to Nation: Alexander von Humboldt and the Formation of Ecuadorian Identity. In: Kutzinski VM, Ette O, Walls LD (eds) Alexander von Humboldt and the Americas. Verlag Walter Frey, Berlin, pp 117–143

Born D (2019) Bearing witness? Polar bears as icons for climate change communication in National Geographic. Environ Commun 13:649–663

Bouguer P (1748) Entretiens sur la cause de l'inclinaison des orbites des planètes, Seconde édition. Ch. A. Jombert, Paris

Bourcier D, van Andel P (2011) La sérendipité. Le hasard heureux, Hermann, Paris

Boussingault J-B (1892–1903) Mémoires de J.-B. Boussingault. 5 tomes, Chamerot et Renouard, Paris

Brady E (2014) Aesthetic value, ethics and climate change. Environ Values 23:551–570

Breyer J, Butcher W (2003) Nothing new under the earth: the geology of Jules Verne's journey to the centre of the earth. Earth Sci Hist 22:36–54

Briffaud S (2006) Le temps du paysage. Alexandre de Humboldt et la géohistoire du sentiment de la nature. In: Blais H, Laboulais I (eds) Géographies plurielles. Les sciences géographiques au moment de l'émergence des sciences humaines (1750–1850). L'Harmattan, Paris, pp 275–301

Britski HA, Figueiredo JL (2019) Peixes do Brasil: aquarelas de Jacques Burkhardt (1865–1866). Edusp, São Paulo

Brodie N, Goodrich C, Swanson FJ (eds) (2016) Forest under story: creative inquiry in an old-growth forest. University of Washington Press, Seattle

Brunet F (2004) Samuel Morse, "père de la photographie américaine". Études photographiques 15: 4–30

Brush A (2017) François Vuilleumier, 1938–2017. Auk 134:776–777

Buffon GLL (1856–1857) Oeuvres Complètes de Buffon—avec des extraits de Daubenton et la classification de Cuvier. A. Delahays, Paris

Burkhardt RW Jr (2001) The leopard in the garden: life in close quarters at the Muséum d'Histoire Naturelle. Isis 98:675–694

Cajori F (1929) History of determinations of the heights of mountains. Isis 12:482–514

Cameron D (1965) Early discoverers XXII: Goethe—discoverer of the Ice Age. J Glaciol 5:751–754

Cañizares-Esguerra J (2006) Nature, empire, and nation: explorations of the history of science in the Iberian world. Stanford University Press, Stanford

Capello E (2018) From imperial pyramids to anticolonial sundials: commemorating and contesting French geodesy in Ecuador. J Hist Geogr 62:37–50

Cárdenas T, Naoki K, Landivar CM et al (2022) Glacier influence on bird assemblages in habitat islands of the high Bolivian Andes. Div Dist 28:242–256

Carlowicz M, Friedl L, Ward K (2018) Earth. NASA, Washington, DC

Carpenter E (1922) Towards democracy. Mitchell Kernerley, New York

Catania KC (2016) Leaping eels electrify threats, supporting Humboldt's account of a battle with horses. Proc Natl Acad Sci USA 113:6979–6984

Cauvy-Fraunié S, Dangles O (2019) A global synthesis of biodiversity responses to glacier retreat. Nat Ecol Evol 3:1675–1685

Cauvy-Fraunié S, Espinosa R, Andino P et al (2014a) Relationships between stream macroinvertebrate communities and new flood-based indices of glacial influence. Freshwat Biol 59:1916–1925

Cauvy-Fraunié S, Andino P, Espinosa R et al (2014b) Glacial flood pulse effects on benthic fauna in equatorial high-Andean streams. Hydrol Proc 28:3008–3017

Cauvy-Fraunié S, Andino P, Espinosa R et al (2015) Temporal scaling of high flow effects on benthic fauna: insights from equatorial glacier-fed streams. Limnol Oceanogr 60:1836–1847

Cauvy-Fraunié S, Andino P, Espinosa R et al (2016) Ecological responses to experimental glacier-runoff reduction in alpine rivers. Nat Comm 7:1–7

Chuine I, Yiou P, Viovy N et al (2004) Grape ripening as a past climate indicator. Nature 432:289–290

Clark R, Lubrich O (eds) (2012) Transatlantic echoes: Alexander von Humboldt in world literature. Berghahn, Oxford

Clavero M, Nores C, Kubersky Piredda S et al (2016) Interdisciplinarity to reconstruct historical introductions: solving the status of cryptogenic crayfish. Biol Rev 91:1036–1049

Coloma L, Duellman W, Almendáriz A et al (2010) Five new (extinct?) species of atelopus (Anura: Bufonidae) from Andean Colombia, Ecuador, and Peru. Zootaxa 2574:1–54

Comstock JH, Comstock AB (1895) A manual for the study of insects. Comstock, Ithaca

Connell JH (1978) Diversity in tropical rain forests and coral reefs. Science 199:1302–1310

Cooper SF (1850) Rural hours. George P Putnam, New York

Cooper SF (2002) Essays on nature and landscape. Johnson R, Patterson D (eds) University of Georgia Press, Athens

Cornell University Proceedings (1885) In memory of Louis Agassiz. Published by order of the Trustees, Ithaca

Cousteau JY, Diolé P (1977) Three adventures: Galapagos, Titicaca, the Blue Holes. Littlehampton Book Services, Worthing

Crespo-Pérez V, Andino P, Espinosa R et al (2016) The altitudinal limit of Leptohyphes Eaton, 1882 and Lachlania Hagen, 1868 (Ephemeroptera: Leptohyphidae, Oligoneuriidae) in Ecuadorian Andes streams: searching for mechanisms. Aquat Insect 37:69–86

Crosland M (1967) The Society of Arcueil—a view of French Science at the time of Napoleon I. Harvard University Press, Cambridge

Dangles O, Mena B (2017) Oso + Páramo. Cumbia, Quito

Dangles O, Rabatel A, Kraemer M et al (2017) Ecosystem sentinels for climate change? Evidence of wetland cover changes over the last 30 years in the tropical Andes. PLoS One 12:e0175814

Darlington PJ (1943) Carabidae of mountains and islands: data on the evolution of isolated faunas, and on atrophy of wings. Ecol Monogr 13:37–61

Darwin C (1839) Journal and remarks (1832–1836), voyage of the adventure and Beagle, vol III. Colburn, London

Darwin C (1840) Journal of researches into the natural history and geology of the countries visited during the voyage of H.M.S. 'Beagle' round the world. H. Colburn, London

Darwin C (1866) On the origin of species by means of natural selection, or the preservation of favoured races in the struggle for life, 4th edn. John Murray, London

Davis SN (1999) Humboldt, Arago, and the temperature of groundwater. Hydrogeol J 7:501–503

De Baldarrago FC, Poma I, Spadaro V (2012) Evaluación etnobotanica de la Yareta (Azorella compacta) en Arequipa (Perú) y sus posibles aplicaciones. Quad Bot Amb Appl 23:15–30

De Botton A (2004) The art of travel. Vintage, New York

Desor E (1844) Excursions et séjours dans les glaciers et les hautes régions des Alpes de M. Agassiz et de ses compagnons de voyage. J.-J. Kissling, Neuchatel

Dettelbach M (1996) Global physics and aesthetic empire. In: Miller D, Reill P (eds) Visions of empire. University of California Press, Los Angeles, pp 258–292

Dettelbach M (2005) The stimulations of travel: Humboldt's physiological construction of the tropics. In: Driver F, Martins L, de Lima Martins L (eds) Tropical visions in an age of empire. University of Chicago Press, Chicago, pp 43–58

De Witt S (1802) Respecting a plan of a meteorological chart, for exhibiting a comparative view of the climates of North America, and the progress of vegetation. Trans Soc Promot Agric Arts Manuf Inst State N Y 1:88–92

Diamond J (2005) Collapse. How societies chose to fail or succeed. Penguin, New York

Dickinson EC (2019) Conflicting options for the first available use of the name Rhea pennata d'Orbigny and the date to be used. Zool Bibliog 5:403–404

Diderot D, d'Alembert J (1751–1780) Encyclopédie, ou Dictionnaire raisonné des sciences, des arts et des métiers. Recueil de planches sur les sciences et les arts. Le Breton, Briasson, David et Durand, Paris

Dollfus-Ausset D (1863–1870) Matériaux pour l'étude des glaciers. F. Savy, Paris

Dorst J (1957) La vie sur les hauts plateaux andins du Pérou. Revue d'Ecologie, Terre et Vie 11:3–50

Drevet P (1997) Le Corps du Monde. Le Seuil, Paris

Drouin J-M, Bensaude-vincent B (1996) Nature for the people. In: Jardine N, Secord JA, Spary EC (eds) Cultures of natural history. Cambridge University Press, Cambridge, pp 408–425

Dry S (2019) Waters of the World. The story of the scientists who unravelled the mysteries of our seas, glaciers, and atmosphere—and made the planet whole. Scribe, Melbourne

Duchicela SA, Cuesta F, Tovar C et al (2021) Microclimatic warming leads to a decrease in species and growth form diversity: insights from a tropical alpine grassland. Front Ecol Evol 9:673655

Dupuys L (2000) Espace et temps dans l'oeuvre de Jules Verne. La Clef d'Argent, Dole

Duviols J-P, Minguet C (1994) Humboldt savant-citoyen du monde. Gallimard, Paris

Dyson F (2009) Birds and frogs. Notices Am Math Soc 56:212–223

Ecklund EH, James SA, Lincoln AE (2012) How academic biologists and physicists view science outreach. PLoS One 7:e36240

Eichler A (1955) Nieve y selva en Ecuador. B. Moritz, Quito

Eisner T (1982) For love of nature: exploration and discovery at biological field stations. BioScience 32:321–326

Eisner T (2003) For love of insects. Harvard University Press, Cambridge

Engel S (2015) The hungry mind: The origins of curiosity in childhood. Harvard University Press, Cambridge

Espinosa R, Andino P, Cauvy-Fraunié S et al (2020) Diversity patterns of aquatic macroinvertebrates in a tropical high-Andean catchment. Rev Biol Trop 68:29–53

Estes JA (2016) Serendipity: an ecologist's quest to understand nature. University of California Press, Oakland

Ette O (2012) TransTropics: Alexander von Humboldt and hemispheric constructions. In: Kutzinski VM, Ette O, Walls LD (eds) Alexander von Humboldt and the Americas. Verlag Walter Frey, Berlin, pp 209–236

Ette O (2014) La pensée Nomade. Alexander von Humboldt ou la Science vivante. In: Savoy B, Blankenstein D (eds) Les frères Humboldt, l'Europe de l'esprit. PSL Research University, Paris, pp 117–130

Ette O, Maier J (2018) Alexander von Humboldt. The complete drawings from the American travel diaries. Prestel, München

Fabre J-H (1925) Souvenirs entomologiques, Études sur l'instinct et les moeurs des insectes. Librairie Delagrave, Paris

Ferreiro LD (2011) Measure of the Earth: the enlightenment expedition that reshaped our world. Basic, New York

Flannery T (2019) Europe: a natural history. Atlantic Monthly Press, New York

Flantua SG, O'dea A, Onstein RE et al (2019) The flickering connectivity system of the north Andean páramos. J Biogeogr 46:1808–1825

Flecker AS (1992) Fish predation and the evolution of invertebrate drift periodicity: evidence from neotropical streams. Ecology 73:438–448

Fleming F (2000) Killing dragons: the conquest of the Alps. Atlantic Monthly Press, New York

Foote JH (2008) Touchdown: the story of the Cornell Bear. Cornell, Ithaca

Forster G (1777) A voyage round the world, in His Britannic Majesty's sloop, resolution, commanded by Capt. James Cook, during the years 1772, 3, 4, and 5. B White, London

Forster JR (1778) Observations made during a voyage round the world, on physical geography, natural history and ethic philosophy. G Robinson, London

Fourier J (1822) Théorie analytique de la chaleur. Firmin Didot, Paris

Fourier J (1827) Mémoire sur la température du globe terrestre et des espaces planétaires. Mémoires de l'Académie royale des sciences de l'Institut de France 7:569–604

Francou B (2013) La première mission géodésique française au Pérou et la détermination de la forme de la Terre (1735–1744). In: Espinosa C, Lomné G (coord) Ecuador y Francia, Diálogos Científicos y Políticos (1735–2013). IFEA-FLACSO, Quito, pp 23–35

Freeman BG, Scholer MN, Ruiz-Gutierrez V et al (2018) Climate change causes upslope shifts and mountaintop extirpations in a tropical bird community. Proc Natl Acad Sci USA 115:11982–11198

Frith CB (2016) Charles Darwin's life with birds: his complete ornithology. Oxford University Press, Oxford

Frochot A (2020) Science, épreuve et passion des glaces chez Daniel Dollfus-Ausset (1797–1870). Romantisme 189:41–51

Frutiger A (2002) The function of the suckers of larval net-winged midges (Diptera: Blephariceridae). Freshwat Biol 47:293–302

Fu C, Wu J, Wang X et al (2004) Patterns of diversity, altitude range and body size among freshwater fishes in the Yangtze River basin, China. Global Ecol Biogeogr 13:543–552

Gade DW (1999) Nature and culture in the Andes. University of Wisconsin Press, Madison

Gade DW (2016) Spell of the Urubamba: anthropogeographical essays on an Andean valley in space and time. Springer, Cham

Geoffroy Saint-Hilaire G, Cuvier F (1824–1842) Histoire naturelle des mammifères avec des figures originales, coloriées, dessinées d'après des animaux vivants. A. Belin/Blaise, Paris

Gervais P (1855) Histoire naturelle des mammifères: avec l'indication de leurs moeurs, et de leurs rapports avec les arts, le commerce et l'agriculture, vol 2. L. Curmer, Paris

Gèze B (1986) La géologie dans les romans de Jules Verne. Travaux du Comité français d'Histoire de la Géologie 2:71–84

Gilles JL, García M, Yucra ES et al (2022) Validating local meteorological forecast knowledge in the Bolivian Altiplano: moving toward the co-production of agricultural forecasts. Clim Dev 1–12

Giono J (1934) Le chant du monde. Gallimard, Paris

Gleason HA (1922) On the relation between species and area. Ecology 3:158–162

Gos C (1928) L'hôtel des Neuchâtelois, un épisode de la conquête des Alpes. Payot, Paris

Gosnell M (2005) Ice: the nature, the history and the uses of an astonishing substance. A. Knof, New York

Gould SJ (1989) Wonderful life: the Burgess Shale and the nature of history. WW Norton, New York

Gould SJ (2000) Church, Humboldt, and Darwin: the tension and harmony of art and science. Frederic Edwin Church. In: Beezley WH, Curcio-Nagy LA (eds) Latin American popular culture: an introduction. Scholarly Resources, Wilmington, pp 27–42

Gould SJ (2011) The hedgehog, the fox, and the magister's pox: mending the gap between science and the humanities. Belknap Press of Harvard University Press, Cambridge

Grattan-Guinness I (1972) Joseph Fourier, 1768–1830: a survey of his life and work. MIT Press, Cambridge

Hairston NG Jr, Likens GE (2009) The legacy of James G. Needham: a century of limnology at Cornell University and the first course on Limnology in the Americas. Limnol Oceanogr Bull 18:30–32

Halloy SR (2002) Variations in community structure and growth rates of high-Andean plants with climatic fluctuations. In: Körner C, Spehn EM (eds) Mountain biodiversity: a global assessment. Parthenon, Nashville, pp 225–237

Hamerlik L, da Silva FL, Jacobsen D (2018) Chironomidae (Insecta: Diptera) of Ecuadorian high altitude streams: a survey and illustrated key. Fla Entomol 101:663–675

Hamy ET (1904) Le Centenaire du retour en Europe d'Alexandre de Humboldt et d'Aimé Goujaud de Bonpland (3 août 1804); Discours prononcé à la séance d'ouverture du XIVème Congrès des Américanistes à Stuttgart, le 18 aout 1904. Burdin et Cie, Angers

Harvey EJ (2020) Alexander von Humboldt and the United States: art, nature, and culture. Smithsonian American Art Museum and Princeton University Press, Lawrenceville

Hay ME (2009) Marine chemical ecology: chemical signals and cues structure marine populations, communities, and ecosystems. Ann Rev Mar Sci 1:193–212

Hazareesingh S (2015) How the French think: an affectionate portrait of an intellectual people. Penguin, London

Heneghan L (2018) Have ecologists lost their senses? Walking and reflection as ecological method. TREE 33:475–478

Herivel J (1975) Joseph Fourier. The man and the physicist. Clarendon, Oxford

Heron-Allen E (1917) Presidential address, 1916–17: Alcide d'Orbigny, his life and his work. J R Microsc Soc 37:1–105

Horn HS (1975) Markovian properties of forest succession. In: Cody ML, Diamond JM (eds) Ecology and evolution of communities. Belknap, Cambridge, pp 196–211

Hornaday WT (1911) The spectacle bear. Bull NY Zool Soc 45:747–748

Horrell M (2019) Feet and wheels to Chimborazo. Mountain Footsteps Press

Hubbard E (1916) Little journey to the home of great scientists, memorial edition. Wise, New York

Hughes D (2022) Picturing ecology: photography and the birth of a new science. Palgrave Macmillan, Singapore

Humboldt A von (1793) Florae fribergensis specimen: plantas cryptogamicas praesertim subterraneas exhibens. Apud Henr. Augustum Rottmann, Berlin

Humboldt A von (1803) Correspondance. Extrait de plusieurs lettres de M. A. de Humboldt. Annales du Muséum National d'histoire naturelle par les professeurs de cet établissement, Tome second. Frères Levrault, Paris, pp 322–337

Humboldt A von (1808) Tableaux de la nature ou, Considérations sur les déserts, sur la physionomie des végétaux, et sur les cataractes de l'Orénoque, Traduit de l'allemand par J.B.B. Eyriès, 2 tomes. F Schoell, Paris

Humboldt A von (1808–1811) Recueil d'observations astronomiques, d'opérations trigonométriques et de mesures barométriques, faites pendant le cours d'un voyage aux regions équinoxiales du Nouveau Continent, depuis 1799 jusqu'en 1804, rédigées et calculées d'après les tables les plus exactes, par Jabbo Oltmanns; ouvrage auquel on a joint des recherches historiques sur la position de plusieurs points importants pour les navigateurs et pour les géographes 2 vol. F Schoell, Treuttel, Würtz, Paris

Humboldt A von (1810) Vues des cordillères, et monumens des peuples indigènes de l'Amérique, F Schoell, Paris

Humboldt A von (ed) (1811–1833) Recueil d'Observations de Zoologie et d'Anatomie comparée, faites dans l'Océan Atlantique, dans l'Intérieur du nouveau Continent et dans la Mer du Sud, pendant les années 1799, 1800, 1801, 1802 et 1803; par Al. de Humboldt et A. Bonpland, 2 vol. F Schoell and Dufour & Cie, J Smith and Gide, Paris

Humboldt A von (1814–1825) Relation historique du Voyage aux régions équinoxiales du Nouveau Continent fait en 1799, 1800, 1801, 1802, 1803 et 1804; par Al. de Humboldt and A. Bonpland, 3 vol. F Schoell, Maze, J Smith and Gide, Paris

Humboldt A von (1846–1858) Cosmos: sketch of a physical description of the universe; Translated under the superintendence of Sabine Edward, 5 Vol. Longman, Brown, Green & John Murray, London

Humboldt A von (1850) Views of nature, or, contemplations on the sublime phenomena of creation. Translated from German by Otté EC, Bohn HG. Henry G Bohn, London

Humboldt A von (1852–1853) Personal narrative of travels to the equinoctial regions of the New continent during the years 1799–1804 by Al. de Humboldt and A. Bonpland. Translated and edited by Thomasina Ross, 3 volumes. Henry G Bohn, London

Humboldt A von (1854) Mélanges de Géologie et de Physique Générale. Tome I. Gide et J. Baudry, Paris

Humboldt A von (1860) Lettres de Alexandre de Humboldt à Varnhagen von Ense. L Held, L Hachette & A Lacroix, Van Meenen and Cie, Genève

Humboldt A von (1865) Correspondance scientifique et littéraire. Ducrocq, Paris

Humboldt A von (1868) Lettres d'Alexandre de Humboldt à Marc-Auguste Pictet (1795–1824). Le Globe Revue genevoise de géographie 7:129–204

Humboldt A von (1905) Lettres américaines d'Alexandre de Humboldt (1798–1807) précédées d'une notice de J-C Delamétherie. et suivies d'un choix de documents en partie inédits publiées avec une introduction et des notes par le Dr E-T Hamy. E Guilmoto, Paris

Humboldt A von (2003) Reise auf dem Rio Magdalena, durch die Anden und durch Mexico, Teil I: Texte. Akademie, Berlin

Humboldt A von, Bonpland A (1808–1809) Plantes équinoxiales : recueillies au Mexique, dans l'île de Cuba, dans les provinces de Caracas, de Cumana et de Barcelone, aux Andes de la Nouvelle-Grenade, de Quito et du Pérou, et sur les bords du Rio-Negro, de l'Orénoque et de la rivière des Amazones, par Al. de Humboldt et A. Bonpland, 2 vol. F Schoell, Paris

Humboldt A von, Bonpland A (2010) Essay on the geography of plants. Edited with an introduction by Stephen T. Jackson ; translated by Sylvie Romanowski. University of Chicago Press, Chicago

Humboldt A von, Bonpland A, Kunth K (1815–1825) Nova genera et species plantarum, vols 1–7. Lutetiae Parisiorum, Paris

Hurlbert SH, Chang CC (1983) Ornitholimnology: effects of grazing by the Andean flamingo (Phoenicoparrus andinus). Proc Natl Acad Sci USA 80:4766–4769

Hutchinson GE (1918) A swimming grasshopper. Entomol Rec J Var 30:138

Hutchinson GE (1933) Limnological studies at high altitudes in Ladak. Nature 132:136–136

Hutchinson GE (1936) The clear mirror. A pattern of life in Goa and in Indian Tibet. Cambridge University Press, Cambridge

Hutchinson GE (1953) The itinerant ivory tower. Yale University Press, New Haven

Hutchinson GE (1963) The naturalist as an art critic. Proc Acad Nat Sci Phila 115:99–111

Hutchinson GE (1978) An introduction to population ecology. Yale University Press, New Haven

Hutchinson GE (1979) The kindly fruits of the Earth. Yale University Press, New Haven

Hutchinson GE (1983) Marginalia: what is science for? Am Sci 71:639–644

Hutchinson GE (1993) A treatise on limnology, vol. 4: The zoobenthos. Wiley, London

Hutchinson GE, Bowen VT (1947) A direct demonstration of the phosphorus cycle in a small lake. Proc Natl Acad Sci USA 33:148

Hutton J (1795) The theory of the earth, with proofs and illustrations. Messrs. Cadell. Jr. and Davies & William Creech, London

Ingold T (2000) The perception of the environment: essays on livelihood, dwelling and skill. Routledge, London

Ingold T (2013) Dreaming of dragons: on the imagination of real life. J R Anthropol Inst 19:734–752

Ingold T, Vergunst JL (eds) (2008) Ways of walking: ethnography and practice on foot. Ashgate, Hampshire

IPCC (2021) Summary for policymakers. In: Masson-Delmotte V, Zhai P, Pirani A et al (eds) Climate Change 2021: the physical science basis. Contribution of Working Group I to the sixth assessment report of the Intergovernmental Panel on Climate Change. Cambridge University Press, Cambridge, pp 3–32

Irmscher C (2013) Louis Agassiz: creator of American science. Houghton Mifflin Harcourt, Boston

Iturralde-Polit P, Dangles O, Burneo SF et al (2017) The effects of climate change on a mega-diverse country: predicted shifts in mammalian species richness and turnover in continental Ecuador. Biotropica 49:821–831

Jackson R (2018) The ascent of John Tyndall. Oxford University Press, Oxford

Jackson R, Jackson N, Brown D (eds) (2020) The poetry of John Tyndall. UCL Press, London

Jacobsen D (2020) The dilemma of altitudinal shifts: caught between high temperature and low oxygen. Front Ecol Environ 18:211–218

Jacobsen D, Dangles O (2017) Ecology of high altitude waters. Oxford University Press, Oxford

Jacobsen D, Milner AM, Brown LE et al (2012) Biodiversity under threat in glacier-fed river systems. Nat Clim Change 2:361–364

Jacobsen D, Laursen SK, Hamerlik L et al (2013) Sacred fish: on beliefs, fieldwork, and freshwater food webs in Tibet. Front Ecol Environ 11:50–51

Jacobsen D, Andino P, Calvez R et al (2014a) Temporal variability in discharge and benthic macroinvertebrate assemblages in a tropical glacier-fed stream. Freshwat Sci 33:32–45

Jacobsen D, Cauvy-Fraunié S, Andino P et al (2014b) Runoff and the longitudinal distribution of macroinvertebrates in a glacier-fed stream: implications for the effects of global warming. Freshwat Biol 59:2038–2050

Jeannel R (1925) L'aptérisme chez les insectes insulaires. C R Acad Sci Paris 1:1222–1224

Jevrejeva S (2001) Severity of winter seasons in the northern Baltic Sea during 1529–1990. Clim Res 17:55–62

Johnson RDO (1912) Notes on the habits of a climbing catfish (Arges marmoratus) from the Republic of Colombia. Ann NY Acad Sci 22:327–333

Jomelli V, Favier V, Rabatel A et al (2009) Fluctuations of glaciers in the tropical Andes over the last millennium and palaeoclimatic implications: a review. Palaeogeogr Palaeoclim Palaeoecol 281:269–228

Jouty S (1989) Comment les savants ont mesuré l'altitude? L'Histoire 126:38–44

Jungblut AD, Hawes I (2017) Using Captain Scott's Discovery specimens to unlock the past: has Antarctic cyanobacterial diversity changed over the last 100 years? Proc R Soc B 284:20170833

Kahneman D (2011) Thinking fast and slow. Farrar, Straus and Giroux, New York

Kelt DA, Lessa EP, Salazar-Bravo J et al (eds) (2007) The quintessential naturalist: honoring the life and legacy of Oliver P. Pearson. Univ Calif Publ Zool 134:1–981

Kessel JJMM van (1994) El zorro en la cosmovisión andina. Revista Chungara 226:33–242

Keynes RD (ed) (1988) Charles Darwin's Beagle diary. Cambridge University Press, Cambridge

Kingsland SE (2010) The beauty of the world: Evelyn Hutchinson's vision of science. In: Skelly DK, Post DM, Smith MD (eds) The art of ecology. Yale University Press, New Haven, pp 1–9

Kohlstedt SG (2005) Nature, not books: scientists and the origins of the nature-study movement in the 1890s. Isis 96:324–352

Körner C (2003) Alpine plant life: functional plant ecology of high mountain ecosystems, 2nd edn. Springer, Berlin

Kurth RT (2016) Susan Fenimore Cooper. New perspectives on her works. iUniverse, Bloomington

Lacépède E de (1804) La Ménagerie du Muséum national d'histoire naturelle, ou Description et histoire des animaux qui y vivent et qui y ont vécu, Tome I. Miger, Paris

Lack HW (2009) Alexander von Humboldt: the botanical exploration of the Americas. Prestel, New York

La Condamine C-M de (1745) Relation abrégée d'un voyage fait dans l'intérieur de l'Amérique méridionale, Depuis la côte de la mer du Sud, jusqu'aux côtes du Brésil & de la Guiane, en descendant la rivière des Amazones. Veuve Pissot, Paris

La Condamine C-M de (1751) Mesure des trois premiers degrés du méridien dans l'hémisphere austral: tirée des observations de mrs. de l'Académie royale des sciences, envoyés par le roi sous l'équateur, Académie Royale des Sciences, Paris

Lansley CM (2018) Charles Darwin's debt to the romantics: how Alexander von Humboldt, Goethe and Wordsworth helped shape Darwin's view of nature. Peter Lang, Pieterlen

Leask NJ (2003) Darwin's second sun: Alexander von Humboldt and the genesis of the voyage of the Beagle. In: Tait T, Small H (eds) Literature, science and psychoanalysis, 1830–1970: essays in honour of Dame Gillian Beer. Oxford University Press, Oxford, pp 13–36

Legré-Zaidline F (2002) Alcide Dessalines d'Orbigny (1802–1857). L'Harmattan, Paris

Leopold A (1949) A Sand County Almanac. Oxford University Press, London. 240p

Lescure J, Bour R, Ineich I, Ohler AM, Ortiz JC (2002) Liste inédite des Reptiles et Amphibiens récoltés par Alcide d'Orbigny en Amérique méridionale. Comptes Rendus Palevol 1:527–532

Levin SA, Marks PL, Paine RT (2013) Resolution of respect: Richard Bruce (Dick) Root. Bull Ecol Soc Am 94:210–215

Llambi LD, Melfo A, Gámez LE et al (2021) Primary succession in the forefield of the last Venezuelan glacier: ecosystem assembly and biotic interactions in an extreme tropical-alpine environment. Front Plant Ecol Evol 9:657755

Lovejoy TE, Bierregaard RO Jr, Rylands AB et al (1986) Edge and other effects of isolation on Amazon forest fragments. In: Soulé ME (ed) Conservation biology: the science of scarcity and diversity. Sinauer, Sunderland, pp 257–285

Loza CB (2005) François Rossignon, un naturalista francés cautivo de las aves de Caupolican (Beni y La Paz, 1833–1847). Bull Inst Fr Études Andines 34:59–80

Lubchenco J (1998) Entering the century of the environment: a new social contract for science. Science 279:491–497

Lubrich O (2019) L'oeuvre de Humboldt. In: Von Humboldt A (ed) Écrits. Classiques Garnier, Paris, pp 25–29

Lubrich O (2022) The Humboldt paradox: science, communication and mythology. In: Falk GC, Strecker MR, Schneider S (eds) Alexander von Humboldt: multiperspective approaches. Springer, Cham, pp 177–194

Mabee C (1943) The American Leonardo: a life of Samuel FB Morse. Alfred A Knopf, New York

Madriñán S, Cortés AJ, Richardson JE (2013) Páramo is the world's fastest evolving and coolest biodiversity hotspot. Front Genet 4:192

Madsen P, Andino P, Espinosa R et al (2015) Altitudinal distribution limits of aquatic macroinvertebrates: an experimental test in a tropical alpine stream. Ecol Entomol 40:629–638

Małecki J, Lovell H, Ewertowski W et al (2018) The glacial landsystem of a tropical glacier: Charquini Sur, Bolivian Andes. Earth Surf Process Landf 43:2584–2602

Mani MS (1968) Ecology and biogeography of high altitude insects. Dr. W Junk NV, The Hague

Manthorne KE (1989) Tropical renaissance: North American artists exploring Latin America, 1839–1879. Smithsonian Institution Press, Washington

Marris E (2007) The escalator effect. Nat Clim Change 1:94–96

Martinez-Conde S (2016) Has contemporary academia outgrown the Carl Sagan effect? J Neurosci 36:2077–2082

Mason K (1987) Abode of snow; A history of Himalayan exploration and mountaineering. Mountaineers, Seattle

Mauboussin MJ (2012) The success equation: untangling skill and luck in business, sports, and investing. Harvard Business Review Press, Cambridge

McAlister E (2017) The secret life of flies. Natural History Museum, Firefly, London

McFarlane R (2003) Mountains of the mind. A history of fasciation. Granta, London

Meineke EK, Classen AT, Sanders NJ et al (2019) Herbarium specimens reveal increasing herbivory over the past century. J Ecol 107:105–117

Meinhardt M (2018) Alexander von Humboldt. How the most famous scientist of the romantic age found the soul of nature. BlueBridge, New York

Melegari D (1895) Journal intime de Benjamin Constant et lettres à sa famille et à ses amis: portraits et autographe. P. Ollendorff, Paris

Meyer H (1907) In den Hochanden von Ecuador: Chimborazo, Cotopaxi etc: Reisen und Studien. Mit 3 farbigen Karten und 138 Abbildungen auf 37 Tafeln. Dietrich Reimer, Berlin

Meyer H (1908) Les Hautes Andes de l'Equateur: Chimborazo, Cotopaxi, etc. Atlas et texte. A. Challamel, Paris

Miller RR (1982) James Orton: a Yankee naturalist in South America, 1867–1877. Proc Am Phil Soc 126:11–25

Minguet C (1969) Alexandre de Humboldt, historien et géographe de l'Amérique espagnole (1799–1804). Maspero, Paris

Mitchell T (1989) Frederic Church's "The icebergs": erratic boulders and time's slow changes. Smithson Stud Am Art 3:3–23

Moret P (1990) Volcanisme et spéciation dans les Andes: à propos de deux nouveaux Dyscolus orophiles [Col. Caraboidea Platyninae]. Bull Soc Entomol France 95:169–174

Moret P (2009) Altitudinal distribution, diversity and endemicity of Carabidae (Coleoptera) in the páramos of Ecuadorian Andes. Ann Soc Entomol France 45:500–510

Moret P (2019) Antisana: the true Humboldtian mountain, Ecology and evolution nature portfolio, retrieved online on May 15th 2023. https://ecoevocommunity.nature.com/posts/53467-antisana-the-true-humboldtian-mountain

Moret P, Muriel P, Jaramillo R et al (2019) Humboldt's tableau physique revisited. Proc Natl Acad Sci USA 116:12889–12894

Morueta-Holme N, Engemann K, Sandoval-Acuña P et al (2015) Strong upslope shifts in Chimborazo's vegetation over two centuries since Humboldt. Proc Natl Acad Sci USA 112:12741–12745

Murienne J, Barragán Á, Manzin S et al (2022) How tectonic, volcanic and climatic processes in Andean 'sky islands' shaped the diversification of endemic ground beetles. J Biogeogr 49:2077–2090

Nabokov V (1973) Strong opinions. McGraw-Hill, New York

Needham JG (1909) Notes on the neuroptera in the collection of the Indian museum. Rec Indian Mus 3:185–210

Needham JG (1921) The new wild life preserve near McLean, N. Y. Sci Mon 12:246–252

Needham PR (1928) A net for capture of stream drift organisms. Ecology 9:339–342

Needham JG (1946) The lengthened shadow of a man and his wife. Sci Mon 62:140–150

Needham JG (1964) Nature study. In: Cornell science leaflet, Nature poetry, 57. NY State College of Agriculture, Ithaca, p 68

Nelken H (1976) Humboldtiana at Harvard Alexander von Humboldt and the United States. Harvard University Press, Cambridge

Nelken H (1980) Alexander von Humboldt. His portraits and their artists, a documentary iconography. Dietrich Reimer, Berlin

Nichols S (2006) Why was Humboldt forgotten in the United States? Geogr Rev 96:399–415

Nico LG (2001) Alexander von Humboldt (1769–1859): contributions to knowledge of new world fishes. In: Aymard G (ed) Alexander von Humboldt: Homenaje al Bicentenario de su llegada a tierras Venezolanas (16 de julio-1799 al 26 de noviembre-1800). BioLlania—special edition 7, Guanare, pp 127–164

Noirtin C, Boiteux P, Guillet P et al (1981) Les simulies, nuisance pour le bétail dans les Vosges: les origines de leur pullulation et les méthodes de lutte. Cahiers Orstom, Série Entomologie Médicale et Parasitologie 19:101–112

Olson R (2009) Don't be such a scientist: talking substance in an age of style. Island Press, Washington

Orbigny A d' (1835–1847) Voyage dans l'Amérique méridionale... exécuté pendant les années 1826, 1827, 1828, 1829, 1830, 1831, 1832 et 1833, 7 t., 21 vol. P. Bertrand & Veuve Levrault, Paris

Orbigny A d' (1839) L'homme américain, considéré sous ses rapport physiologiques et moraux, 2 t. Pitois-Levrault & Levrault, Paris

Orlove BS, Chiang JC, Cane MA (2002) Ethnoclimatology in the Andes: a cross-disciplinary study uncovers a scientific basis for the scheme Andean potato farmers traditionally use to predict the coming rains. Am Sci 90:428–435

Orlove B, Wiegandt E, Luckman BH (eds) (2008) Darkening peaks: glacier retreat, science, and society. University of California Press, Los Angeles

Ortiz JD, Jackson R (2022) Understanding Eunice Foote's 1856 experiments: heat absorption by atmospheric gases. Notes Rec 76:67–84

Orton J (1871) XXI. On the condors and humming-birds of the Equatorial Andes. J Nat Hist 8:185–192

Päßler U (2018) Im freien Spiel dynamischer Kräfte. Pflanzengeographische Schriften, Manuskripte und Korrespondenzen Alexander von Humboldts. In: Edition Humboldt digital, hg. v. Ottmar Ette. Berlin-Brandenburgische Akademie der Wissenschaften, Berlin. Version 3 vom 14.09.2018

Paine RT (1966) Food web complexity and species diversity. Am Nat 100:65–75

Paisley S, Saunders NJ (2010) A god forsaken: the sacred bear in Andean iconography and cosmology. World Archaeol 42:245–260

Pastoureau M (2007) L'Ours. Histoire d'un roi déchu. Seuil, Paris

Pearson OP, Bradford DF (1976) Thermoregulation of lizards and toads at high altitudes in Peru. Copeia 1976:155–170

Pérez Arias JD (2022) Amnesia: tras las huellas de unos fotógrafos olvidados. . . en busca de un país perdido. Mariscal Editions, Quito

Peyton B (1980) Ecology, distribution, and food habits of spectacled bears, Tremarctos ornatus, in Peru. J Mammal 61:639–652

Pörtner HO, Roberts DC, Masson-Delmotte V et al (2019) The ocean and cryosphere in a changing climate. IPCC special report on the ocean and cryosphere in a changing climate, 1155. Cambridge University Press, Cambridge

Pouchet F-A (1872) L'Univers. Les infiniments grands et les infiniments petits. Librairie Hachette et Cie, Paris

Primack RB, Miller-Rushing AJ (2012) Uncovering, collecting, and analyzing records to investigate the ecological impacts of climate change: a template from Thoreau's Concord. BioScience 62:170–181

Prothero DR (2015) The story of life in 25 Fossils. Columbia University Press, New York

Provençal JM, Humboldt A de (1808) Recherches sur la respiration des poisons. Mémoires de Physique et de Chimie de la Société d'Arcueil 2:360–404

Quammen D (1997) The song of the dodo: island biogeography in an age of extinctions. Scribner, New York

Quenta-Herrera E, Daza A, Lazzaro X et al (2021) Aquatic biota responses to temperature in a high Andean geothermal stream. Freshwat Biol 66:1889–1900

Raguso RA (2011) For love of Eisner. J Chem Ecol 37:546–547

Raimondi A (1862) Apuntes sobre la provincia litoral de Loreto. M.D. Cortes, Lima

Ralph CP (1978) Observations on Azorella compacta (Umbelliferae), a tropical Andean cushion plant. Biotropica 10:62–67

Ramond de Carbonnières LF (1792) Voyage et observations faites dans les Pyrénées. Dumoulin, Liège

Ramond de Carbonnières LF (1931) Carnets pyrénéens 2 vol. Editions de l'Echauguette, Lourdes

Réaumur RAF (1734–1742) Mémoires pour server à l'histoire des insects, 6 volumes. Imprimerie Royale, Paris

Recht R (1989) La lettre de Humboldt—du jardinière paysager au daguerreotype. Christian Bourgeois, Paris

Reider KE, Schmidt SK (2021) Vicuña dung gardens at the edge of the cryosphere. Ecology 102(2): e03228

Reider KE, Larson DJ, Barnes BM et al (2021) Thermal adaptations to extreme freeze–thaw cycles in the high tropical Andes. Biotropica 53:296–306

Reinhard J, Ceruti MC (2010) Inca rituals and sacred mountains: a study of the world's highest archaeological sites. Cotsen Institute of Archaeology at UCL, Los Angeles

Reiss W, Stübel A (1873) Alturas tomadas en la Republica del Ecuador en los años 1871, 1872, 1873. Guzman, Quito

Rémy F (2019a) Les voyages polaires de Jules Verne. Cybergeo Eur J Geogr. http://journals.openedition.org/cybergeo/32455

Rémy F (2019b) Rouge… comme neige. Pour la science 504:80–84

Renner SS, Otto R, Martín-Esquivel JL et al (2023a) Vegetation change on Mt. Teide, the Atlantic's highest volcano, inferred by incorporating the data underlying Humboldt's Tableau Physique des Iles Canaries. J Biogeogr 50:251–261

Renner SS, Päßler U, Moret P (2023b) "My reputation is at stake." Humboldt's mountain plant geography in the making (1803–1825). J Hist Biol. https://doi.org/10.1007/s10739-023-09705-z1

Rincón AD, Soibelzon LH (2007) The fossil record of the short-faced bears (Ursidae, Tremarctinae) from Venezuela. Systematic, biogeographic, and paleoecological implications. Neues Jahrb fur Geol Palaeontol Abh 244:287–298

Rivet P (1933) D'Orbigny, Ethnologue. In: Publication du Muséum d'Histoire Naturelle, 3, Commémoration du voyage d'Alcide d'Orbigny en Amérique du Sud 1826–1833. Masson & Cie, Paris, pp 15–26

Rosero P, Crespo-Pérez V, Espinosa R et al (2021) Multi-taxa colonisation along the foreland of a vanishing equatorial glacier. Ecography 44:1010–1021

Rosero-López D, Walter T, De Bievre B et al (2020) Design of a paired-weir system for experimental manipulation of environmental flows in streams. J Ecohydraulics 1–8

Rosero-López D, Walter T, Flecker AS et al (2022) A whole-ecosystem experiment reveals flow-induced shifts in a stream community. Commun Biol 5:420

Roux J-C (2004) La modernité de l'oeuvre d'Alcide d'Orbigny: Un précurseur des droits de l'homme et de la géographie du développement Outre-mers. Revue d'Historie 91:115–142

Sachs A (2006) The Humboldt current: nineteenth-century exploration and the roots of American environmentalism. Penguin, New York

Sagan D (2016) Fictional realism: scaling the twin peaks of art and science. In: Blackwell SH, Johnson K (eds) Fine lines: Vladimir Nabokov's scientific art. Yale University Press, New Haven, pp 243–250

Salzer L, Nöbauer A (2021) (Auf) Humboldts Spuren. Eine bauforscherische Untersuchung der "Casa Humboldt" am Antisana in Ecuador. HiN-Alexander von Humboldt im Netz. Internationale Zeitschrift für Humboldt-Studien 22:65–82

Saussure H-B de (1779) Voyages dans les Alpes, 4 vol. S Fauche, Neuchatel

Schaumann C (2020) Peak pursuit. The emergence of mountaineering in the nineteenth century. Yale University Press, New Haven

Scheffer M (2014) The forgotten half of scientific thinking. Proc Natl Ac Sci USA 111:6119–6119

Scheuchzer JJ (1723) Ouresiphoítes Helveticus, sive itinera per Helvetiæ alpinas regiones facta annis MDCCII. MDCCIII. MDCCIV. MDCCV. MDCCVI. MDCCVII. MDCCIX. MDCCX. MDCCCXI, 4 vol. P. vander Aa, Lugduni Batavorum

Schifko G (2010) Jules Vernes literarische Thematisierung der Kanarischen Inseln als Hommage an Alexander von Humboldt. Alexander von Humboldt im Netz. Int Rev Humboldtian Stud 11:65–70

Schippers MP, Ramirez O, Arana M et al (2012) Increase in carbohydrate utilization in high-altitude Andean mice. Curr Biol 22:2350–2354

Schlett J (2015) A not too greatly changed eden. Cornell University Press, Ithaca

Shepherd N (2014) The living mountain. Canongate, Edinburgh

Simberloff DS, Wilson EO (1969) Experimental zoogeography of islands: the colonization of empty islands. Ecology 50:278–296

Slack NG (2010) G. Evelyn Hutchinson and the invention of modern ecology. Yale University Press, New Haven

Soibelzon LH, Schubert BW (2011) The largest known bear, *Arctotherium angustidens*, from the early Pleistocene pampean region of Argentina: with a discussion of size and diet trends in bears. J Paleontol 85:69–75

Stafford BM (1984) Voyage into substance: art, science, nature, and the illustrated travel account, 1760–1840. MIT Press, Cambridge

Statzner B (1988) Growth and Reynolds number of lotic macroinvertebrates: a problem for adaptation of shape to drag. Oikos 51:84–87

Steele W (1987) Eisner on nature's designs. Cornell Alumni News 90:16–21

Steinbauer MJ, Grytnes JA, Jurasinski G et al (2018) Accelerated increase in plant species richness on mountain summits is linked to warming. Nature 556:231–234

Steinitz-Kannan M, López C, Jacobsen D et al (2020) History of limnology in Ecuador: a foundation for a growing field in the country. Hydrobiologia 847:4191–4206

Steleanu A von (1959) Alexander von Humbold und die Bedeutung seines wissenschaftlichen Werkes für die hydrobiology. In: Nielsen N (ed) Alexander von Humboldt; Gedenkschrift zur 100. Akademie, Berlin, pp 425–444

Stensrud AB (2020) Sentient springs and sources of life: Water, climate change and world-making practices in the Andes. In: Ray C (ed) Sacred waters, a cross-cultural compendium of hallowed springs and holy wells. Routledge, London, pp 368–377

Storz JF, Quiroga-Carmona M, Opazo JC et al (2020) Discovery of the world's highest-dwelling mammal. Proc Natl Acad Sci USA 117:18169–18171

Stucchi M, Torres D, Soibelzon L (2002) Los parientes desaparecidos del oso frontino. Sociedad de Ciencias Naturales La Salle. Caracas, Venezuela. Revista Natura 120:10–16

Sulloway FJ (1982) Darwin and his finches: the evolution of a legend. J Hist Biol 15:1–53

Terra H de (1955) The life and times of Alexander von Humboldt 1769–1859. A. Knopf, New York

Terra H de, Hutchinson GE (1934) Evidence of recent climatic changes shown by Tibetan highland lakes. Geogr J 84:311–320

Terrall M (2014) Catching Nature in the Act: Réaumur and the Practice of Natural History in the Eighteenth Century. The University of Chicago Press, Chicago

The Cornell Lab of Ornithology, Vyn G (2015) Living bird: 100 years of listening to nature. Mountaineers, Seattle

Torres D (2021) Oso andino: animal y mito. Fundación AndígenA, Merida

Trystam F (1979) Le procès des étoiles. Seghers, Paris

Tschudi J-J von (1847) Travels in Peru, translation by Thomasina Ross. D. Bogue, London

Tyndall J (1860) The glaciers of the alps; being a narrative of excursions and ascents, an account of the origin and phenomena of glaciers and an exposition of the physical principles to which they are related. J. Murray, London

Tyndall J (1863) Heat considered as a mode of motion. Longman, Green, London

Urai AE, Kelly C (2023) Point of view: rethinking academia in a time of climate crisis. eLife 12: e84991

Velasco J de (1841–1844) Historia del reino de Quito en la America meridional, Año de 1789, Quito, 3 t. Suarez de Valdés, Quito

Vénec-Peyré M-T (2002) Alcide d'Orbigny et les foraminifères. In: Taquet P (ed) Alcide d'Orbigny, du nouveau monde... au passé du monde. Muséum national d'Histoire naturelle & Nathan, Paris, pp 77–88

Verberk WC, Bilton DT, Calosi P et al (2011) Oxygen supply in aquatic ectotherms: partial pressure and solubility together explain biodiversity and size patterns. Ecology 92:1565–1572

Verne J (1867) Voyage au centre de la Terre. J Hetzel, Paris

Verne J (1874) The fur country: or, seventy degrees north latitude. J Osgood and Cie, Boston

Verne J (1875) L'île mystérieuse. Hetzel et Cie, Paris

Verne J (1876) The voyages and adventures of captain Hatteras. J Osgood and Cie, Boston

Verne J (2002) The Mighty Orinoco. Wesleyan University Press, Middletown

Vimos DJ, Encalada AC, Ríos-Touma B et al (2015) Effects of exotic trout on benthic communities in high-Andean tropical streams. Freshwat Sci 34:770–783

Vogel S (1988) Life's devices, the physical world of plants and animals. Princeton University Press, Princeton

Vosoughi S, Roy D, Aral S (2018) The spread of true and false news online. Science 359:1146–1151

Vuilleumier F (2004) In memoriam: Jean Dorst, 1924–2001. Auk 121:1289–1290

Vuilleumier F, Monasterio M (eds) (1986) High altitude tropical biogeography. Oxford University Press, Oxford

Wallace A (1853a) Travels on the Amazon and Rio Negro: with an account of the native tribes, and observations on the climate, geology, and natural history of the Amazon valley. Reeve, London

Wallace AR (1853b) On the Rio Negro. J R Geogr Soc 23:212–217

Wallace AR (1905) My life: a record of events and opinions, 2 vol. Chapman and Hall, London

Wallace AR (1911) The world of life; a manifestation of creative power, directive mind and ultimate purpose. Moffat, Yard, New York

Walls LD (1995) Seeing new worlds: Henry David Thoreau and nineteenth-century natural science. University of Wisconsin Press, Madison

Walls LD (2006) The search for Humboldt. Geogr Rev 96:473–477

Walls LD (2009) The passage to cosmos. University Chicago Press, Chicago

Walls LD (2016) O pioneer. Am Sci 104:118–119

Wesenberg-Lund C (1920–1921) Contributions to the biology of the Danish Culicidae. D. Kgl. danske videnskabernes selskabs skrifter. Naturvidenski. og mathematisk, vol 8. Host AF, Copenhagen

Wharton DA (2002) Life at the limits. Cambridge University Press, Cambridge

Whitaker R (2004) The mapmaker's wife: a true tale of love, murder, and survival in the Amazon. Transworld, London

Whymper E (1892) Travels amongst the great Andes of the Equator. John Lehmann, London

Wiggins GB, Richardson JS (1989) Biosystematics of Eocosmoecus, a new Nearctic caddisfly genus (Trichoptera: Limnephilidae, Dicosmoecinae). J N Am Benthol Soc 8:355–369

Williams G (2015) Naturalists at sea: scientific travellers from Dampier to Darwin. Yale University Press, New Haven

Wilson E (2003) The spiritual history of ice: Romanticism, science and the imagination. Palgrave Macmillan, London

Winthrop T (1869) Companion to heart of the Andes. Appelton, New York

Wollaston TV (1854) Insecta Maderensia; being an account of the insects of the islands of the Madeiran group. 13 pls. Van Voorst, London

Woodward G, Dybkjaer JB, Olafsson JS et al (2010) Sentinel systems on the razor's edge: effects of warming on Arctic geothermal stream ecosystems. Glob Change Biol 16:1979–1991

Wright AH (1946) Cornell's educational pioneers. Jean Louis Rodolphe Agassiz. Cornell Alumni News 48:388–389

Wulf A (2015) The invention of nature: Alexander von Humboldt's new world. Alfred Knopf, New York

Yépez A (2022) Sacred mountains and their oracles in the equatorial glacial volcanic arc. J Glacial Archaeol 6:33–45

Zerefos CS, Tetsis P, Kazantzidis A et al (2014) Further evidence of important environmental information content in red-to-green ratios as depicted in paintings by great masters. Atmos Chem Phys 14:2987–3015

Zhong Mengual E (2021) Apprendre à Voir. Le point de vue du vivant. Actes Sud, Arles

Index

© The Author(s), under exclusive license to Springer Nature Switzerland AG 2023   255
O. Dangles, *Climate Change on Mountains*,
https://doi.org/10.1007/978-3-031-39528-4